WITHDRAWN

Finite-Element Modelling of Unbounded Media

Finite-Element Modelling of Unbounded Media

John P. Wolf and Chongmin Song
Swiss Federal Institute of Technology, Lausanne, Switzerland

JOHN WILEY & SONS
Chichester · New York · Brisbane · Toronto · Singapore

Copyright © 1996 by John Wiley & Sons Ltd,
Baffins Lane, Chichester,
West Sussex PO19 1UD, England

Reprinted May 1997

National   01243 779777
International (+44) 1243 779777
e-mail (for orders and customer service enquiries): cs-books@wiley.co.uk
Visit our Home page on http://www.wiley.co.uk
or http://www.wiley.com

All Rights Reserved. No part of this book may be reproduced, stored in a retrieval system, or transmitted in any form or by any means, electronic, mechanical, photocopying, recording or otherwise, except under the terms of the Copyright Designs and Patents Act 1988 or under the terms of a licence issued by the Copyright Licensing Agency, 90 Tottenham Court Road, London, UK W1P 9HE, without the permission in writing of the publisher.

Other Wiley Editorial Offices

John Wiley & Sons, Inc., 605 Third Avenue,
New York, NY 10158-0012, USA

Jacaranda Wiley Ltd, 33 Park Road, Milton,
Queensland 4064, Australia

John Wiley & Sons (Canada) Ltd, 22 Worcester Road,
Rexdale, Ontario M9W 1L1, Canada

John Wiley & Sons (Asia) Pte Ltd, 2 Clementi Loop #02-01,
Jin Xing Distripark, Singapore 0512

Cover illustration: Mixing red and blue produces purple. Analogously, assembling a finite-element cell and an unbounded medium yields another unbounded medium – the concept of the consistent infinitesimal finite-element cell method (Figure 5-1).

British Library Cataloguing in Publication Data

A catalogue record for this book is available from the British Library.

ISBN 0 471 96134 5

Produced from camera-ready copy supplied by the authors using LaTeX
Printed and bound in Great Britain by Bookcraft (Bath) Ltd
This book is printed on acid-free paper responsibly manufactured from sustainable forestation, for which at least two trees are planted for each one used for paper production.

Contents

FOREWORD xiii

PREFACE xv

1 INTRODUCTION 1
   1.1 STATEMENT OF PROBLEM 1
   1.2 RADIATION CONDITION 2
   1.3 SUBSTRUCTURE METHOD AND DIRECT METHOD 4
   1.4 STATE OF THE ART 7
       1.4.1 Rigorous Modelling 8
       1.4.2 Approximate Modelling 9
   1.5 OVERVIEW OF NOVEL METHODS 10
       1.5.1 Similarity-based Formulation 11
       1.5.2 Damping-solvent Extraction Method 16
       1.5.3 Doubly-asymptotic Multi-directional Transmitting Boundary . 19
   1.6 ORGANIZATION OF TEXT 20

I SIMILARITY-BASED FORMULATION FOR UNIT-IMPULSE RESPONSE AND DYNAMIC STIFFNESS   **23**

2 DISPLACEMENT, VELOCITY AND ACCELERATION UNIT-IMPULSE RESPONSE WITH DYNAMIC STIFFNESS AND RATIONAL APPROXIMATION 31
   2.1 INTERACTION FORCE-DISPLACEMENT RELATIONSHIP . . . 31
   2.2 INTERACTION FORCE-ACCELERATION RELATIONSHIP . . . 34
   2.3 INTERACTION FORCE-VELOCITY RELATIONSHIP 37
   2.4 RATIONAL APPROXIMATION AND REALIZATION 40

3 DYNAMIC STIFFNESS AND UNIT-IMPULSE RESPONSE AT SIMILAR STRUCTURE-MEDIUM INTERFACES OF UNBOUNDED MEDIUM 43
3.1 FREQUENCY DOMAIN 44
3.2 TIME DOMAIN 46

4 FORECASTING METHOD 49
4.1 CONCEPT BASED ON WAVE PROPAGATION 49
4.2 FUNDAMENTAL EQUATIONS 52
4.2.1 Interaction Force-acceleration Relationship of Unbounded Medium 52
4.2.2 Equations of Motion of Finite-element Region 53
4.3 TIME DISCRETIZATION AND NUMERICAL ALGORITHM 54
4.4 ACCURACY 58
4.4.1 Semi-infinite Rod on Elastic Foundation 58
4.4.2 Out-of-plane Motion of Semi-infinite Wedge 59
4.4.3 In-plane Motion of Semi-infinite Wedge 60
4.4.4 Strip Foundation with Rectangular Cross-section Embedded in Half-plane 61

5 CONSISTENT INFINITESIMAL FINITE-ELEMENT CELL METHOD FOR WAVE PROPAGATION 65
5.1 CONCEPT BASED ON INFINITESIMAL CELL WIDTH 65
5.2 FUNDAMENTAL EQUATIONS 67
5.2.1 Coefficient Matrices of Finite-element Cell for Three-dimensional Vector Wave Equation 68
5.2.2 Summary of Coefficient Matrices for Three-dimensional Scalar Wave Equation 76
5.2.3 Summary of Coefficient Matrices for Two-dimensional Vector Wave Equation 78
5.2.4 Summary of Coefficient Matrices for Two-dimensional Scalar Wave Equation 80
5.2.5 Summary of Coefficient Matrices for Axisymmetric Vector Wave Equation 81
5.2.6 Summary of Coefficient Matrices for Axisymmetric Scalar Wave Equation 87
5.2.7 Assemblage of Finite-element Cell and Unbounded Medium 89
5.2.8 Consistent Infinitesimal Finite-element Cell Equation in Frequency Domain 91
5.2.9 Consistent Infinitesimal Finite-element Cell Equation in Time Domain 93

		5.2.10	Two-dimensional Layered Unbounded Medium as a Special Case 94

- 5.2.10 Two-dimensional Layered Unbounded Medium as a Special Case 94
- 5.2.11 Variation of Material Properties in Radial Direction 99
- 5.2.12 Insight on Radiation Damping 102

5.3 TIME DISCRETIZATION 104
- 5.3.1 First Time Step 105
- 5.3.2 nth Time Step 106

5.4 ACCURACY 107
- 5.4.1 Spherical Cavity Embedded in Full-space 107
- 5.4.2 Spherical Cavity Embedded in Full-space with Varying Material Properties in Radial Direction 112
- 5.4.3 Out-of-plane Motion of Circular Cavity Embedded in Full-plane with Varying Shear Modulus in Radial Direction 114
- 5.4.4 Out-of-plane Motion of Semi-infinite Layer of Constant Depth 120
- 5.4.5 Out-of-plane Motion of Semi-infinite Wedge 121
- 5.4.6 In-plane Motion of Semi-infinite Wedge 121
- 5.4.7 Circular Cavity Embedded in Full-plane 121
- 5.4.8 Out-of-plane Motion of Strip Foundation with Rectangular Cross-section Embedded in Half-plane 124
- 5.4.9 In-plane Motion of Strip Foundation with Rectangular Cross-section Embedded in Half-plane 125
- 5.4.10 Cylinder Embedded in Half-space for Vector Wave Equation . 128
- 5.4.11 Prism Embedded in Half-space for Scalar Wave Equation ... 129
- 5.4.12 Prism Embedded in Half-space for Vector Wave Equation .. 132

6 CONSISTENT INFINITESIMAL FINITE-ELEMENT CELL METHOD FOR INCOMPRESSIBLE ELASTICITY 137

6.1 FUNDAMENTAL EQUATIONS 138
- 6.1.1 Coefficient Matrices of Finite-element Cell 138
- 6.1.2 Consistent Infinitesimal Finite-element Cell Equation 139

6.2 ACCURACY 144
- 6.2.1 Spherical Cavity Embedded in Full-space 144
- 6.2.2 In-plane Motion of Semi-infinite Layer of Constant Depth .. 145
- 6.2.3 Circular Cavity Embedded in Full-plane 147
- 6.2.4 Prism Embedded in Half-space 147

7 CONSISTENT INFINITESIMAL FINITE-ELEMENT CELL METHOD IN FREQUENCY DOMAIN 151

7.1 OVERVIEW 151
7.2 ASYMPTOTIC EXPANSION FOR HIGH FREQUENCY 152
7.3 DYNAMIC CONDENSATION AND SUBSTRUCTURE DELETION METHODS 155

	7.4	ACCURACY 158
	7.4.1	Spherical Cavity Embedded in Full-space 158
	7.4.2	Circular Cavity Embedded in Full-plane 162
	7.4.3	Prism Embedded in Half-space 162

8 CONSISTENT INFINITESIMAL FINITE-ELEMENT CELL METHOD FOR STATICS 165

- 8.1 SUMMARY OF FUNDAMENTAL EQUATIONS 165
- 8.2 DISPLACEMENT AND STRAIN AT INTERNAL POINT 167
- 8.3 ACCURACY 171
  - 8.3.1 Spherical Cavity Embedded in Full-space 171
  - 8.3.2 Semi-infinite Wedge 172
  - 8.3.3 Hemispherical Foundation Embedded in Half-space 173
  - 8.3.4 Prism Embedded in Half-space 173
  - 8.3.5 Tunnel in Inhomogeneous Transversely Isotropic Unbounded Rock 176

9 CONSISTENT INFINITESIMAL FINITE-ELEMENT CELL METHOD FOR DIFFUSION 179

- 9.1 FUNDAMENTAL EQUATIONS 180
  - 9.1.1 Definition of Problem 180
  - 9.1.2 Unit-impulse Response Matrices 181
  - 9.1.3 Dynamic Stiffness at Similar Structure-medium Interfaces .. 182
  - 9.1.4 Coefficient Matrices of Finite-element Cell 184
  - 9.1.5 Assemblage of Finite-element Cell and Unbounded Medium . 185
  - 9.1.6 Consistent Infinitesimal Finite-element Cell Equation in Frequency Domain 186
  - 9.1.7 Consistent Infinitesimal Finite-element Cell Equation in Time Domain 187
- 9.2 TIME DISCRETIZATION 188
  - 9.2.1 First Time Step 188
  - 9.2.2 nth Time Step 188
- 9.3 DYNAMIC STIFFNESS 189
- 9.4 ACCURACY 190
  - 9.4.1 Spherical Cavity Embedded in Full-space 190
  - 9.4.2 Semi-infinite Wedge 194
  - 9.4.3 Prism Embedded in Inhomogeneous Half-space 198

10 CONSISTENT INFINITESIMAL FINITE-ELEMENT CELL METHOD APPLIED TO BOUNDED MEDIUM 201

- 10.1 SUMMARY OF FUNDAMENTAL EQUATIONS 202
- 10.2 STATICS 204

CONTENTS ix

      10.2.1 Static-stiffness Matrix 204
      10.2.2 Displacement and Strain at Internal Point 206
  10.3 MASS MATRIX 207
  10.4 ACCURACY 208
      10.4.1 Solid Sphere 208
      10.4.2 Cantilever 212
      10.4.3 Edge-cracked Plate 213

II DAMPING-SOLVENT EXTRACTION FOR DYNAMIC STIFFNESS AND INTERACTION FORCE   217

11 FUNDAMENTALS OF DAMPING-SOLVENT EXTRACTION METHOD 223
  11.1 CONCEPT 223
  11.2 EFFECT OF DAMPING ON DYNAMIC STIFFNESS 225
  11.3 EXTRACTION OF EFFECT OF DAMPING 228

12 IMPLEMENTATION, VERIFICATION AND ACCURACY OF DAMPING-SOLVENT EXTRACTION METHOD 231
  12.1 IMPLEMENTATION IN FREQUENCY DOMAIN 231
  12.2 IMPLEMENTATION IN TIME DOMAIN 233
  12.3 FLEXIBILITY FORMULATION 235
  12.4 ANALYTICAL VERIFICATION 237
  12.5 ACCURACY IN FREQUENCY DOMAIN 246
      12.5.1 Out-of-plane Motion of Semi-infinite Layer of Constant Depth 246
      12.5.2 In-plane Motion of Semi-infinite Wedge 248
      12.5.3 Strip Foundation with Rectangular Cross-section Embedded in Half-plane 250
  12.6 ACCURACY IN TIME DOMAIN 251

III DOUBLY-ASYMPTOTIC MULTI-DIRECTIONAL TRANSMITTING BOUNDARY   257

13 CONCEPT AND NUMERICAL IMPLEMENTATION OF DOUBLY-ASYMPTOTIC MULTI-DIRECTIONAL TRANSMITTING BOUNDARY 263
  13.1 CONCEPT 263
  13.2 NUMERICAL IMPLEMENTATION 267

14 ACCURACY AND MODELLING PROCEDURE OF DOUBLY-ASYMPTOTIC MULTI-DIRECTIONAL TRANSMITTING BOUNDARY . . . 271
14.1 PERFORMANCE AND COMPARISON WITH OTHER TRANSMITTING BOUNDARIES . . . 271
  14.1.1 Semi-infinite Rod on Elastic Foundation . . . 271
  14.1.2 In-plane Motion of Semi-infinite Layer of Constant Depth . . 273
  14.1.3 In-plane Motion of Semi-infinite Wedge . . . 274
14.2 DECISION FLOWCHART OF MODELLING PROCEDURE . . . 276

Appendix A BENCHMARK EXAMPLES . . . 281
A.1 SPHERICAL CAVITY EMBEDDED IN FULL-SPACE . . . 282
A.2 SEMI-INFINITE ROD ON ELASTIC FOUNDATION . . . 289
A.3 SEMI-INFINITE LAYER OF CONSTANT DEPTH . . . 294
  A.3.1 Out-of-plane Motion . . . 294
  A.3.2 In-plane Motion . . . 301
A.4 SEMI-INFINITE WEDGE . . . 302
  A.4.1 Out-of-plane Motion . . . 302
  A.4.2 In-plane Motion . . . 307
A.5 IN-PLANE MOTION OF CIRCULAR CAVITY . . . 307
A.6 IN-PLANE MOTION OF STRIP FOUNDATION WITH RECTANGULAR CROSS-SECTION EMBEDDED IN HALF-PLANE . . . 314
A.7 PRISM EMBEDDED IN HALF-SPACE FOR VECTOR WAVE EQUATION . . . 314
A.8 TRANSVERSELY ISOTROPIC MATERIAL . . . 314

Appendix B DESCRIPTION OF COMPUTER PROGRAMME *SIMILAR* FOR CONSISTENT INFINITESIMAL FINITE-ELEMENT CELL METHOD . . . 317

REFERENCES . . . 319

INDEX . . . 325

Le silence éternel de ces espaces infinis m'effraie.

(The eternal silence of these infinite spaces frightens me.)

Blaise Pascal   1623–1662

FOREWORD

It is now widely recognized that the dynamic interaction between a structure and the supporting soil may have a significant effect on the response of the structure to earthquake excitation, wind loads, and other dynamic disturbances. At the present time, analysis techniques of varying degrees of accuracy exist to model the structure and a portion of the surrounding soil where nonlinearities may occur. One of the major remaining difficulties for a complete analysis of the full soil-structure interaction problem is the need to consider the dynamic response of the (practically) unbounded region of soil beyond the immediate vicinity of the structure. In the last twenty years significant advances have been made in this area, most notably the development of procedures based on boundary-integral or boundary-element methods for unbounded media. Nevertheless, a major need still exists for alternative approaches, particularly for procedures that could be implemented within the context of finite-element analyses. These procedures would be more familiar to structural analysts, and would allow greater flexibility in terms of the geometrical and material characterization of the unbounded medium.

The present book by John P. Wolf and Chongmin Song responds exactly to this need for finite-element based procedures to address the problems of unbounded media as they arise in several fields, including earthquake and geotechnical engineering, geophysics, and acoustics. The authors present in full detail three approaches, two of which are suitable for substructure analyses of the structure-unbounded medium problem and one approach applicable to direct analyses of the complete problem. This last approach (Part III) involves the development of a transmitting boundary which is exact in the low-frequency limit and at high frequencies for plane waves impinging on the transmitting boundary at selected angles. This effort represents a synthesis and a significant advancement over earlier transmitting boundaries suggested by Lysmer (1970), Underwood and Geers (1981), Liao and Wong (1984) and Higdon (1986).

The second approach (Part II) involves consideration of an artificially bounded region in the vicinity of the structure. A large fictitious damping is prescribed over this region to minimize reflections from the artificial boundary. A simple procedure is then used to approximately remove the unwanted effects of the artificial damping.

A major part of the book (Part I) is devoted to the presentation of a procedure which takes full advantage of conditions resulting from an assumed geometrical similarity. The most general of two variants of this approach leads to a boundary finite-element procedure in which only the interface between the structure and the unbounded medium needs to be discretized. The procedure is applicable to problems of wave propagation in compressible and incompressible elastic media, and to static and diffusion problems. The only limitations arise from the geometrical requirements imposed to obtain similarity. This approach represents a major accomplishment which is many logical steps away from an initial seminal suggestion by Dasgupta (1982).

All developments are presented in a clear and detailed fashion, and the resulting algorithms are carefully tested in a number of benchmark cases for which exact or reliable solutions are known. The material in this book and the accompanying computer program will do much to facilitate and make more accessible the analysis of interaction problems involving a finite structure and an unbounded medium. As I anticipate the impact of this book, I cannot help but think of the significant advances that have been made since the pioneering contributions sixty years ago of Reissner (1936), who considered the dynamic response of a circular foundation supported on an elastic half-space, and of Sezawa and Kanai (1935, 1936), who considered the seismic response of a cylindrical elastic structure supported on an elastic hemispherical foundation resting on an elastic half-space.

J. Enrique Luco
University of California, San Diego

PREFACE

In many fields of engineering a *bounded structure* with finite dimensions interacts dynamically with an adjacent *unbounded medium* with infinite dimensions. In wave propagation, examples are soil-structure interaction, fluid-structure interaction and many aspects of acoustics, electromagnetism and geophysics, and in diffusion, heat conduction and consolidation of soil can be mentioned. Statics as a special case of the time-dependent problem is important. In this *dynamic unbounded medium-structure-interaction analysis* the numerical modelling of the structure with, for instance, finite elements is well developed, but that of the unbounded medium has not progressed sufficiently to solve practical problems accurately and efficiently.

The unbounded medium is represented by a boundary condition on its interface with the structure, which yields a model with a finite number of degrees of freedom. This *boundary condition* describes the force-motion relationship, whose *rigorous* form is *global in space and time*. The *boundary-element method* is the only general procedure for constructing such a rigorous representation. This method differs significantly from the familiar finite-element method used to model the structure. The boundary-element method requires a fundamental solution, which is not always available and can be very complicated, leading to singular integrals. The *consistent boundary* is also rigorous but can only be applied to a layered unbounded medium with parallel interfaces of the layers. An *approximate* form of the force-motion relationship can also be specified which is constructed to be *local in space and time* to reduce the computational effort. The *transmitting boundaries* provide such approximate representations. They are easily incorporated in the finite-element method but often exhibit inadequate accuracy. The same applies to the *infinite element*. The book presents three novel finite-element based methods to model the unbounded medium, which are characterized as follows

1. The *consistent infinitesimal finite-element cell method* is a *boundary finite-element procedure*, as it requires the spatial discretization of the boundary only and is based on finite-element assemblage. A reduction of the spatial dimension by one thus results as in the boundary-element method but without using a fundamental solution. The consistent infinitesimal finite-element cell method converges to the exact solution in the finite-element sense. The advantages of the boundary-element and finite-element methods are thus combined. Similarity

is used in the derivation. Conditions on boundaries extending to infinity such as on free surfaces and on interfaces between different materials compatible with similarity are incorporated automatically without any spatial discretization. Anisotropic material is processed straightforwardly without increasing the computational effort. As another implementation based on finite elements and similarity the *forecasting method* is discussed. Only the banded symmetric system of equations is solved in this procedure which makes use of the time delay of wave propagation.

2. The *damping-solvent extraction method* permits an efficient analysis of the unbounded medium by modelling only part of the unbounded medium adjacent to the structure with finite elements: *artificial damping as a solvent* is first *introduced* to reduce the amplitudes of the reflected waves and then *extracted*.

3. The *doubly-asymptotic multi-directional transmitting boundary* is rigorous for the low-frequency limit and the high-frequency limit for plane waves propagating at preselected angles.

Wave propagation in an unbounded medium is addressed in the derivation of the novel methods. The consistent infinitesimal finite-element cell method is applied to diffusion and statics in an unbounded medium and to a bounded medium.

The book commences with an overview in Chapter 1. The three novel methods are developed in three parts. Part I addresses the similarity-based formulation to calculate the unit-impulse response in the time domain and the dynamic stiffness in the frequency domain to be used in the force-motion relationship. Chapter 2 discusses the unit-impulse response and dynamic stiffness. Chapter 3 develops the relationships of the dynamic stiffness and unit-impulse response at similar structure-medium interfaces of an unbounded medium. Chapter 4 addresses the forecasting method for wave propagation in the time domain. Chapter 5 derives the consistent infinitesimal finite-element cell method and applies it to the wave propagation in the time domain. Chapter 6 presents the extension to incompressible elasticity. Chapter 7 examines the solution of the consistent infinitesimal finite-element cell equation in the frequency domain. Chapter 8 presents the consistent infinitesimal finite-element cell method for statics and Chapter 9 for diffusion in the time and frequency domains. Chapter 10 applies the consistent infinitesimal finite-element cell method to a bounded medium. Part II examines the damping-solvent extraction method. Chapter 11 discusses its fundamentals and Chapter 12 its implementation, verification and accuracy. Part III develops the doubly-asymptotic multi-directional transmitting boundary. Chapter 13 addresses its concept and numerical implementation and Chapter 14 its accuracy and modelling procedure. Appendix A describes the benchmark examples used throughout the book. Appendix B contains a description of the computer programme *SIMILAR* for the consistent infinitesimal finite-element cell method.

As a prerequisite basic knowledge of the finite-element method and of dynamic analysis is necessary. The derivations are presented step by step in great detail and are thus self-contained. The same benchmark examples are solved by the novel methods demonstrating their high accuracy and versatility. The computer programme *SIMILAR* for the conSistent Infinitesimal finite-element cell Method - a fInite-eLement boundARy is available on disk. This programme was used to calculate the examples in the book. The source code can be incorporated into a general finite-element programme and forms the basis for developing more advanced versions.

This book should appeal to numerical analysts and software developers in many fields of engineering such as engineering mechanics, earthquake engineering, acoustics, electromagnetism and computational mathematics. It serves as the text of an advanced course on finite elements and dynamic soil-structure interaction taught at the Swiss Federal Institute of Technology in Lausanne.

The financial support for the development of the finite-element based methods for modelling unbounded media provided by the Division of Safety of Dams of the Swiss Federal Office for Water Resources and the Swiss National Science Foundation is gratefully acknowledged.

John P. Wolf
Chongmin Song
Swiss Federal Institute of Technology

1

INTRODUCTION

1.1 STATEMENT OF PROBLEM

An *infinite* or *semi-infinite medium* has to be addressed in many fields of engineering and physical science. Often, a *bounded structure* is also present together with this *unbounded medium*. The common finite surface is called the *structure-medium interface*. When this system is excited by a time-varying load, the structure interacts dynamically with the unbounded medium. To determine the resulting response, a so-called *dynamic unbounded medium-structure-interaction* analysis is performed.

In wave propagation, examples are soil-structure interaction, fluid-structure interaction and many aspects of geophysics, acoustics and electromagnetism. In diffusion, heat conduction and settlement of a structure caused by consolidation of the soil can be mentioned. Statics as a special case of the time-dependent problem is common.

Dynamic unbounded medium-structure-interaction analysis can be characterized by considering soil-structure interaction. A typical example as encountered in the analysis of nuclear power plants is shown in Figure 1-1. The reactor building denoted as the actual structure is embedded in soft soil, which is the unbounded medium. The latter is divided into two parts: the irregular bounded medium adjacent to the structure and the remaining regular unbounded medium extending to infinity. The structure and the irregular bounded medium form the generalized structure which can exhibit nonlinear behaviour. The *regular unbounded medium must remain linear*. The generalized structure interacts with the regular unbounded medium at the generalized structure-medium interface. (The size of the irregular bounded medium can vary significantly. It can be very large in geophysical applications and can vanish in certain soil-structure interaction cases.) For the sake of simplicity, the generalized structure is called the *structure*, the regular unbounded medium the *unbounded medium*, and the generalized structure-medium interface the *structure-medium interface*, through which the dynamic interaction occurs.

The loading (Figure 1-1) can be introduced into the dynamic system through the unbounded medium. For soil-structure interaction these incident waves are generated from earthquakes, underground explosions, or the passage of vehicles distant from

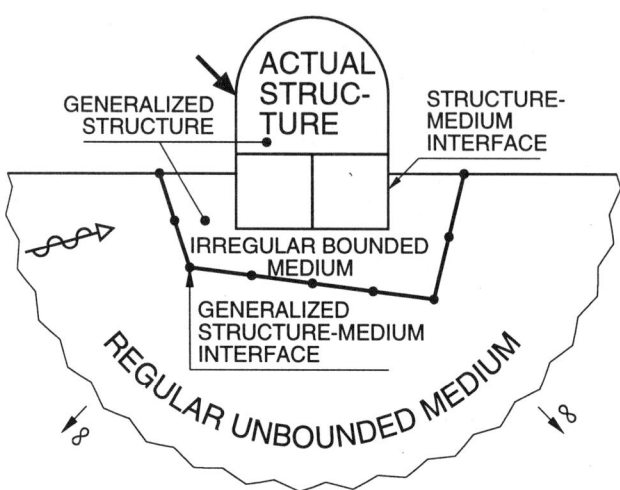

Figure 1-1 Problem definition of dynamic unbounded medium-structure-interaction analysis

the structure. Alternatively, the loading can act directly on the structure, arising from machines, impacts, vehicles moving within the structure, etc.

Loads acting on the structure lead to waves propagating from the structure-medium interface towards infinity in a soil-structure-interaction analysis. As the amplitudes of all body and surface waves in three dimensions decay owing to geometric spreading, the nonlinearity will be limited to the irregular bounded medium when chosen sufficiently large. The remaining regular unbounded medium will thus behave linearly.

Numerical modelling of an unbounded medium is a difficult task. The boundary condition at infinity, often called the *radiation condition*, states that no energy be radiated from infinity towards the structure. For a wave-propagation problem for each increase in time an additional part of the unbounded medium (adjacent to the wave front) which was at rest is excited and this requires energy. As far as the structure's response is concerned, this corresponds to *radiation damping*.

In this book emphasis is placed on methods of analysis working in the time domain. This is not only the natural approach for dynamics considering the sequence of developments from one time to the next, but also permits a straightforward implementation of the nonlinearity of the structure.

1.2 RADIATION CONDITION

For the unbounded medium the solution must be *unique*, as is the one realized in nature. The governing equations of elasto-dynamics and the boundary conditions

1.2 RADIATION CONDITION

on the structure-medium interface and the free surface, if present, of the unbounded medium are not sufficient to define a unique solution. Conceptually, in addition, a boundary at a finite distance from the structure-medium interface which does not intersect the latter and the free surface could be introduced, leading to a solution which satisfies the boundary conditions on the structure-medium interface and the free surface, but which differs from that for the unbounded medium. Another condition is thus necessary. In this section only the conclusions are stated. For a detailed proof which is quite involved the reader is referred to References [18] and [47].

In the *time domain* the additional *condition for uniqueness* [18] states that *outside the domain of influence the displacement vanishes*. The domain of influence at a specific time is defined as that part of the unbounded medium enclosed at this instant by the wave front propagating with the largest wave velocity, the dilatational-wave velocity. Thus, for waves starting at the structure-medium interface, at each point in time there will be an adjacent domain of influence and outside up to infinity the medium is at rest.

For a finite time the response of the unbounded medium can be determined from that of a sufficiently large (larger than the domain of influence) bounded medium. A straightforward application is the modelling of an unbounded medium by the so-called extended mesh whose outer boundary lies outside the domain of influence.

For each increase in time, an additional part of the unbounded medium which is at rest adjacent to the wave front is incorporated in the domain of influence. This requires energy which is supplied through the structure-medium interface. As far as the structure's response is concerned, this corresponds to *radiation damping*.

The uniqueness theorem can be formulated to calculate the response of the unbounded medium at the initial time. The direction perpendicular to the infinitesimal area dA of the structure-medium interface is addressed. The unbounded medium is initially at rest. After applying the load per unit area p during the first infinitesimal time dt the wave front is at the distance $c_p dt$ (dilatational-wave velocity c_p) and the domain of influence equals $c_p dt dA$. The law of conservation of momentum is applied for the first infinitesimal time dt. The initial momentum vanishes. The momentum (mass times velocity) at dt equals $\rho c_p dt dA \dot{u}$ (mass density ρ, velocity $\dot{u}$ at dt). The law is formulated as

$$p dA dt = \rho c_p dt dA \dot{u} \tag{1.1}$$

which yields

$$p = \rho c_p \dot{u} \tag{1.2}$$

The initial response perpendicular to the structure-medium interface of the unbounded medium is thus modelled by a dashpot with the coefficient per unit area ρc_p which is called the *impedance*. Analogously, the initial response in the tangential directions is described by dashpots with the same coefficient ρc_s (shear-wave velocity c_s).

In the *frequency domain*, the *condition for uniqueness*, called the *radiation condition*, is formulated at an infinite distance from the structure-medium interface [47]. The condition of vanishing displacement amplitude at infinity is not sufficient. The radiation condition states that *no energy is radiated from infinity towards the structure-medium interface*. The unbounded medium thus acts as an *energy sink* and never as an energy source.

For the scalar wave equation, the radiation condition of an infinite medium is formulated with the radial coordinate r as

$$\lim_{r \to \infty} r^{\frac{s-1}{2}} \left(u(\omega),_r + \frac{i\omega}{c} u(\omega) \right) = 0 \tag{1.3}$$

with the displacement amplitude $u(\omega)$, wave velocity c and the spatial dimension s ($= 2$ or $= 3$). Note that the expression in the parenthesis of Eq. 1.3 can be interpreted as Eq. 1.2 using $c = c_p = \sqrt{(\lambda + 2G)/\rho}$ and $p(\omega) = -(\lambda + 2G)u(\omega),_r$ (Lamé constants λ, G).

The most striking feature in an unbounded medium, which is never encountered in a bounded medium, is the presence of radiation damping in a frequency domain analysis of an elastic system. Mathematically, the corresponding dynamic stiffness at the structure-medium interface describing the relationship between the amplitudes of the displacements and those of the forces is complex. Physical insight can be gained by examining the elastic restoring force and the inertial force of the medium as the radial coordinate increases towards infinity. The elastic restoring force tends to maintain the original position of the medium, and the inertial force favours movement away from the original position which transmits energy to the adjacent medium. When the inertial force dominates over the elastic restoring force, radiation damping will occur. This is the case commonly encountered. When the restoring force dominates, no radiation damping will arise. Thus, (elastic) unbounded media exist where no radiation damping appears. The inertial force is proportional to the square of the excitation frequency. When the elastic restoring force and the inertial force are the same function of the radial coordinate, for sufficiently small and large frequencies radiation damping will vanish and be present, respectively. The frequency at which radiation first occurs is called the *cut-off frequency*. These important concepts are explained in depth in Section 5.2.12.

1.3 SUBSTRUCTURE METHOD AND DIRECT METHOD

The bounded structure of finite dimensions can be modelled using the finite-element method or some other domain procedure introducing a finite number of degrees of freedom. This part of the analysis is well understood, permitting even strong nonlinearities to be processed.

In contrast, the unbounded medium of infinite dimensions cannot be modelled

1.3 SUBSTRUCTURE METHOD AND DIRECT METHOD

with (a finite number of) finite elements without special treatment. A surface which encloses the structure called the *interaction horizon* is chosen. The numerical dynamic model encompasses those nodes which all lie within or on the horizon. On the interaction horizon a boundary condition is formulated representing the unbounded medium exterior to this boundary. A certain arbitrariness exists when selecting the location of the interaction horizon, which actually has no physical significance. Two extremes are possible. First, the interaction horizon can coincide with the structure-medium interface leading to the *substructure method* (Figure 1-2a). In this case, in

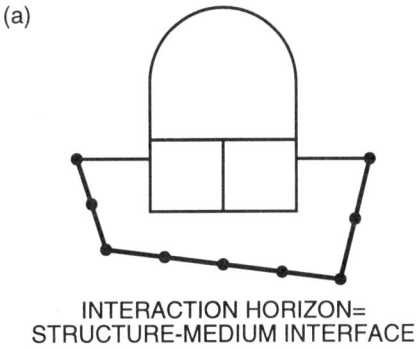

(a) INTERACTION HORIZON= STRUCTURE-MEDIUM INTERFACE

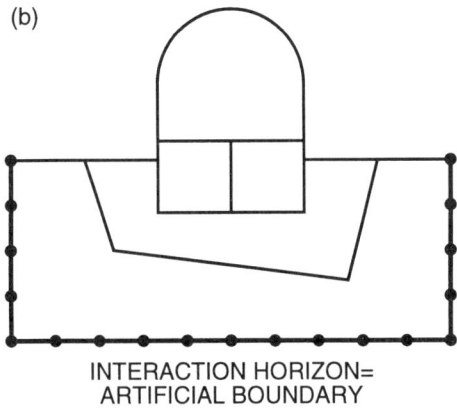

(b) INTERACTION HORIZON= ARTIFICIAL BOUNDARY

Figure 1-2 Dynamic model of unbounded medium-structure-interaction analysis: (a) substructure method; (b) direct method

general, the *rigorous boundary condition* has to be enforced to achieve sufficient accuracy. Second, part of the unbounded medium adjacent to the structure-medium interface is modelled with finite elements up to an artificial boundary. The latter must be introduced, as it is impossible to cover the unbounded medium with a finite number of finite elements with bounded dimensions. The artificial boundary is identical with the interaction horizon (Figure 1-2b). This results in the *direct method*. An *approximate* and thus simpler *boundary condition* can then be applied. In general, accurate results in the structure are obtained for a sufficiently large finite-element mesh.

The boundary condition formulated in the degrees of freedom of the nodes on the interaction horizon is equal to the *interaction force R - displacement u relationship* of the unbounded medium on the exterior (Figure 1-3).

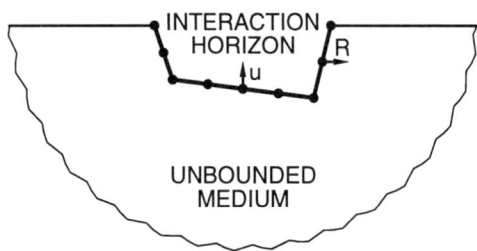

Figure 1-3 Interaction force-displacement relationship of unbounded medium formulated as boundary condition on interaction horizon

In a time-domain analysis the rigorous form of this stiffness relationship is *global in space and time*. The interaction force of a specific degree of freedom at a specific time depends on the displacements corresponding to all degrees of freedom at all previous times from the start of the excitation onwards. This follows from the fact that an enforced unit-impulse displacement in a degree of freedom will yield interaction forces in all degrees of freedom (spatial coupling) also at later time (temporal coupling). As a consequence, the interaction-force vector of all degrees of freedom in the nodes on the interaction horizon at a specific time t $\{R(t)\}$ is equal to the convolution integral of the *unit-impulse response matrix of the unbounded medium* $[S^\infty(t)]$ and the corresponding displacement vector $\{u(t)\}$

$$\{R(t)\} = \int_0^t [S^\infty(t-\tau)]\{u(\tau)\}d\tau \qquad (1.4)$$

The superscript ∞ is used to denote the unbounded medium. At this stage of the development, Eq. 1.4 should be regarded as a formal expression only. The detailed derivation and discussion are deferred to Section 2.1.

In a frequency-domain analysis the interaction force-displacement relationship with the corresponding amplitudes is written for a frequency ω as

$$\{R(\omega)\} = [S^\infty(\omega)]\{u(\omega)\} \qquad (1.5)$$

The fully coupled matrix $[S^\infty(\omega)]$ represents the *dynamic-stiffness matrix*.

The rigorous form of the interaction force-displacement relationship is used in the substructure method of the unbounded medium-structure-interaction analysis, providing the same accuracy as for the discretized structure. Assembling the interaction force-displacement relationship of the unbounded medium with the equations of motion of the structure leads to the basic equations of the total dynamic system [58].

When the loading is introduced into the dynamic system through the unbounded medium as in the case of waves from earthquakes, the so-called *scattered motion* can be of interest. The scattered motion is defined as the displacements in the nodes on the structure-medium interface of the unbounded medium only (Figure 1-3), i.e. without the presence of the structure, caused by the free-field motion of the incident waves. As is well known (see e.g. Reference [58]), the scattered motion is determined using standard matrix operations from the free-field response using the unit-impulse response matrix $[S^\infty(t)]$ or the dynamic-stiffness matrix $[S^\infty(\omega)]$. In this book only the methods for calculating $[S^\infty(t)]$ and $[S^\infty(\omega)]$ are discussed.

In order to reduce the computational effort arising from the global coupling of the rigorous form of the interaction force-displacement relationship of the unbounded medium, a simple but approximate form can be constructed. This stiffness relationship in a time-domain analysis is *local in space and time*. It uses information only from the specific node or its nearby region of the finite-element mesh at a specific time or, at most, during a limited past time. This form is applied in the direct method of the unbounded medium-structure-interaction analysis, where it is often called a *transmitting boundary*. As one would intuitively expect, results of sufficient accuracy in the structure can be determined when the distance of the structure-medium interface from the artificial boundary is chosen sufficiently large.

The direct method of analysis permits a transient excitation to be processed directly in the time domain without calculating convolution integrals.

1.4 STATE OF THE ART

To be able to evaluate the novel finite-element based modelling of the unbounded medium addressed in this book, it is appropriate to briefly review the state of the art. Not all methods which have been developed can be discussed. An accurate historical review of the development of dynamic unbounded medium-structure-interaction analysis lies outside the scope of this book. Preference is given to those methods which are widely applied, especially in soil-structure-interaction analysis. The results of these procedures are also used for comparison with those calculated by the novel methods of the book. In addition, a selection of historical references as well as some review papers and books are mentioned.

In order to classify the methods, rigorous and approximate modelling procedures of the unbounded medium are addressed separately in the following. In dynamic soil-structure interaction, review papers of general interest include [31], [29], [43], [59], and the text books [57], [58] can be mentioned. Books addressing many fields of applications also exist [17], [7].

1.4.1 Rigorous Modelling

The established rigorous procedures used in the substructure method of analysis (Section 1.3) are addressed. To model the unbounded medium exhibiting dynamic characteristics which differ from those of the bounded structure, the interaction force-displacement relationship in the degrees of freedom of the nodes on the structure-medium interface is established. This relationship represents the significant dynamic features of the unbounded medium located on the exterior of this interface. In particular, the radiation condition must be satisfied which enforces that no energy is radiated from infinity towards the structure (Section 1.2). This radiation condition can be incorporated straightforwardly in certain analytical solutions. The latter can be used as a fundamental solution to formulate the *boundary-integral equation*, which in discretized form is called the *boundary-element method*. A review paper [6] and several books [9], [34], [15], [4] are available. A detailed description will thus not be given here. It is sufficient to state that only the boundary has to be discretized as the governing differential equations are satisfied exactly. This leads to a *reduction of the spatial dimension by one*. Compared with a domain procedure, e.g. the finite-element method, the mesh generation is simpler and the resulting system of equations is smaller. One of the most suitable problems to be solved with the boundary-element method is the dynamic modelling of an unbounded medium, as the *radiation condition is satisfied automatically* as part of the fundamental solution. The method is exact, i.e. no error remains as the size of the boundary elements becomes infinitesimal. Unfortunately, the fundamental solution is not always available. For instance, for certain anisotropic materials the fundamental solution of even the full-space is not known. Except for the homogeneous isotropic full-space, the fundamental solution in elasto-dynamics is very complicated, making the boundary-element method less attractive. For instance, the fundamental solution for a homogeneous isotropic half-space with a free surface involves infinite integrals of Bessel functions. As a further disadvantage, the (non-symmetric) coefficient matrices of the final system of equations are difficult to evaluate: singularities and special functions (not encountered in the finite-element method) arise. Only a finite part of the boundaries which extend to infinity such as the free surface and the interfaces between two different materials can be discretized. This truncation results in errors.

Although the boundary-element method is regarded as the most powerful procedure for modelling the unbounded medium in many areas of engineering and science, it requires a strong analytical and numerical background, which the engineer who is

1.4 STATE OF THE ART

familiar with finite elements has to acquire.

As another rigorous procedure the *consistent boundary*, also called the thin-layer method, is discussed [30], [33], [55], [23], [24]. The method which is formulated in the frequency domain applies only to a horizontally layered unbounded medium. It makes use of the exact displacement functions in the horizontal direction that satisfy the radiation condition and an expansion in the vertical direction consistent with that used for finite elements. The consistent-boundary formulation is *exact in the horizontal direction and converges to the exact solution in the finite-element sense in the vertical direction*. It is based on the finite-element methodology and thus does not require a fundamental solution. This method is well suited to process horizontal layers with material properties varying in the vertical direction. The boundary conditions on the free surface and on the interfaces between adjacent layers are rigorously satisfied with the same computational effort as for a homogeneous medium. Many extensions exist, e.g. the formulation for a poro-elastic unbounded medium is available [8].

1.4.2 Approximate Modelling

In the direct method of analysis (Section 1.3) results of acceptable accuracy in the structure can be obtained by using an approximate boundary condition on the artificial boundary when chosen sufficiently far away from the structure-medium interface. The radiation condition which should be enforced at infinity is formulated approximately on the artificial boundary in such a way that a highly absorbing boundary condition results. This *transmitting boundary* is, in general, local in space and time and is formulated directly in the time domain.

Obviously, many possibilities exist to construct approximate solutions. In most cases the transmitting boundaries are based on the theory of wave propagation, enforcing *outgoing plane waves*. They appear to be vastly different from each other in their conceptual basis, mathematical formulation and numerical implementation. However, in the limit represented by the continuum, most transmitting boundaries are equivalent, corresponding to the same kind of differential equations but with varying coefficients [22].

The most popular transmitting boundaries are mentioned in the following. The *viscous boundary* [32] consists of dashpots which absorb plane waves propagating perpendicularly to the artificial boundary. The *superposition boundary* [46], [13] averages the solutions from two sets of boundary conditions corresponding to symmetry and anti-symmetry which eliminates the reflected waves for a single boundary. The *paraxial boundary* [16], [12] constructs a differential equation which favours outgoing waves. The *doubly-asymptotic boundary* [54] uses viscous dashpots and coupled static springs and is asymptotically exact at low frequencies and at high frequencies for waves propagating perpendicularly to the artificial boundary. Besides this first-order doubly-asymptotic boundary a second-order formulation also exists [37]. The *extrapolation boundary* [27] calculates the displacements at the artificial boundary by extrapolating the data in the interior nodes at earlier times which requires an estimate of the propagation velocity. The *multi-directional boundary* [25], [19], [20] constructs a differential equation which absorbs plane waves at preselected angles.

The unbounded medium can also be modelled approximately with *infinite elements* [7]. This represents an extension of the finite-element method. Decay functions representing the wave propagation towards infinity are used as the shape functions of the displacements. The decay rate and the phase velocity must be specified in the frequency domain. The infinite-element method is not exact in the finite-element sense. Thus, for an infinitesimal element size, an error caused by the modelling of the unbounded medium remains. An analysis directly in the time domain is not feasible, as the shape functions cannot represent the displacements which in the time domain can exhibit spatially any pattern.

1.5 OVERVIEW OF NOVEL METHODS

As is apparent from the state-of-the-art reviews (Section 1.4), the finite-element modelling of the unbounded medium has not progressed sufficiently to analyse general dynamic medium-structure interaction as encountered in practice accurately and efficiently. The structure is routinely discretized with finite elements and this is regarded as the appropriate method for modelling nonlinear behaviour working in the time domain. Engineers and scientists are familiar with the finite-element method. It is desirable to develop *finite-element based procedures for the unbounded medium* and this is the aim of this book. Three such methods are addressed:

1. The *forecasting method* and the *consistent infinitesimal finite-element cell method*, both based on similarity, which converge to the exact solution in the finite-element sense and are used in the substructure method in the time or frequency domain.
2. The *damping-solvent extraction method* which is highly accurate and is used in the substructure method in the frequency or time domain.
3. The *doubly-asymptotic multi-directional transmitting boundary* which is highly accurate and is applied in the direct method in the time domain.

A brief superficial description with the characteristic features follows, which cannot be fully understood at this stage. Detailed discussions, derivations and examples can be found in the following chapters.

1.5 OVERVIEW OF NOVEL METHODS

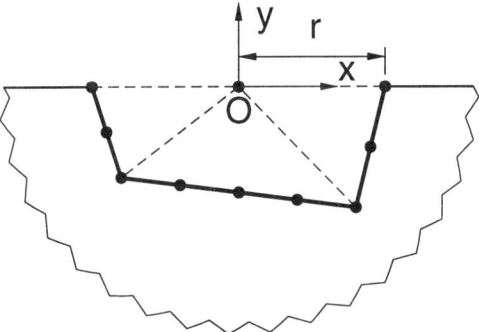

Figure 1-4 Unbounded medium with structure-medium interface defined by characteristic length and similarity centre

1.5.1 Similarity-based Formulation

In the substructure method of analysis (Section 1.3, Figure 1-2a) the dynamic properties of the unbounded medium are represented in the time domain by the unit-impulse response matrix and in the frequency domain by the dynamic-stiffness matrix corresponding to the degrees of freedom in the nodes on the structure-medium interface (Figure 1-4). These two matrices permit the calculation of the interaction forces of the unbounded medium, as specified in Eqs. 1.4 and 1.5.

To explain the concept, *similarity* is addressed in the context of an unbounded medium. A two-dimensional homogeneous half-plane with a discretized structure-medium interface is used as an example (Figure 1-4). A similarity centre denoted as O is selected outside the unbounded medium whereby the two free surfaces have to pass through it. The origin of the coordinates x, y coincides with the similarity centre. To construct a similar structure-medium interface, all coordinates of the nodes on the structure-medium interface are multiplied by a non-zero constant, the similarity factor. Thus, the characteristic length r can be used to define the structure-medium interface (other choices for the characteristic length are possible which are all proportional to r). It will be proven that starting from the unit-impulse response matrix (or the dynamic-stiffness matrix) at a structure-medium interface of characteristic length r that at any other similar structure-medium interface of the same unbounded medium can be calculated via a dimensionless variable. Thus, a relationship linking the unit-impulse response matrices (or the dynamic-stiffness matrices) at two similar structure-medium interfaces of the same unbounded medium can be formulated. It will be demonstrated that another relationship follows from the standard finite-element assemblage at the two structure-medium interfaces. The unit-impulse response matrix (or the dynamic-stiffness matrix) can then be obtained from these two relationships. *This similarity-*

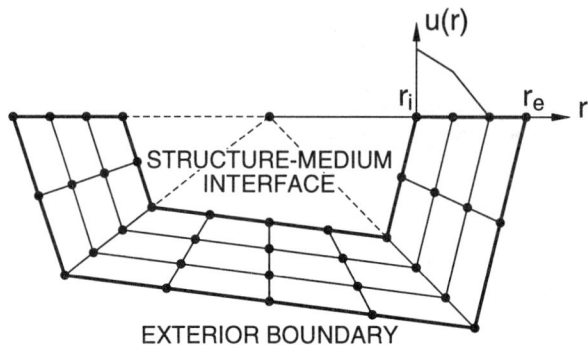

Figure 1-5 Finite-element region between similar interior and exterior boundaries used in forecasting method

based formulation is a stand-alone finite-element method capable of capturing the radiation condition at infinity without using analytical solutions.

It is worth mentioning that the similarity-based formulation can also calculate an unbounded medium whose material properties vary as a power function of the radial direction r measured from the similarity centre to infinity.

In the pioneering paper of the so-called *cloning* algorithm [14] to determine the dynamic-stiffness matrix, an average value of the two characteristic lengths is used to calculate the dimensionless variable when formulating the similarity-based relationship. In the general case of varying characteristic lengths incorrect results are obtained. In the *generalized cloning* method [71] the variation of the characteristic length is considered yielding a rigorous procedure.

For the unit-impulse response matrix two implementations exist: the forecasting method and the consistent infinitesimal finite-element cell method. The concepts of these two methods are discussed, and a brief historical review is included.

In the *forecasting method* a finite-element region adjacent to the structure-medium interface consisting of 2 or 3 rows only in the radial direction is introduced (Figure 1-5). Its exterior boundary is chosen to be similar to its interior boundary which coincides with the structure-medium interface. The forecasting method is first explained for the beginning of the analysis starting from time $t = 0$. As illustrated in Figure 1-5, it takes time for a wave to propagate from the interior boundary to the exterior boundary (time delay). The boundary condition at the exterior boundary which is at rest is formulated straightforwardly. This permits the unit-impulse response matrix at the interior boundary to be calculated before the wave reaches the exterior boundary. The *unit-impulse response matrix at the exterior boundary is forecasted from the unit-impulse response matrix at the interior boundary.* It then provides the consistent representation of the unbounded medium with an interface characterized by r_e, and thus the boundary condition at the exterior boundary of

1.5 OVERVIEW OF NOVEL METHODS

the finite-element region for times after the wave reaches the exterior boundary. The procedure also works for later times. Conceptually, the time is subdivided into time segments, which are shorter than the propagation time from the interior boundary to the exterior boundary. Figure 1-5 thus still applies to the latest time segment. Only the standard banded symmetric system of equations of the finite-element method is solved. No approximation in modelling the unbounded medium other than that of the finite-element method is introduced. Therefore, the forecasting method *converges to the exact solution in the finite-element sense*. As use is made of the time delay resulting from the finite wave velocity, the procedure can only be used for wave-propagation problems [49].

In the *consistent infinitesimal finite-element cell method* one row of finite elements in the radial direction, called the finite-element cell, is introduced adjacent to the structure-medium interface in the derivation (Figure 1-6). Again its exterior boundary is chosen to be similar to its interior boundary which coincides with the structure-medium interface. The unit-impulse response matrix at the exterior boundary is calculated from that at the interior boundary using the relationship based on similarity. The *limit of infinitesimal cell width is performed analytically* in the relationship of the standard finite-element assemblage. Substituting the relationship based on similarity results in the consistent infinitesimal finite-element cell equation which can be used to calculate the unit-impulse response matrix starting from $t = 0$. The same concept can also be applied to calculate the dynamic-stiffness matrix at decreasing frequencies ω starting from a very large ω.

In an actual calculation the discretization with surface finite elements is limited to the structure-medium interface (Figure 1-7). This results in a *reduction of the spatial dimension by one* as in the boundary-element method. But the consistent infinitesimal finite-element cell method is based solely on the finite-element formulation and does not require a fundamental solution. Any finite-element discretization can be applied. The coefficient matrices of the consistent infinitesimal finite-element cell equation depend on the geometry of the structure-medium interface and on the material properties of the unbounded medium. The procedure is *exact in the radial direction and converges to the exact solution in the finite-element sense in the circumferential directions*. As the discretization and the unknowns are only introduced on the boundary, the consistent infinitesimal finite-element cell method can be regarded as a boundary-element method based on finite elements called the *boundary finite-element method*.

The consistent-boundary method [55] for analysing a two-dimensional horizontally layered medium (in the frequency domain) is a special case of the consistent infinitesimal finite-element cell method (in the time domain). The characteristic lengths of the interior and exterior boundaries are the same for the layered medium. This corresponds to placing the centre of similarity at infinity.

The analytical limit of the infinitesimal cell width is performed for one-dimensional

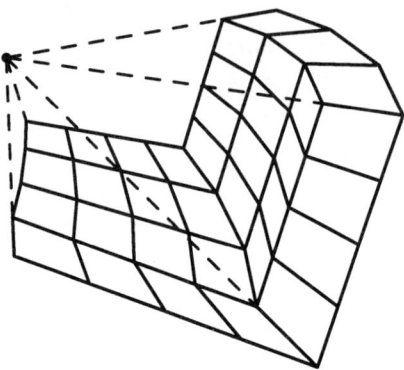

Figure 1-6 Cell of finite elements with similar interior and exterior boundaries used in derivation of consistent infinitesimal finite-element cell method

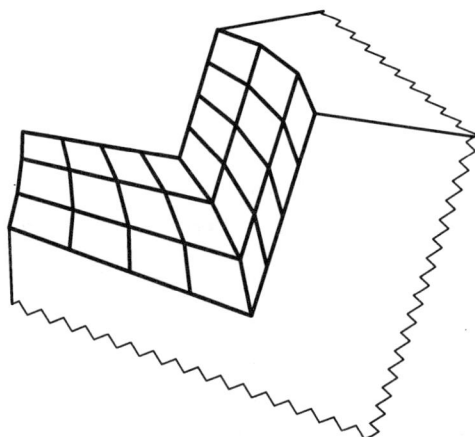

Figure 1-7 Finite-element discretization of structure-medium interface used in application of consistent infinitesimal finite-element cell method

wave propagation in Reference [71]. In the two- and three-dimensional cases the formulation is derived for the scalar wave equation [50], [70], the compressible vector wave equation [68], [51] and for the incompressible vector wave equation [53]. The static problem follows as a special case of wave propagation [69]. The diffusion equation is also addressed [52].

Common features of the two similarity-based methods are outlined and compared with the properties of the boundary-element method.

The similarity-based formulations satisfy the radiation condition which is also the case for the boundary-element method.

As the similarity-based procedures use polynomial shape functions as in the

1.5 OVERVIEW OF NOVEL METHODS

conventional finite-element method, the integrations are simple to perform. This is in contrast to the integrations of singular functions occurring in the fundamental solutions of the boundary-element method.

The unit-impulse response matrix and the dynamic-stiffness matrix resulting from the similarity-based formulations are symmetric. This does not apply to the results of the conventional boundary-element method where special and more complicated formulations are necessary to enforce symmetry. Symmetry is required to perform an unbounded medium-structure-interaction analysis consistently where the structure is modelled with finite elements.

In an unbounded medium-structure-interaction analysis, the unit-impulse response matrix or the dynamic-stiffness matrix has to be calculated. When only the unbounded medium is to be analysed, the boundary-element method permits a loading case to be processed directly without first determining the unit-impulse response matrix or the dynamic-stiffness matrix.

For the sake of simplicity, a homogeneous unbounded medium is assumed to explain similarity in Figure 1-4. However, the relationship based on similarity is also applicable to the inhomogeneous case compatible with similarity. An example with inhomogeneities in shear modulus G, Poisson's ratio ν and mass density ρ is shown in Figure 1-8. Obviously, the other relationship describing the assemblage of

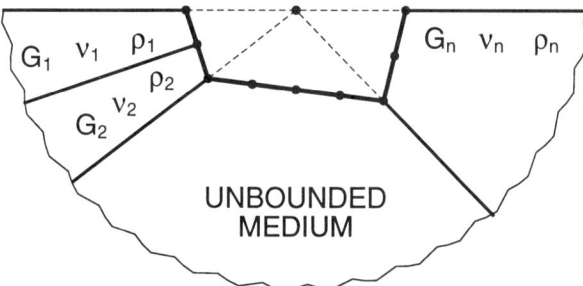

Figure 1-8 Unbounded medium with material inhomogeneity satisfying similarity

finite elements is also valid for inhomogeneities. Thus, the forecasting method and the consistent infinitesimal finite-element cell method which are based on these two relationships can be used. The boundary conditions at the interfaces and at the free surface are satisfied rigorously and automatically without any further discretization. Material inhomogeneities which satisfy similarity and the boundary condition at the free surface can thus be processed without any increase in the number of degrees of freedom. In contrast, in the boundary-element method, the interfaces and the free surface must be discretized, which increases the computational effort. In addition, on these boundaries extending to infinity, truncation of the discretization has to be introduced in the boundary-element method and this leads to errors. These are difficult to evaluate.

The two finite-element implementations based on similarity can also process materials with anisotropic behaviour and with properties varying in the radial direction as a power function straightforwardly, whereby the computational effort is hardly affected. In contrast, the fundamental solution of the boundary-element method is very complicated in the orthotropic case and does not exist in the general anisotropic case.

The consistent infinitesimal finite-element method and the boundary-element method lead to a reduction of the spatial dimension by one in the spatial discretization. As the discretization is limited to the structure-medium interface, the stresses on this boundary are calculated in both methods directly and more accurately than in the standard finite-element method.

In the forecasting method the banded symmetric equations of finite elements are solved while in the boundary-element method a non-symmetric fully populated coefficient matrix (of the order of the number of degrees of freedom on the structure-medium interface) of the equations occurs. As will be discussed, in the consistent infinitesimal finite-element cell method a quadratic matrix equation of the order of the number of degrees of freedom is solved once.

Both the forecasting method and the consistent infinitesimal finite-element cell method require similarity of the unbounded medium, which is not always satisfied in practice. An example with a layered unbounded medium (Figure 1-9a) and another with strongly inclined interfaces between the parts of a half-plane (Figure 1-10a) are used for illustration. By moving the structure-medium interface outwards, an approximate representation of the unbounded medium satisfying similarity (Figures 1-9b and 1-10b) can be constructed. The farther away the structure-medium interface is chosen, the better the approximation becomes. This procedure should be compared with that of truncating the discretization of the interfaces extending to infinity as used in the boundary-element method.

The consistent infinitesimal finite-element cell method can be applied to a *bounded medium* to determine the static-stiffness and mass matrices with the degrees of freedom defined on the boundary.

1.5.2 Damping-solvent Extraction Method

In the substructure method of analysis (Section 1.3, Figure 1-2a) the dynamic properties of the unbounded medium are represented in the frequency domain by the dynamic-stiffness matrix defined at the structure-medium interface.

The concept of the *damping-solvent extraction method* [48] is best explained starting with a familiar process. To extract salt from the ground, water can be injected into the ground. The salt dissolves in the water which can then be pumped to the surface. After evaporation of the water, the salt remains. To calculate the dynamic-

1.5 OVERVIEW OF NOVEL METHODS

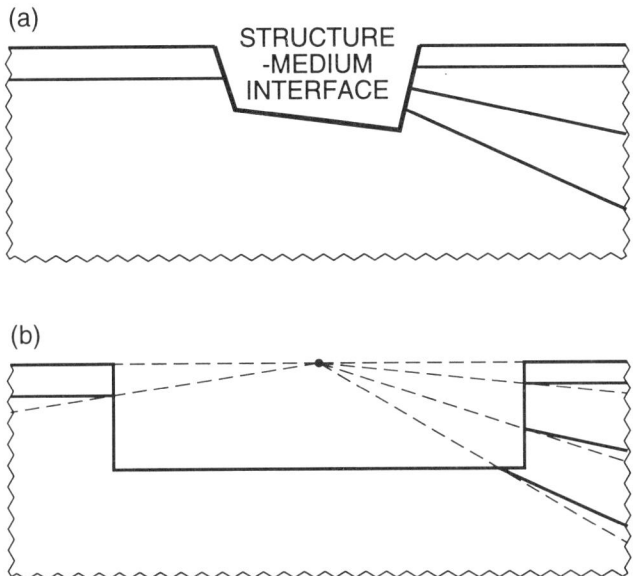

Figure 1-9 (a) Original dynamic unbounded medium-structure-interaction problem with layers; (b) Approximate representation enforcing similarity of unbounded medium with structure-medium interface moved outwards

stiffness matrix in the frequency domain of the undamped unbounded medium, damping is used as a solvent. Three steps are involved in the procedure.

In the *first step*, a *finite region of the unbounded medium* adjacent to the structure, a bounded medium, is modelled with *finite elements* (Figure 1-11), whereby *damping which is not present in the actual medium is introduced artificially as a solvent*. The effect of this damping consists of reducing the amplitudes of the outgoing waves f propagating from the structure-medium interface towards the outer boundary and after reflection, diminishing the amplitudes of the reflected waves g, resulting in negligible amplitudes when reaching the structure-medium interface. The damping acting as a solvent thus leads to the structure-medium interface's motion depending only on the outgoing waves f. The dynamic-stiffness matrix of the artificially damped bounded medium follows straightforwardly by eliminating all degrees of freedom which are not located on the structure-medium interface.

In the *second step*, the dynamic-stiffness matrix of the artificially damped bounded medium determined in the first step is assumed to be equal to the *dynamic-stiffness matrix of the unbounded medium with the same introduced artificial damping*. (The same is also assumed to apply for their first derivatives with respect to frequency).

In the *third step*, the influence of the *introduced artificial damping*, the solvent, on the dynamic-stiffness matrix *is extracted*. As will be demonstrated, this elimination of

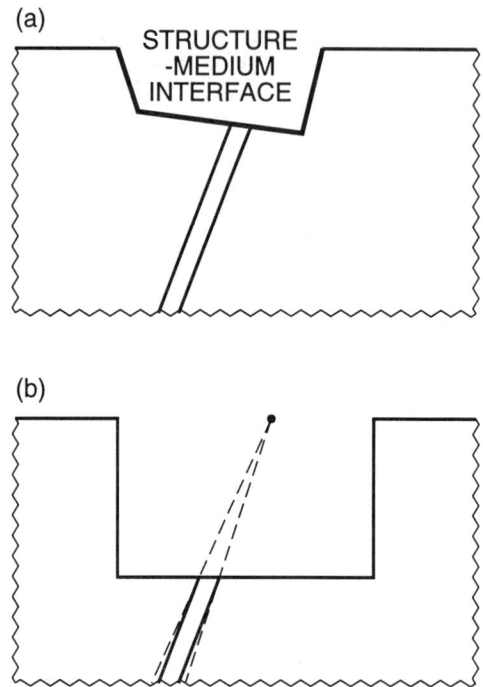

Figure 1-10 (a) Original dynamic unbounded medium-structure-interaction problem with inclined interfaces; (b) Approximate representation enforcing similarity of unbounded medium with structure-medium interface moved outwards

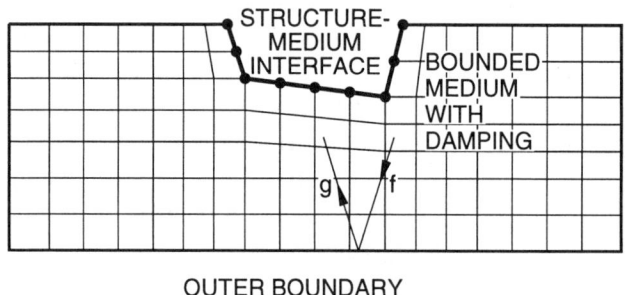

Figure 1-11 Finite-element discretization of bounded medium in first step of damping-solvent extraction method

1.5 OVERVIEW OF NOVEL METHODS

the damping solvent can be performed for each element of the matrix independently of the others and for each frequency using a Taylor expansion, resulting in a negligible computational effort. This results in an approximation of the dynamic-stiffness matrix of the undamped unbounded medium.

The damping-solvent extraction method permits an efficient calculation of the dynamic-stiffness matrix of an unbounded medium by analysing the adjacent bounded medium only, which exhibits the same dynamic characteristics as the (bounded) structure. It must also be stressed that the dynamic-stiffness matrix at any specific frequency of interest can be calculated without processing the full range of frequencies.

The damping-solvent extraction method can also be applied directly in the time domain. It contrast to the similarity-based methods where unit-impulse response functions have to be first determined, a transient response can be calculated directly without processing a convolution integral.

1.5.3 Doubly-asymptotic Multi-directional Transmitting Boundary

In the direct method of analysis in the time domain (Section 1.3, Figure 1-2b), to be able to represent that part of the medium on the outside of the artificial boundary which is not explicitly modelled, a highly absorbing boundary which is local in space and time called transmitting boundary is needed.

The *multi-directional boundary* [20] absorbs plane waves propagating outwardly at preselected angles. The assumed propagation of plane waves does not cover the static case and evanescent waves such as exponentially decaying displacement patterns which do not propagate. Satisfactory behaviour for these cases is achieved using the *doubly-asymptotic boundary* [54], which is rigorous for the low-frequency limit, i.e. for the static case, and for the high-frequency limit of waves propagating perpendicularly to the artificial boundary.

To construct the *doubly-asymptotic multi-directional transmitting boundary* [67], the multi-directional outward plane wave boundary condition is formulated for the interaction forces (and not for the displacements), whereby the contributions of the two limits covered rigorously by the doubly-asymptotic boundary are subtracted beforehand. The resulting boundary condition is discretized with finite differences leading to an explicit formulation, which is straightforwardly implemented in the finite-element method.

The doubly-asymptotic multi-directional transmitting boundary combines the advantages of the doubly-asymptotic and multi-directional formulations. It is *rigorous for the low-frequency limit and the high-frequency limit in the wave-propagation direction perpendicular to the artificial boundary and at all the preselected angles. It is highly accurate for plane waves at intermediate frequencies and at other angles.*

1.6 ORGANIZATION OF TEXT

The following chapters are divided into three parts which can be read independently from one another.

Part I addresses the rigorous procedures based on similarity for calculating the unit-impulse response and the dynamic stiffness of the unbounded medium for use in the substructure method of dynamic unbounded medium-structure-interaction analysis in the time and frequency domains. The contents of the various chapters are illustrated in the flow chart (Figure 1-12). For a specific problem the arrows indicate which chapters should be consulted. Chapter 2 discusses the relationship of the response functions to unit-impulses of displacement, velocity and acceleration in the time domain and the corresponding dynamic stiffness in the frequency domain. Chapter 3 addresses the relationship of the dynamic stiffnesses and unit-impulse responses at similar structure-medium interfaces of an unbounded medium. Chapter 4 describes the forecasting method for a time-domain analysis of wave propagation. The remaining chapters of Part I explain the consistent infinitesimal finite-element cell method. Chapter 5 is devoted to the calculation of the unit-impulse response for wave propagation. Chapter 6 presents the extension to incompressible elasticity. Chapter 7 covers the frequency-domain analysis. Chapter 8 addresses the statics. Chapter 9 deals with diffusion. Chapter10 applies the consistent infinitesimal finite-element cell method to a bounded medium. Chapters 2 to 10 on the similarity-based formulation make up the largest part of the text.

Part II discusses the approximate damping-solvent extraction method for calculating the dynamic stiffness of the unbounded medium for use in the substructure method in the frequency domain and for determining the interaction forces in the time domain for a transient excitation. Chapter 11 describes the concept. Chapter 12 presents the verifications, the implementation in the frequency and time domains and demonstrates the high accuracy.

Part III develops the doubly-asymptotic multi-directional transmitting boundary for use in the direct method in the time domain. Chapter 13 explains the concept and the numerical implementation. Chapter 14 evaluates the performance compared with that of other transmitting boundaries.

Appendix A describes the benchmark examples with exact solutions which are used for comparison throughout the text.

Appendix B contains the description of a computer programme of the consistent infinitesimal finite-element cell method which is available on a floppy disk.

1.6 ORGANIZATION OF TEXT

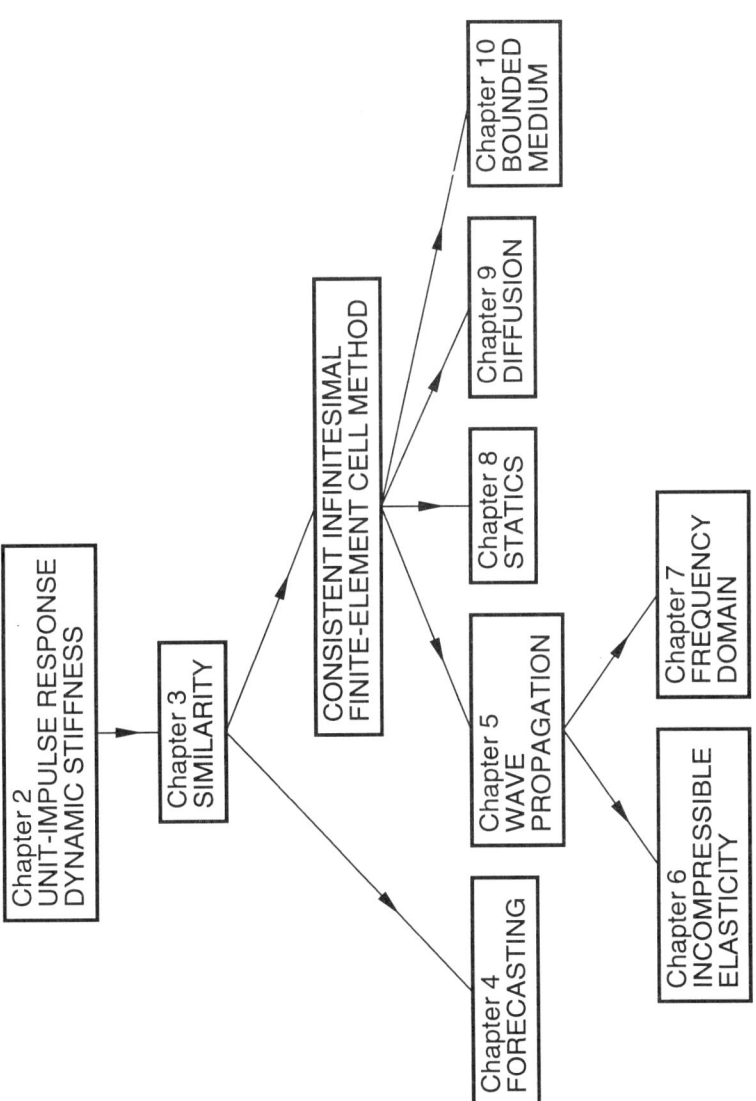

Figure 1-12 Content of chapters related to similarity-based formulation

Part I

SIMILARITY-BASED FORMULATION FOR UNIT-IMPULSE RESPONSE AND DYNAMIC STIFFNESS

Le fini s'anéantit en présence de l'infini …

(The finite vanishes in the presence of the infinite …)

Blaise Pascal   1623–1662

To calculate the interaction force-displacement relationship of the degrees of freedom in the nodes on the structure-medium interface of the unbounded medium for use in the substructure method of dynamic medium-structure-interaction analysis, the rigorous formulation based on similarity and the finite-element method is developed. Two implementations exist: the forecasting method and the consistent infinitesimal finite-element cell method, which are both based on the same concept.

The common concept is illustrated in Figure I-1 for the two-dimensional case. The

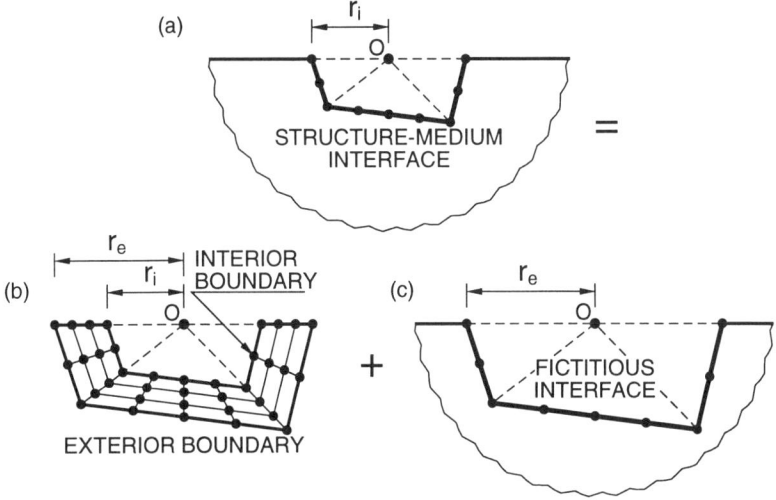

Figure I-1 Concept of similarity-based formulation

unbounded medium with the structure-medium interface is indicated in Figure I-1a. The nodes on the structure-medium interface coincide with those of the finite-element discretization of the structure (Figure 1-1). To make use of similarity, another *similar* fictitious interface is conceptually introduced by multiplying the coordinates of the structure-medium interface referred to the similarity centre O by a similarity factor (see discussion in connection with Figure 1-4 in Section 1.5.1). The similar interfaces are defined by the characteristic length r, with r_i for the structure-medium interface and r_e for the fictitious interface (Figure I-1c). The similarity factor thus equals r_e/r_i. The bounded region between these two interfaces is discretized with finite elements (Figure I-1b) with its *interior boundary* (subscript i for interior) coinciding with the structure-medium interface and its (similar) *exterior* boundary (subscript e for exterior) with the fictitious interface. The remaining part of the unbounded medium which has not been discretized with finite elements can be interpreted as the same unbounded medium but with the fictitious interface representing another structure-medium interface (Figure I-1c). The latter system is denoted as the unbounded medium with the interface characterized by r_e, called the unbounded medium with r_e

for short. Analogously, the unbounded medium with the structure-medium interface characterized by r_i is denoted as the unbounded medium with r_i. Obviously, adding the bounded region to the unbounded medium with r_e results in the similar unbounded medium with r_i. This concept can be applied to their dynamic-stiffness matrices in the frequency domain. Assembling the dynamic-stiffness matrix of the bounded region discretized with finite elements (which is straightforwardly determined from the static-stiffness matrix and mass matrix) and the unknown dynamic-stiffness matrix of the unbounded medium with r_e results in the unknown dynamic-stiffness matrix of the unbounded medium with r_i. This *assemblage enforcing compatibility and equilibrium* results in one set of equations linking the dynamic-stiffness matrices of the unbounded medium with r_i and of unbounded medium with r_e. The other set of equations follows from *similarity* of the unbounded medium with r_i and unbounded medium with r_e, as will be demonstrated. From these two sets of equations the *dynamic-stiffness matrices of the unbounded medium with r_i and of unbounded medium with r_e can be expressed as a function of the static-stiffness matrix and the mass matrix of the bounded region*. Analogously, this applies also to the unit-impulse response matrices. The implementation of this concept in the forecasting method is different from that in the consistent infinitesimal finite-element cell method.

In Part I the following topics are discussed (see also Figure 1-12).

To calculate the interaction forces on the structure-medium interface of the unbounded medium in the time domain, the unit-impulse response matrix must be determined. The response matrix to a unit impulse of displacement, velocity or acceleration can be used. The relationships between these different unit-impulse response matrices and between the corresponding dynamic-stiffness matrices are discussed in Chapter 2.

Using dimensional analysis the relationships of the dynamic-stiffness matrices and the unit-impulse response matrices at similar structure-medium interfaces of an unbounded medium are derived in Chapter 3. Dimensionless frequency and time are introduced.

Chapters 2 and 3 are the prerequisites to understanding the forecasting method and the consistent infinitesimal finite-element cell method.

The forecasting method is developed for wave propagation analysed in the time domain in Chapter 4.

The consistent infinitesimal finite-element cell method is explained in the frequency and time domains in Sections 5.1 and 5.2 using wave propagation which serves also as the basis of Chapters 6 to 10. Wave propagation in the time domain is addressed in Sections 5.3 and 5.4.

The consistent infinitesimal finite-element cell method in the time domain is extended to wave propagation in incompressible elasticity by performing the limit of Poisson's ratio $\rightarrow 0.5$ analytically in Chapter 6.

The consistent infinitesimal finite-element cell method is applied to wave propagation in the frequency domain in Chapter 7. Methods working with a finite cell width called dynamic condensation and substructure deletion are also discussed.

The consistent infinitesimal finite-element cell method for statics is then examined in Chapter 8 as a special case which leads to a significant simplification.

The consistent infinitesimal finite-element cell method is formulated in Chapter 9 for diffusion, addressing time-domain and frequency-domain analyses.

Finally, the consistent infinitesimal finite-element cell method is applied to a bounded medium to determine the static-stiffness and mass matrices in Chapter 10.

2

DISPLACEMENT, VELOCITY AND ACCELERATION UNIT-IMPULSE RESPONSE WITH DYNAMIC STIFFNESS AND RATIONAL APPROXIMATION

In the substructure method of analysing dynamic unbounded medium-structure interaction, the coupling of the unbounded medium and the structure is performed via the interaction force-motion relationship in the degrees of freedom of the nodes on the structure-medium interface. As will become apparent, the interaction forces can be expressed as a function of the displacements, velocities or accelerations. As in conventional static structural analysis a displacement-based formulation is used, this approach is first addressed in the dynamic analysis of an unbounded medium.

The forecasting method and the consistent infinitesimal finite-element cell method are formulated using the response to a unit-impulse of acceleration, i.e. the so-called acceleration unit-impulse response. Thus, only the interaction force-acceleration relationship of Section 2.2 is actually applied.

This chapter is based on Reference [66].

2.1 INTERACTION FORCE-DISPLACEMENT RELATIONSHIP

The interaction force-displacement relationship of the unbounded medium is addressed (Figure 1-3). It is written in the frequency domain as

$$\{R(\omega)\} = [S^\infty(\omega)]\{u(\omega)\} \tag{2.1}$$

with the displacement amplitudes $\{u(\omega)\}$ and interaction force amplitudes $\{R(\omega)\}$. $[S^\infty(\omega)]$ is the *displacement dynamic-stiffness matrix* in the frequency domain. The superscript ∞ denotes the unbounded medium. Often, $[S^\infty(\omega)]$ is just called the dynamic-stiffness matrix. The interaction force-displacement relationship in the time domain is formulated as

$$\{R(t)\} = \int_0^t [S^\infty(t-\tau)]\{u(\tau)\}d\tau \tag{2.2}$$

for an unbounded medium that is assumed to be initially at rest

$$\{u(t=0)\} = 0 \tag{2.3a}$$
$$\{\dot{u}(t=0)\} = 0 \tag{2.3b}$$

Eq. 2.2 follows from Eq. 2.1 mathematically, as a product in the frequency domain corresponds to a convolution integral in the time domain. $[S^\infty(t)]$ is the *displacement unit-impulse response matrix* in the time domain. $[S^\infty(t)]$ and $[S^\infty(\omega)]$ form a Fourier transform pair

$$[S^\infty(t)] = \frac{1}{2\pi}\int_{-\infty}^{+\infty}[S^\infty(\omega)]e^{i\omega t}d\omega \tag{2.4}$$

Eq. 2.2 can also be derived physically from the definition of the unit-impulse response matrix $[S^\infty(t)]$. The contribution of the infinitesimal pulse $\{u(\tau)\}d\tau$ acting at time τ to the interaction forces $\{dR(t)\}$ at time t ($t > \tau$) depends on the displacement unit-impulse response matrix evaluated for the time difference $t-\tau$, $[S^\infty(t-\tau)]$, resulting in

$$\{dR(t)\} = [S^\infty(t-\tau)]\{u(\tau)\}d\tau \tag{2.5}$$

For the total time history, the interaction forces $\{R(t)\}$ follow as the integral of Eq. 2.5 over τ for $0 \leq \tau \leq t$, corresponding to a convolution integral of the displacement unit-impulse response matrix $[S^\infty(t)]$ and the displacement $\{u(t)\}$, as expressed in Eq. 2.2. This is, of course, analogous to the familiar Duhamel integral introduced in elementary structural dynamics in the context of a flexibility formulation.

Eq. 2.4 should be regarded as a formal result only, as $[S^\infty(\omega)]$ is not square integrable. To actually calculate $[S^\infty(t)]$, $[S^\infty(\omega)]$ is decomposed into the *singular* part $[S_s^\infty(\omega)]$, which is equal to its asymptotic value for $\omega \to \infty$, and the remaining *regular* part $[S_r^\infty(\omega)]$

$$[S^\infty(\omega)] = [S_s^\infty(\omega)] + [S_r^\infty(\omega)] \tag{2.6}$$

For compressible elasticity two matrices in the singular part appear

$$[S_s^\infty(\omega)] = i\omega[C_\infty] + [K_\infty] \tag{2.7}$$

The subscript ∞ on the right-hand side denotes the limit $\omega \to \infty$. The constant matrix $[C_\infty]$ corresponds to dashpot coefficients and the constant matrix $[K_\infty]$ to spring coefficients. The regular part $[S_r^\infty(\omega)]$ is square integrable.

2.1 INTERACTION FORCE-DISPLACEMENT RELATIONSHIP

The inverse Fourier transform (Eq. 2.4) of Eq. 2.6 with Eq. 2.7 results in the displacement unit-impulse response matrix

$$[S^\infty(t)] = [C_\infty]\dot\delta(t) + [K_\infty]\delta(t) + [S_r^\infty(t)] \tag{2.8}$$

where

$$[S_r^\infty(t)] = \frac{1}{2\pi}\int_{-\infty}^{+\infty}[S_r^\infty(\omega)]e^{i\omega t}d\omega \tag{2.9}$$

$\delta(t)$ represents the Dirac-delta function.

Substituting Eq. 2.8 in Eq. 2.2 leads to the interaction force-displacement relationship

$$\{R(t)\} = [C_\infty]\{\dot u(t)\} + [K_\infty]\{u(t)\} + \int_0^t [S_r^\infty(t-\tau)]\{u(\tau)\}d\tau \tag{2.10}$$

The first two terms on the right-hand side of Eq. 2.10 (corresponding to the singular part) represent the *instantaneous* response and the convolution integral (corresponding to the regular part) represents the *lingering* response. $[S^\infty(t)]$ in Eq. 2.8 is a formal expression containing singular functions. Eq. 2.10 instead of Eq. 2.2 is applied to calculate the interaction forces numerically.

For the initial time, the first term $[C_\infty]\{\dot u(t)\}$ in Eq. 2.10 dominates, which should be compared with Eq. 1.2 for the scalar case. Thus, $[C_\infty]$ can also be formulated based on the law of conservation of momentum. The force per unit area perpendicular to the structure-medium interface equals the product of the mass density ρ and the dilatational wave velocity c_p, multiplied by the normal velocity component. The forces per unit area tangential to the structure-medium interface equal ρc_s (shear-wave velocity c_s) multiplied by the tangential velocity components in the corresponding directions. The coefficients in the banded matrix $[C_\infty]$ follow from the shape functions of the finite-element discretization applying the principle of virtual work. Note that the high-frequency limit of the dynamic-stiffness matrix (Eq. 2.7) $i\omega[C_\infty]$ corresponds to the initial time response.

Eq. 2.10 permits an analysis of unbounded medium-structure interaction to be performed directly in the time domain. In the similarity-based procedures the singular part complicates the formulation and can introduce numerical difficulties in the forecasting method, as the singularity corresponding to a unit impulse in displacement cannot be discretized in time efficiently. As will be demonstrated, a formulation based on a convolution with the acceleration and the corresponding acceleration unit-impulse response is possible, which eliminates the numerical difficulties. In addition, no singular part in the acceleration unit-impulse response is present. This alternative formulation is derived in the next Section 2.2.

2.2 INTERACTION FORCE-ACCELERATION RELATIONSHIP

Using the same physical derivation as in connection with Eq. 2.5 but applied to the acceleration unit-impulse response, the interaction force-acceleration relationship is formulated as

$$\{R(t)\} = \int_0^t [M^\infty(t-\tau)]\{\ddot{u}(\tau)\}d\tau \tag{2.11}$$

$[M^\infty(t)]$ is the *acceleration unit-impulse response matrix* in the time domain. Transforming Eq. 2.11 to the frequency domain and with

$$\{\ddot{u}(\omega)\} = (i\omega)^2\{u(\omega)\} \tag{2.12}$$

yields

$$\{R(\omega)\} = [M^\infty(\omega)](i\omega)^2\{u(\omega)\} \tag{2.13}$$

$[M^\infty(\omega)]$ is denoted as the *acceleration dynamic-stiffness matrix* in the frequency domain. $[M^\infty(t)]$ and $[M^\infty(\omega)]$ form a Fourier transform pair. Comparing Eqs. 2.1 and 2.13 leads to the relationship between the acceleration and displacement dynamic-stiffness matrices

$$[M^\infty(\omega)] = \frac{[S^\infty(\omega)]}{(i\omega)^2} \tag{2.14}$$

The behaviour of $[M^\infty(\omega)]$ for large ω is examined. As the highest order in $i\omega$ in the singular part $[S_s^\infty(\omega)]$ is one (Eq. 2.7), the division by $(i\omega)^2$ in Eq. 2.14 results in no singular part being present in $[M^\infty(\omega)]$.

The relationship between the acceleration and displacement unit-impulse response matrices is established in the time domain. The starting point is the interaction force-displacement relationship (Eq. 2.10) of the unbounded medium which is initially at rest ($\{u(t=0)\} = \{\dot{u}(t=0)\} = 0$). The velocity vector in the first term on the right-hand side results in

$$\{\dot{u}(t)\} = \int_0^t \{\ddot{u}(\tau)\}d\tau \tag{2.15a}$$

Integration by parts is applied to the displacement vector in the second term to allow introduction of the acceleration vector

$$\{u(t)\} = \int_0^t \{\dot{u}(\tau)\}d\tau = t\{\dot{u}(t)\} - \int_0^t \tau\{\ddot{u}(\tau)\}d\tau$$
$$= \int_0^t (t-\tau)\{\ddot{u}(\tau)\}d\tau \tag{2.15b}$$

2.2 INTERACTION FORCE-ACCELERATION RELATIONSHIP

The third term leads to

$$\int_0^t [S_r^\infty(t-\tau)]\{u(\tau)\}d\tau = -\int_0^t d\left(\int_0^{t-\tau}[S_r^\infty(\tau')]d\tau'\right)\{u(\tau)\}d\tau$$

$$= -\left(\int_0^{t-\tau}[S_r^\infty(\tau')]d\tau'\right)\{u(\tau)\}\bigg|_0^t + \int_0^t \left(\int_0^{t-\tau}[S_r^\infty(\tau')]d\tau'\right)\{\dot u(\tau)\}d\tau$$

$$= -\int_0^t d\left(\int_0^{t-\tau}\int_0^{\tau'}[S_r^\infty(\tau'')]d\tau''d\tau'\right)\{\dot u(\tau)\}d\tau$$

$$= \int_0^t \left(\int_0^{t-\tau}\int_0^{\tau'}[S_r^\infty(\tau'')]d\tau''d\tau'\right)\{\ddot u(\tau)\}d\tau$$

$$= \int_0^t [M_S^\infty(t-\tau)]\{\ddot u(\tau)\}d\tau \qquad (2.15c)$$

where the abbreviation

$$[M_S^\infty(t)] = \int_0^t \int_0^\tau [S_r^\infty(\tau')]d\tau'd\tau \qquad (2.16)$$

is introduced. Eq. 2.10 is thus written as

$$\{R(t)\} = \int_0^t ([C_\infty] + [K_\infty](t-\tau) + [M_S^\infty(t-\tau)])\{\ddot u(\tau)\}d\tau \qquad (2.17)$$

From comparing Eqs. 2.11 and 2.17, it follows that $[M^\infty(t)]$ is equal to the expression in parenthesis in Eq. 2.17

$$[M^\infty(t)] = [C_\infty]H(t) + [K_\infty]tH(t) + [M_S^\infty(t)] \qquad (2.18)$$

with the Heaviside-step function $H(t)$. For $t < 0$, $[M^\infty(t)] = 0$ applies. At $t = 0$ discontinuities in the function and its derivative occur. For $t = 0^+$ (i.e. to the right of the discontinuity) $[M_S^\infty(t=0)]$ and $[\dot M_S^\infty(t=0)]$ are zero (Eq. 2.16), as $[S_r^\infty(t)]$ is finite. For $t = 0^+$

$$[M^\infty(t=0)] = [C_\infty] \qquad (2.19a)$$
$$[\dot M^\infty(t=0)] = [K_\infty] \qquad (2.19b)$$

holds. Figure 2-1 illustrates Eq. 2.18 for the scalar case. Eq. 2.18 can also be derived via the frequency domain. Substituting Eqs. 2.6 and 2.7 in Eq. 2.14 and determining the inverse Fourier transform yields $[M^\infty(t)]$.

Substituting Eq. 2.16 in Eq. 2.18 leads to

$$[M^\infty(t)] = [C_\infty]H(t) + [K_\infty]tH(t) + \int_0^t \int_0^\tau [S_r^\infty(\tau')]d\tau'd\tau \qquad (2.20)$$

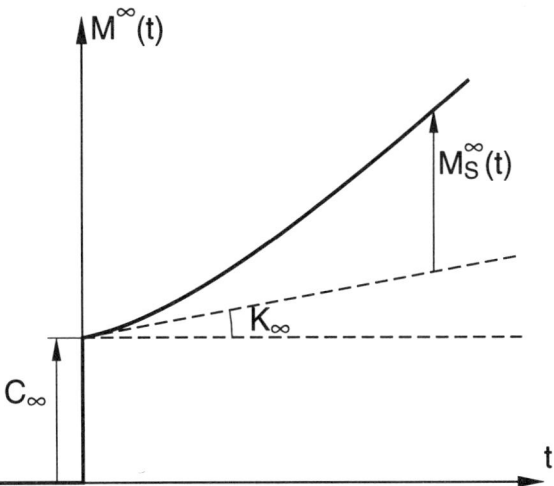

Figure 2-1 Acceleration unit-impulse response coefficient with discontinuities in function and first derivative at $t = 0$

which links the unit-impulse response matrix for displacement $[S^\infty(t)]$ (Eq. 2.8) to that for acceleration $[M^\infty(t)]$. Eq. 2.20 also follows from integrating Eq. 2.8 twice

$$[M^\infty(t)] = \int_0^t \int_0^\tau [S^\infty(\tau')]d\tau'd\tau \qquad (2.21)$$

The acceleration unit-impulse response matrix $[M^\infty(t)]$ is finite (with no singular part). $[M^\infty(t)]$ contains all information necessary to calculate the singular and regular parts of $[S^\infty(t)]$. $[C_\infty]$ and $[K_\infty]$ follow from Eq. 2.19, which then permits $[M_S^\infty(t)]$ to be calculated from Eq. 2.18. The regular part of the displacement unit-impulse response matrix follows from Eq. 2.16 as

$$[S_r^\infty(t)] = [\ddot{M}_S^\infty(t)] \qquad (2.22)$$

To evaluate the accuracy of the methods, comparisons with the results in the literature which are often presented in the frequency domain are performed. For this purpose, the dynamic-stiffness matrix $[S^\infty(\omega)]$ is determined directly from the acceleration unit-impulse response matrix $[M^\infty(t)]$. To achieve this, $[M^\infty(t)]$ is decomposed as (Figure 2-2)

$$[M^\infty(t)] = [C]H(t) + [K]tH(t) + [M_f^\infty(t)] \qquad (2.23)$$

with $[K]$ and $[C]$ chosen in such a way as to achieve $[M_f^\infty(t \to \infty)] = 0$ (using data of the last few time stations). The Fourier transformation of Eq. 2.23 yields

$$[M^\infty(\omega)] = \frac{1}{i\omega}[C] + \frac{1}{(i\omega)^2}[K] + [M_f^\infty(\omega)] \qquad (2.24)$$

2.3 INTERACTION FORCE-VELOCITY RELATIONSHIP

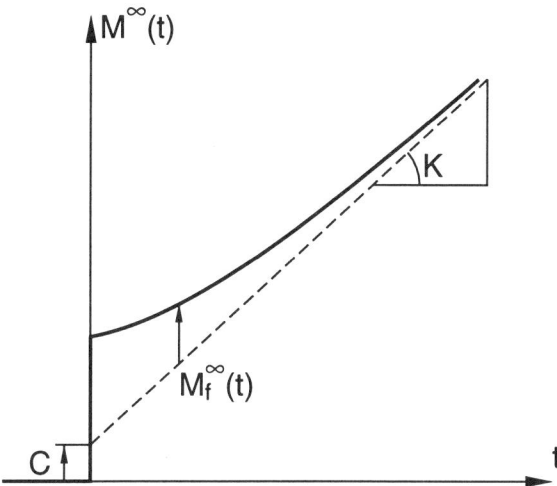

Figure 2-2 Decomposition of acceleration unit-impulse response coefficient for Fourier transformation

where $[M_f^\infty(\omega)]$ is evaluated numerically from $[M_f^\infty(t)]$. To convert $[M^\infty(\omega)]$ to the dynamic-stiffness matrix $[S^\infty(\omega)]$, a multiplication with $(i\omega)^2$ is performed (Eq. 2.14), yielding

$$[S^\infty(\omega)] = i\omega[C] + [K] + (i\omega)^2[M_f^\infty(\omega)] \qquad (2.25)$$

It follows from Eq. 2.25 that $[K]$ and $[C]$ are the static-stiffness matrix $[K^\infty]$ and the matrix of dashpot coefficients at $\omega = 0$ of the unbounded medium.

Both the forecasting method and the consistent infinitesimal finite-element cell method are formulated for the acceleration unit-impulse response matrix.

As an example, Appendix A.1 contains the displacement and acceleration unit-impulse response coefficients for the spherical cavity embedded in full-space with symmetric waves (Eqs. A.1.23a, A.1.23b, A.1.27 and A.1.48). They can be used to illustrate the derivation and to verify the final relationship.

2.3 INTERACTION FORCE-VELOCITY RELATIONSHIP

For the sake of completeness, the velocity unit-impulse response is also addressed. The interaction force-velocity relationship is formulated as

$$\{R(t)\} = \int_0^t [V^\infty(t-\tau)]\{\dot{u}(\tau)\}d\tau \qquad (2.26)$$

$[V^\infty(t)]$ is the *velocity unit-impulse response matrix* in the time domain. Transforming Eq. 2.26 to the frequency domain and with

$$\{\dot{u}(\omega)\} = i\omega\{u(\omega)\} \tag{2.27}$$

yields

$$\{R(\omega)\} = [V^\infty(\omega)]i\omega\{u(\omega)\} \tag{2.28}$$

$[V^\infty(\omega)]$ is denoted as the *velocity dynamic-stiffness matrix* in the frequency domain. $[V^\infty(t)]$ and $[V^\infty(\omega)]$ form a Fourier transform pair. Comparing Eq. 2.28 with Eqs. 2.1 and 2.13 leads to the relationship between the velocity, displacement and acceleration dynamic-stiffness matrices

$$[V^\infty(\omega)] = \frac{[S^\infty(\omega)]}{i\omega} = i\omega[M^\infty(\omega)] \tag{2.29}$$

To discuss the asymptotic value of $[V^\infty(\omega)]$ for $\omega \to \infty$ and thus the singular part, Eqs. 2.6 and 2.7 are substituted in Eq. 2.29, yielding

$$[V^\infty(\omega)] = [C_\infty] + [V_r^\infty(\omega)] \tag{2.30}$$

with the regular part

$$[V_r^\infty(\omega)] = \frac{[K_\infty]}{i\omega} + \frac{[S_r^\infty(\omega)]}{i\omega} \tag{2.31}$$

The singular part of $[V^\infty(\omega)]$ consists of one matrix $[C_\infty]$.

To establish the relationship between the velocity and displacement unit-impulse response matrices in the time domain, integration by parts is performed in Eq. 2.10

$$\{R(t)\} = [C_\infty]\int_0^t \delta(t-\tau)\{\dot{u}(\tau)\}d\tau + [K_\infty]\int_0^t \{\dot{u}(\tau)\}d\tau$$
$$+ \int_0^t \left(\int_0^{t-\tau}[S_r^\infty(\tau')]d\tau'\right)\{\dot{u}(\tau)\}d\tau \tag{2.32}$$

The last term on the right-hand side is an intermediate result of Eq. 2.15c. From comparing Eqs. 2.26 and 2.32, it follows that the coefficient matrix of $\{\dot{u}(\tau)\}$ in Eq. 2.32 is equal to $[V^\infty(t-\tau)]$. This can be formulated as

$$[V^\infty(t)] = [C_\infty]\delta(t) + [V_r^\infty(t)] \tag{2.33}$$

where the regular part equals

$$[V_r^\infty(t)] = [K_\infty]H(t) + \int_0^t [S_r^\infty(\tau)]d\tau \tag{2.34}$$

2.3 INTERACTION FORCE-VELOCITY RELATIONSHIP

For $t < 0$, $[V_r^\infty(t)] = 0$ applies. At $t = 0$, a discontinuity in the function occurs. For $t = 0^+$

$$[V_r^\infty(t=0)] = [K_\infty] \tag{2.35}$$

holds. Eq. 2.32 is then written as

$$\{R(t)\} = [C_\infty]\{\dot{u}(t)\} + \int_0^t [V_r^\infty(t-\tau)]\{\dot{u}(\tau)\}d\tau \tag{2.36}$$

Eqs. 2.33 and 2.34 can also be formulated directly from the corresponding relations in the frequency domain, Eqs. 2.30 and 2.31.

Substituting Eq. 2.34 in Eq. 2.33 leads to

$$[V^\infty(t)] = [C_\infty]\delta(t) + [K_\infty]H(t) + \int_0^t [S_r^\infty(\tau)]d\tau \tag{2.37}$$

which links the unit-impulse response matrix for displacement $[S^\infty(t)]$ (Eq. 2.8) to that for velocity $[V^\infty(t)]$. Obviously,

$$[V^\infty(t)] = \int_0^t [S^\infty(\tau)]d\tau \tag{2.38}$$

applies. $[V^\infty(t)]$ contains all information to calculate the singular and regular parts of $[S^\infty(t)]$. From Eq. 2.33 $[C_\infty]$ and $[V_r^\infty(t)]$ follow. $[K_\infty]$ is determined from Eq. 2.35. Differentiating Eq. 2.34 yields for $t \geq 0^+$

$$[S_r^\infty(t)] = [\dot{V}_r^\infty(t)] \tag{2.39}$$

To link the unit-impulse response matrix for velocity to that for acceleration, Eq. 2.20 is differentiated

$$[\dot{M}^\infty(t)] = [C_\infty]\delta(t) + [K_\infty]H(t) + \int_0^t [S_r^\infty(\tau)]d\tau \tag{2.40}$$

The right-hand side is equal to $[V^\infty(t)]$ (Eq. 2.37), leading to

$$[V^\infty(t)] = [\dot{M}^\infty(t)] \tag{2.41}$$

The singular part of $[V^\infty(t)]$ (Eq. 2.33), i.e. $[C_\infty]$, follows from Eq. 2.19a. Its regular part $[V_r^\infty(t)]$ is calculated using Eq. 2.41 for $t \geq 0^+$

$$[V_r^\infty(t)] = [\dot{M}^\infty(t)] \qquad t \geq 0^+ \tag{2.42}$$

To calculate the unit-impulse response matrix for acceleration from that for velocity

$$[M^\infty(t)] = \int_0^t [V^\infty(\tau)]d\tau = [C_\infty]H(t) + \int_0^t [V_r^\infty(\tau)]d\tau \tag{2.43}$$

applies, which follows from integrating Eq. 2.33.

2.4 RATIONAL APPROXIMATION AND REALIZATION

The interaction force-motion relationship (Eqs. 2.10, 2.11 or 2.36) involves a convolution integral of the unit-impulse response and the motion. This integral has to be recalculated from time zero onwards for each time station in an analysis of a transient. In this rigorous evaluation of the convolution integrals a large computational effort (proportional to the *square* of the number of time stations) and storage requirement result, which makes it unrealistic to perform large practical unbounded medium-structure-interaction analyses with many degrees of freedom on the structure-medium interface.

To reduce the computational effort the concepts of *linear system theory* can be applied. A rigorous treatment of this field, which is routinely applied in signal processing and control theory, lies outside the scope of this text. Only certain concepts are discussed superficially, and the final equations are formulated omitting all derivations. Formulations in dynamic unbounded medium-structure-interaction analysis are described in References [60], [61], [38].

The interaction force-displacement relationship discussed in Section 2.1 is addressed with the formulation in the frequency domain specified in Eq. 2.1 and in the time domain in Eq. 2.10. In general, the (regular part of the displacement) unit-impulse response matrix $[S_r^\infty(t_j)]$ at distinct time stations t_j, or the displacement dynamic-stiffness matrix $[S^\infty(\omega_j)]$ at distinct frequencies ω_j is known, and these are the two starting points of the procedures. All methods approximate $[S^\infty(\omega_j)]$ by a rational function in $i\omega$ (corresponding to a dynamic system with a finite number of degrees of freedom), denoted as the *rational approximation*. This permits in the continuous-time formulation a system of linear differential equations or in the discrete-time formulation explicit finite-difference or recursive equations to be established, called the *realization*. The solution of the system of differential equations or the processing of the finite-difference equations describing the model of the unbounded medium results in reduced computational effort (proportional to the number of time stations) and storage requirement, as in a dynamic analysis of a structure.

Starting from either $[S_r^\infty(t_j)]$ or $[S^\infty(\omega_j)]$ a large number of procedures working in continuous or discrete time exist. They either address the whole matrices or scalar elements which can e.g. be calculated by applying transformations.

Performing a rational approximation and a realization the rigorous interaction force-displacement relationship (Eq. 2.10)

$$\{R(t)\} = [C_\infty]\{\dot{u}(t)\} + [K_\infty]\{u(t)\} + \int_0^t [S_r^\infty(t-\tau)]\{u(\tau)\}d\tau \qquad (2.44)$$

leads to the following *continuous-time formulations*. The system of linear differential

2.4 RATIONAL APPROXIMATION AND REALIZATION

equations of first order equals

$$\{\dot{x}(t)\} = [A]\{x(t)\} + [B]\{u(t)\} \tag{2.45a}$$
$$\{R(t)\} = [C]\{x(t)\} + [C_\infty]\{\dot{u}(t)\} + [K_\infty]\{u(t)\} \tag{2.45b}$$

with the state variables (internal variables) $\{x(t)\}$ and the time-independent matrices $[A]$, $[B]$, $[C]$. The latter are calculated from e.g. $[S_r^\infty(t_j)]$. This non-symmetric system can be transformed to a system of first-order differential equations with the symmetric damping and static-stiffness matrices $[\bar{C}]$ and $[\bar{K}]$

$$[\bar{C}]\left\{\begin{array}{c}\{\dot{u}(t)\}\\\{\dot{x}(t)\}\end{array}\right\} + [\bar{K}]\left\{\begin{array}{c}\{u(t)\}\\\{x(t)\}\end{array}\right\} = \left\{\begin{array}{c}\{R(t)\}\\0\end{array}\right\} \tag{2.46}$$

or to a system of second-order differential equations with the symmetric mass, damping and static-stiffness matrices $[M]$, $[C]$ and $[K]$

$$[M]\left\{\begin{array}{c}\{\ddot{u}(t)\}\\\{\ddot{w}(t)\}\end{array}\right\} + [C]\left\{\begin{array}{c}\{\dot{u}(t)\}\\\{\dot{w}(t)\}\end{array}\right\} + [K]\left\{\begin{array}{c}\{u(t)\}\\\{w(t)\}\end{array}\right\} = \left\{\begin{array}{c}\{R(t)\}\\0\end{array}\right\} \tag{2.47}$$

Typically, the order of the internal variable vector $\{w\}$ is half of that of $\{x\}$. Eq. 2.47 for the model of the unbounded medium is in the same form as the equation of motion in finite-element analysis of a structure with a finite number of degrees of freedom.

Alternatively, the following *discrete-time formulations* are obtained. The explicit finite-difference equations of first order at time station n equal

$$\{x\}_{n+1} = [\tilde{A}]\{x\}_n + [\tilde{B}]\{u\}_n \tag{2.48a}$$
$$\{R\}_n = [\tilde{C}]\{x\}_n + [\tilde{D}]\{u\}_n + [C_\infty]\{\dot{u}\}_n + [K_\infty]\{u\}_n \tag{2.48b}$$

with time-station-independent matrices $[\tilde{A}]$, $[\tilde{B}]$, $[\tilde{C}]$, $[\tilde{D}]$. Alternatively, *recursive equations* expressing the interaction forces $\{R\}_n$ at time station n as a function of the past few interaction forces $\{R\}_{n-j}$ and displacements $\{u\}_{n-j}$ are written as

$$\{R_r\}_n = \sum_{j=1}^M [a]_j\{R\}_{n-j} + \sum_{j=0}^M [b]_j\{u\}_{n-j} \tag{2.49a}$$
$$\{R\}_n = \{R_r\}_n + [C_\infty]\{\dot{u}\}_n + [K_\infty]\{u\}_n \tag{2.49b}$$

with the time-station-independent matrices $[a]_j$, $[b]_j$. M will be larger than or equal to the order of $\{x\}$.

Analogous formulations follow for the interaction force-velocity relationship and the interaction force-acceleration relationship. The latter (Eq. 2.11) is written rigorously after substituting Eq. 2.23 as

$$\{R(t)\} = [C_0]\{\dot{u}(t)\} + [K^\infty]\{u(t)\} + \int_0^t [M_f^\infty(t-\tau)]\{\ddot{u}(\tau)\}d\tau \tag{2.50}$$

$[K]$ and $[C]$ in Eq. 2.23 are replaced by the static-stiffness matrix $[K^\infty]$ and the matrix of dashpot coefficients $[C_0]$ at $\omega = 0$. As an example, the continuous-time formulation can be written as

$$\{\dot{x}(t)\} = [A]\{x(t)\} + [B]\{\ddot{u}(t)\} \tag{2.51a}$$

$$\{R(t)\} = [C]\{x(t)\} + [C_0]\{\dot{u}(t)\} + [K^\infty]\{u(t)\} \tag{2.51b}$$

and the discrete-time formulation as

$$\{\dot{x}\}_{n+1} = [\tilde{A}]\{x\}_n + [\tilde{B}]\{\ddot{u}\}_n \tag{2.52a}$$

$$\{R\}_n = [\tilde{C}]\{x\}_n + [\tilde{D}]\{\ddot{u}\}_n + [C_0]\{\dot{u}\}_n + [K^\infty]\{u\}_n \tag{2.52b}$$

All coefficient matrices of the realization depend on the rational approximation of $[M_f^\infty(t_j)]$.

3

DYNAMIC STIFFNESS AND UNIT-IMPULSE RESPONSE AT SIMILAR STRUCTURE-MEDIUM INTERFACES OF UNBOUNDED MEDIUM

Both the forecasting method and the consistent infinitesimal finite-element cell method are based on addressing *similar* structure-medium interfaces of the unbounded medium. The construction of such similar interfaces is discussed in Section 1.5.1 (Figure 1-4). The only independent variable which fully defines a specific structure-

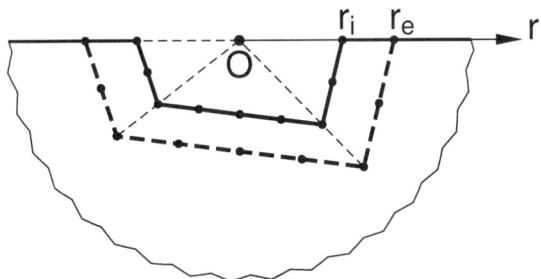

Figure 3-1 Similar structure-medium interfaces of unbounded medium defined by characteristic length

medium interface is the characteristic length r, which is regarded as a variable in the following derivation (Figure 3-1). The dependence of the dynamic-stiffness matrix and the unit-impulse response matrix on the characteristic length r is established using dimensional analysis.

3.1 FREQUENCY DOMAIN

The general form of the interaction force-displacement relationship in the frequency domain involving the dynamic-stiffness matrix is discussed

$$\{R(\omega)\} = [S^\infty(r,\omega)]\{u(\omega)\} \tag{3.1}$$

A *dimensional analysis* is performed to identify the independent dimensionless variables of which $[S^\infty(r,\omega)]$ is a function. Besides Poisson's ratio ν (which is already dimensionless), the characteristic length r, the shear modulus G, the mass density ρ and the frequency ω are sufficient to determine $[S^\infty(r,\omega)]$. G and ρ are not indicated in the argument of $[S^\infty(r,\omega)]$ in Eq. 3.1, as they are constant in the radial direction in the unbounded medium. A bracket denotes the dimension; $[L]$, $[M]$ and $[T]$ are the dimensions of length, mass and time. The dimensions of the variables are

$$[[S^\infty]] = [L]^{s-3}[M][T]^{-2} \tag{3.2a}$$
$$[r] = [L] \tag{3.2b}$$
$$[G] = [L]^{-1}[M][T]^{-2} \tag{3.2c}$$
$$[\rho] = [L]^{-3}[M] \tag{3.2d}$$
$$[\omega] = [T]^{-1} \tag{3.2e}$$

with the spatial dimension s ($= 2$ or $= 3$). The product of all these variables raised to unknown powers n_i ($i = 1, 2, \ldots, 5$) must be dimensionless

$$[[S^\infty]]^{n_1}[r]^{n_2}[G]^{n_3}[\rho]^{n_4}[\omega]^{n_5} = [L]^{(s-3)n_1+n_2-n_3-3n_4}[M]^{n_1+n_3+n_4}[T]^{-2n_1-2n_3-n_5} \tag{3.3}$$

This results in

$$(s-3)n_1 + n_2 - n_3 - 3n_4 = 0 \tag{3.4a}$$
$$n_1 + n_3 + n_4 = 0 \tag{3.4b}$$
$$-2n_1 - 2n_3 - n_5 = 0 \tag{3.4c}$$

The rank of the coefficient matrix of the system of 3 equations (Eq. 3.4) with 5 unknowns equals 3, which permits the two unknowns n_1, n_5 to be chosen arbitrarily. For $n_1 = 1$, $n_5 = 0$, the other unknowns are $n_2 = 2 - s$, $n_3 = -1$, $n_4 = 0$, yielding the first dimensionless variable $[S^\infty]r^{2-s}G^{-1}$. For $n_1 = 0$, $n_5 = 1$, the other unknowns are $n_2 = 1$, $n_3 = -0.5$, $n_4 = 0.5$, resulting in the second dimensionless variable $rG^{-0.5}\rho^{0.5}\omega$, which is the *dimensionless frequency*

$$a_0 = \frac{\omega r}{c_s} \tag{3.5}$$

3.1 FREQUENCY DOMAIN

with the shear-wave velocity $c_s = \sqrt{G/\rho}$. Note that a_0 varies with r. The first dimensionless variable $[S^\infty] r^{2-s} G^{-1}$ will be a function $[\overline{S}^\infty]$ of the second dimensionless variable a_0 (and of ν) yielding

$$[S^\infty(r,\omega)] = G r^{s-2} [\overline{S}^\infty(a_0)] \tag{3.6}$$

The arbitrary function $[\overline{S}^\infty(a_0)]$ cannot be determined based on dimensional analysis. ω and r do not appear explicitly in $[\overline{S}^\infty(a_0)]$, but only as a product in a_0 (Eq. 3.5).

As an example, the dynamic-stiffness coefficient of the spherical cavity of radius r embedded in a full-space with symmetric waves equals (Appendix A.1, Eq. A.1.16)

$$S^\infty(r,\omega) = 16\pi G r \left(1 + \frac{1-\nu}{2(1-2\nu)} \frac{(ia_0)^2}{1+ia_0} \right) \tag{3.7}$$

The dimensionless frequency $a_0 = \omega r/c_p$ is defined with respect to the dilatational-wave velocity c_p and not c_s ($c_p = \sqrt{2(1-\nu)/(1-2\nu)}\, c_s$). The spatial dimension s equals 3.

The specific form of Eq. 3.6 permits similar structure-medium interfaces to be addressed. The dynamic-stiffness matrices of the unbounded medium at two such structure-medium interfaces characterized by r_i (interior) and r_e (exterior) are examined (Figure 3-1). Assuming that $[S^\infty(r_i,\omega)]$ is known, the dynamic-stiffness matrix at r_e can be calculated for the same a_0, which corresponds to the frequency $(r_i/r_e)\omega$ at the exterior interface ($a_0 = \omega r_i/c_s = (r_i/r_e)\omega r_e/c_s$). Formulating Eq. 3.6 for $r = r_i$ with ω and for $r = r_e$ with $(r_i/r_e)\omega$ yields after eliminating the common term $[\overline{S}^\infty(a_0)]$

$$[S^\infty(r_e, \frac{r_i}{r_e}\omega)] = \left(\frac{r_e}{r_i}\right)^{s-2} [S^\infty(r_i,\omega)] \tag{3.8}$$

This presents one form of the relationship between the dynamic-stiffness matrices of similar structure-medium interfaces of an unbounded medium. It will be used to interpret the results of the dynamic condensation and substructure deletion methods (Section 7.3).

In another form to be used in the consistent infinitesimal finite-element cell method in the frequency domain, the variation of the dynamic-stiffness matrix as a function of the characteristic length r and of the excitation frequency ω is studied.

Eq. 3.6 is addressed. $[\overline{S}^\infty(a_0)]$ is a function of one independent variable, i.e. a_0. For constant ω, $[\overline{S}^\infty(a_0)]$ is a function of r, or just as valid, for constant r, $[\overline{S}^\infty(a_0)]$ is a function of ω. The same change in a_0 can be achieved by varying the values of either r or ω with the other fixed. The derivative thus follows for a constant ω varying r as

$$[\overline{S}^\infty(a_0)]_{,a_0} = \frac{c_s}{\omega} [\overline{S}^\infty(a_0)]_{,r} \tag{3.9a}$$

and the same result is calculated for a constant r but varying ω as

$$[\overline{S}^\infty(a_0)]_{,a_0} = \frac{c_s}{r}[\overline{S}^\infty(a_0)]_{,\omega} \tag{3.9b}$$

Setting the two right-hand sides of Eq. 3.9 equal yields

$$r[\overline{S}^\infty(a_0)]_{,r} = \omega[\overline{S}^\infty(a_0)]_{,\omega} \tag{3.10}$$

The partial derivative of $[\overline{S}^\infty(a_0)]$ with respect to r can thus be replaced by that with respect to ω for similar interfaces. The partial derivatives with respect to r and ω are calculated from Eq. 3.6 as

$$[\overline{S}^\infty(a_0)]_{,r} = \frac{1}{Gr^{s-2}}\left(-\frac{s-2}{r}[S^\infty(r,\omega)] + [S^\infty(r,\omega)]_{,r}\right) \tag{3.11a}$$

$$[\overline{S}^\infty(a_0)]_{,\omega} = \frac{1}{Gr^{s-2}}[S^\infty(r,\omega)]_{,\omega} \tag{3.11b}$$

Substituting Eq. 3.11 in Eq. 3.10 leads to

$$r[S^\infty(r,\omega)]_{,r} = (s-2)[S^\infty(r,\omega)] + \omega[S^\infty(r,\omega)]_{,\omega} \tag{3.12}$$

This Eq. 3.12 will be used in the derivation of the consistent infinitesimal finite-element cell equation to eliminate the partial derivative $[S^\infty(r,\omega)]_{,r}$ in the relationship expressing the finite-element assemblage (Section 5.2.8).

It is straightforwardly verified that Eq. 3.7 satisfies Eq. 3.12.

3.2 TIME DOMAIN

As the acceleration unit-impulse response matrix plays a central role in the similarity-based formulations in the time domain, it is appropriate to examine its characteristics at similar structure-medium interfaces of an unbounded medium starting from dimensional analysis.

The interaction force-acceleration relationship in the time domain (Eq. 2.11) involving the acceleration unit-impulse response matrix $[M^\infty(r,t)]$ is formulated as

$$\{R(t)\} = \int_0^t [M^\infty(r, t-\tau)]\{\ddot{u}(\tau)\}d\tau \tag{3.13}$$

$[M^\infty(r,t)]$ is a function of v, r, G, ρ and time t which have the following dimensions

$$[[M^\infty]] = [L]^{s-3}[M][T]^{-1} \tag{3.14a}$$
$$[r] = [L] \tag{3.14b}$$
$$[G] = [L]^{-1}[M][T]^{-2} \tag{3.14c}$$
$$[\rho] = [L]^{-3}[M] \tag{3.14d}$$
$$[t] = [T] \tag{3.14e}$$

3.2 TIME DOMAIN

with the symbols defined in Section 3.1. The following product must be dimensionless

$$[[M^\infty]]^{n_1}[r]^{n_2}[G]^{n_3}[\rho]^{n_4}[t]^{n_5} = [L]^{(s-3)n_1+n_2-n_3-3n_4}[M]^{n_1+n_3+n_4}[T]^{-n_1-2n_3+n_5} \quad (3.15)$$

This yields

$$(s-3)n_1 + n_2 - n_3 - 3n_4 = 0 \quad (3.16a)$$
$$n_1 + n_3 + n_4 = 0 \quad (3.16b)$$
$$-n_1 - 2n_3 + n_5 = 0 \quad (3.16c)$$

n_1 and n_5 are chosen arbitrarily. For $n_1 = 1$, $n_5 = 0$, the other unknowns are $n_2 = 1-s$, $n_3 = -0.5$, $n_4 = -0.5$, leading to the first dimensionless variable $[M^\infty]r^{1-s}G^{-0.5}\rho^{-0.5} = [M^\infty]r^{1-s}G^{-1}c_s$. For $n_1 = 0$, $n_5 = 1$, the values are $n_2 = -1$, $n_3 = 0.5$, $n_4 = -0.5$, resulting in the second dimensionless variable $r^{-1}G^{0.5}\rho^{-0.5}t$, which is the *dimensionless time*

$$\bar{t} = \frac{c_s}{r}t \quad (3.17)$$

Note that $\bar{t}$ varies with r. The first dimensionless variable will be a function $[\overline{M}^\infty]$ of the second dimensionless variable (and of ν) yielding

$$[M^\infty(r,t)] = G\frac{r^{s-1}}{c_s}[\overline{M}^\infty(\bar{t})] \quad (3.18)$$

t and r do not appear explicitly in $[\overline{M}^\infty(\bar{t})]$, but only as a ratio in $\bar{t}$.

Again the spherical cavity is used as an example. From Appendix A.1, Eq. A.1.49, the acceleration unit-impulse response coefficient equals

$$M^\infty(r,t) = 16\pi G \frac{r^2}{c_p}\left(\bar{t} + \frac{1-\nu}{2(1-2\nu)}e^{-\bar{t}}H(\bar{t})\right) \quad (3.19)$$

c_p instead of c_s is introduced which also affects the dimensionless time $\bar{t} = tc_p/r$.

Eq. 3.18 permits various similar structure-medium interfaces to be addressed. As will become apparent in the forecasting algorithm, the acceleration unit-impulse response matrices of the unbounded medium at two similar structure-medium interfaces characterized by r_i (interior) and r_e (exterior) are examined (Figure 3-1). To be able to calculate $[M^\infty(r_e,t)]$, the unit-impulse response matrix at r_i for the same dimensionless time $\bar{t}$ can be used, which corresponds to the time $(r_i/r_e)t$ at the interior boundary ($\bar{t} = tc_s/r_e = (r_i/r_e)tc_s/r_i$). Formulating Eq. 3.18 for $r = r_i$ with $(r_i/r_e)t$ and for $r = r_e$ with t yields after eliminating the common term $[\overline{M}^\infty(\bar{t})]$

$$[M^\infty(r_e,t)] = \left(\frac{r_e}{r_i}\right)^{s-1}[M^\infty(r_i,\frac{r_i}{r_e}t)] \quad (3.20)$$

For the sake of completeness the other form (which could be used in the derivation in the time domain of the consistent infinitesimal finite-element cell method) can be derived analogously as the corresponding relationship in the frequency domain (Eqs. 3.9 to 3.12)

$$r[M^\infty(r,t)]_{,r} = (s-1)[M^\infty(r,t)] + t[\dot{M}^\infty(r,t)] \tag{3.21}$$

As an alternative of the derivation in the time domain, the results in the frequency domain in Section 3.1 can be transformed to the time domain after converting $[S^\infty(\omega)]$ to $[M^\infty(\omega)]$, applying Eq. 2.14. Eqs. 3.6, 3.8 and 3.12 in the frequency domain correspond to Eqs. 3.18, 3.20 and 3.21 in the time domain, respectively.

4

FORECASTING METHOD

This chapter is based on Reference [49].

The characteristics and the concept of the forecasting method are described in Section 4.1. The fundamental equations in continuous time are derived in Section 4.2, followed by the time discretization and the numerical algorithm in Section 4.3. The accuracy of the forecasting method is evaluated in Section 4.4.

4.1 CONCEPT BASED ON WAVE PROPAGATION

The forecasting method is a straightforward implementation based on similarity and finite elements to calculate the unit-impulse response matrix of the unbounded medium. Only a small finite-element region is introduced. As in the standard finite-element method applied to a bounded structure, the banded symmetric system of equations of the finite-element mesh is solved at each time step. The forecasting method is based on *wave propagation with a finite velocity*. This results in a *time delay* occurring at different locations for the same event, as is routinely used in weather and flood forecasting. It can thus be used for wave-propagation problems analysed in the time domain. The forecasting method cannot be applied, however, to problems where the total unbounded medium is excited at the same time (instantaneous response), such as in incompressible elasto-dynamics and diffusion in the time domain, statics and harmonic excitation (frequency domain). All the advantages of the similarity-based finite-element modelling of the unbounded medium apply for the forecasting method, such as the automatic and rigorous consideration of the free surface and interfaces between material inhomogeneities (when compatible with similarity), straightforwardly processing anisotropy and satisfying the radiation condition without introducing a fundamental solution (Section 1.5.1). No approximation other than that of the finite-element method is introduced. The forecasting method thus converges to the exact solution in the finite-element sense.

Without addressing any details, the concept of forecasting can be described as follows. The unit-impulse response matrix at the structure-medium interface of the unbounded medium is to be calculated. Conceptually, another similar interface is

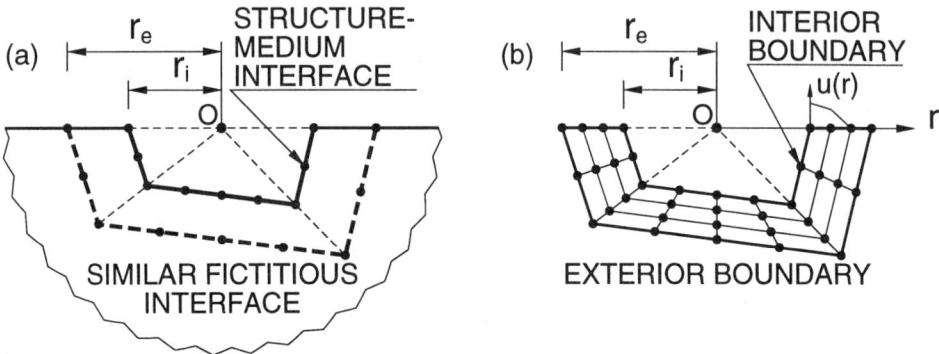

Figure 4-1 Concept of forecasting method with part of unbounded medium discretized by finite elements

introduced denoted as the fictitious interface (Figure 4-1a). The similar interfaces of the same unbounded medium are defined by the characteristic lengths r (r_i for the structure-medium interface, r_e for the fictitious interface). The unit-impulse response matrix at the fictitious interface can be *forecasted* from that at the structure-medium interface using the relationship based on similarity (Section 3.2, Eq. 3.20).

The region between the structure-medium interface and the fictitious interface is discretized with finite elements (Figure 4-1b). Its interior boundary thus coincides with the structure-medium interface and its exterior boundary with the fictitious interface. The arrangement of the nodes within the finite-element region (with the exception of those on the two boundaries) does not have to correspond to similarity as in the case of Figure 4-1b.

The unit-impulse response matrix at the structure-medium interface of the unbounded medium is calculated using the finite-element region of Figure 4-1b with the appropriate boundary condition at the exterior boundary representing the remaining part of the unbounded medium. This boundary condition is introduced using the forecasted unit-impulse response matrix at the fictitious boundary of the unbounded medium, formulating compatibility and equilibrium at the exterior boundary by the standard assemblage of finite elements.

The algorithm is first explained for the beginning of the analysis starting from $t = 0$. A unit impulse is applied to the interior boundary. As illustrated in Figure 4-1b it takes time for a wave to propagate from the interior boundary to the exterior boundary. Before the wave reaches the exterior boundary, the boundary is at rest. The uniqueness theorem for the solution of the unbounded medium discussed in Section 1.2 is thus satisfied. Such a boundary condition can be expressed as either vanishing surface tractions (free boundary) or zero displacements (fixed boundary). This permits the unit-impulse response matrix at the interior boundary, the structure-medium interface,

4.1 CONCEPT BASED ON WAVE PROPAGATION

to be calculated up to the time when the wave arrives at the exterior boundary. For an accurate result, the wave must be represented well, which requires 2 or 3 rows of finite elements in the radial direction. From the relationship based on similarity the *unit-impulse response matrix at the exterior boundary*, the fictitious interface, *can be calculated from that at the interior boundary*. This is done using Eq. 3.20. As $r_e \geq r_i$, the time involved at the interior boundary $(r_i/r_e)t$ is always shorter than (or equal to) that at the exterior boundary t. It is thus always possible to calculate the unit-impulse response matrix at the exterior boundary up to at least the same time as it is available at the interior boundary. The *forecasted unit-impulse response matrix at the exterior boundary is incorporated* in the equation of motion of the finite-element region *to provide the boundary condition at times larger than t*. The procedure also works for later times. Conceptually, the time is subdivided into time segments which are shorter than the propagation time from the interior to the exterior boundary. Figure 4-1b thus still applies to the latest time segment. The contributions of the previous time segments are calculated, evaluating convolution integrals as is shown below.

The key feature of the forecasting method can also be verified mathematically addressing the interaction force-acceleration relationship at the exterior boundary (Eq. 2.11), formulated at time t_n

$$\{R_e(t_n)\} = \int_0^{t_n} [M_e^\infty(t_n - \tau)]\{\ddot{u}_e(\tau)\}d\tau \tag{4.1}$$

The unit-impulse response matrix at the interior boundary is known up to t_m. Using forecasting (Eq. 3.20), that at the exterior boundary can be determined up to $(r_e/r_i)t_m$. The most stringent case $r_e = r_i$ is discussed where the unit-impulse response matrix at the exterior boundary is known only up to t_m. The wave reaches the exterior boundary at t_0, i.e. $\{\ddot{u}_e(t)\} = 0$ for $t \leq t_0$. t_0 equals the shortest distance between the interior and exterior boundaries divided by the dilatational-wave velocity c_p. Eq. 4.1 is written as

$$\begin{aligned}\{R_e(t_n)\} &= \int_{t_0}^{t_n} [M_e^\infty(t_n - \tau)]\{\ddot{u}_e(\tau)\}d\tau \\ &= \int_0^{t_n - t_0} [M_e^\infty(t_n - t_0 - \tau)]\{\ddot{u}_e(t_0 + \tau)\}d\tau\end{aligned} \tag{4.2}$$

It follows that $[M_e^\infty(t)]$ is only required for $0 \leq t \leq t_n - t_0$. t_0 can thus always be chosen such that $t_n - t_0 < t_m$ applies. Thus, the boundary condition at the exterior boundary at t_n can always be formulated from Eq. 4.2 as a linear equation in $\{R_e(t_n)\}$ and $\{\ddot{u}_e(t_n)\}$, which enables the unit-impulse response matrix at the interior boundary at time t_n to be determined. The relationship based on similarity then provides $[M_e^\infty(t)]$ up to t_n for use at later times.

4.2 FUNDAMENTAL EQUATIONS

4.2.1 Interaction Force-acceleration Relationship of Unbounded Medium

The unbounded medium is assumed to be initially at rest. The interaction force-acceleration relationship at the structure-medium interface defined by the characteristic length r of the unbounded medium (Figure 1-4) is formulated as (Eq. 2.11)

$$\{R(r,t)\} = \int_0^t [M^\infty(r, t-\tau)]\{\ddot{u}(r,\tau)\}d\tau \tag{4.3}$$

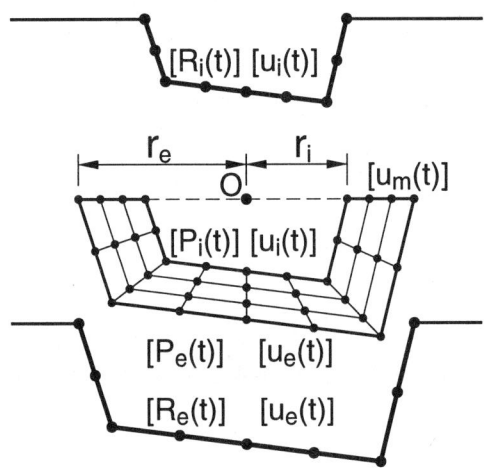

Figure 4-2 Finite-element region between interior and exterior boundaries and corresponding interfaces of unbounded medium

To determine the acceleration unit-impulse response matrix at the structure-medium interface coinciding with the interior boundary i of the finite-element region $[M_i^\infty(t)] = [M^\infty(r_i, t)]$, the unit-impulse acceleration matrix is enforced at the same location (Figure 4-2)

$$[\ddot{u}_i(t)] = c\delta(t)[I] \tag{4.4}$$

which corresponds to

$$[u_i(t)] = ctH(t)[I] \tag{4.5a}$$
$$[\dot{u}_i(t)] = cH(t)[I] \tag{4.5b}$$

$[I]$ denotes the unit matrix. The arbitrary constant c with the dimension length divided by time is introduced to provide the correct dimension for $[\ddot{u}_i(t)]$.

4.2 FUNDAMENTAL EQUATIONS

Note that the acceleration matrix in Eq. 4.4 represents a combination of independent acceleration vectors. This input yields from Eq. 4.3

$$[R_i(t)] = c[M_i^\infty(t)] \tag{4.6}$$

For the fictitious interface coinciding with the exterior boundary e of the finite-element region $[M_e^\infty(t)] = [M^\infty(r_e,t)]$, Eq. 4.3 results in

$$[R_e(t)] = \int_0^t [M_e^\infty(t-\tau)][\ddot{u}_e(\tau)]d\tau \tag{4.7}$$

As discussed in Section 3.2, the unit-impulse response matrices at similar interfaces are related as (Eq. 3.20)

$$[M_e^\infty(t)] = \left(\frac{r_e}{r_i}\right)^{s-1} [M_i^\infty(\frac{r_i}{r_e}t)] \tag{4.8}$$

s denotes the spatial dimension.

4.2.2 Equations of Motion of Finite-element Region

Between the interior and exterior boundaries satisfying similarity a finite-element discretization is performed (Figure 4-2). In principle, the arrangement of the nodes within the finite-element region does not have to correspond to similarity as is the case in the figure. The equations of motion are formulated as

$$[M][\ddot{u}(t)] + [K][u(t)] = [P(t)] \tag{4.9}$$

with the mass matrix $[M]$, the static-stiffness matrix $[K]$ and the nodal-force matrix $[P(t)]$. Partitioning into the nodes lying on the interior and exterior boundaries and the remaining nodes (index m for "middle") yields

$$[u(t)] = \begin{bmatrix} [u_i(t)] \\ [u_m(t)] \\ [u_e(t)] \end{bmatrix} \tag{4.10a}$$

$$[P(t)] = \begin{bmatrix} [P_i(t)] \\ 0 \\ [P_e(t)] \end{bmatrix} \tag{4.10b}$$

$$[M] = \begin{bmatrix} [M_{ii}] & [M_{im}] & 0 \\ [M_{mi}] & [M_{mm}] & [M_{me}] \\ 0 & [M_{em}] & [M_{ee}] \end{bmatrix} \quad (4.11a)$$

$$[K] = \begin{bmatrix} [K_{ii}] & [K_{im}] & 0 \\ [K_{mi}] & [K_{mm}] & [K_{me}] \\ 0 & [K_{em}] & [K_{ee}] \end{bmatrix} \quad (4.11b)$$

Equilibrium at the interior and exterior boundaries relates the interaction forces at the structure-medium interface and at the fictitious interface of the unbounded medium to the forces of the finite-element region

$$[P_i(t)] = [R_i(t)] \quad (4.12a)$$
$$[P_e(t)] = -[R_e(t)] \quad (4.12b)$$

Eqs. 4.6, 4.7, 4.8, 4.9 and 4.12 permit the unit-impulse response matrix $[M_i^\infty(t)]$ to be calculated. To achieve this numerically, a time discretization is performed enabling the forecasting algorithm to be formulated.

4.3 TIME DISCRETIZATION AND NUMERICAL ALGORITHM

In principle, any scheme of time discretization can be applied. For instance, the Newmark time integrator with the parameters β and γ based on a predictor-corrector scheme for the nth time step yields

$$[u]_n = [\tilde{u}]_n + \beta \Delta t^2 [\ddot{u}]_n \quad (4.13a)$$
$$[\dot{u}]_n = [\tilde{\dot{u}}]_n + \gamma \Delta t [\ddot{u}]_n \quad (4.13b)$$

where the predicted values equal

$$[\tilde{u}]_n = [u]_{n-1} + \Delta t [\dot{u}]_{n-1} + (0.5 - \beta) \Delta t^2 [\ddot{u}]_{n-1} \quad (4.14a)$$
$$[\tilde{\dot{u}}]_n = [\dot{u}]_{n-1} + (1 - \gamma) \Delta t [\ddot{u}]_{n-1} \quad (4.14b)$$

The motion up to the $(n-1)$th time step is known. Substituting Eq. 4.13a in Eq. 4.9 leads to

$$[D][\ddot{u}]_n = [P]_n - [K][\tilde{u}]_n \quad (4.15)$$

where

$$[D] = [M] + \beta \Delta t^2 [K] \quad (4.16)$$

4.3 TIME DISCRETIZATION AND NUMERICAL ALGORITHM

At the interior boundary, a unit-impulse acceleration matrix (Eq. 4.4) is enforced as already discussed for the continuous-time formulation. This corresponds to the following condition

$$[u_i(t=0)] = 0 \tag{4.17a}$$
$$[\dot{u}_i(t=0)] = 0 \tag{4.17b}$$
$$[\ddot{u}_i(t)] = c\delta(t)[I] \tag{4.17c}$$

An equivalent condition which avoids the δ-function is formulated immediately after applying the unit-impulse acceleration matrix, i.e. at $t = 0^+$, as

$$[u_i(t=0)] = 0 \tag{4.18a}$$
$$[\dot{u}_i(t=0)] = c[I] \tag{4.18b}$$
$$[\ddot{u}_i(t)] = 0 \tag{4.18c}$$

The motion in Eq. 4.18 is written in discretized form compatible with the Newmark time integrator as

$$[u_i]_0 = 0 \tag{4.19a}$$
$$[\dot{u}_i]_0 = c[I] \tag{4.19b}$$
$$[\ddot{u}_i]_n = 0 \quad n = 0, 1, \ldots \tag{4.19c}$$

Enforcing Eq. 4.19, it follows from Eqs. 4.13 and 4.14 that the discretized $\{u_i\}_n$ and $\{\dot{u}_i\}_n$ satisfy Eq. 4.5 exactly.

At the exterior boundary, the boundary condition consists of the interaction force-acceleration relationship of the unbounded medium formulated at the same location (Eq. 4.7). It is discretized assuming for each time step j ($t_{j-1} \leq t < t_j = j\Delta t$) a constant $[M_e^\infty(t)]$, which is approximated by its value in the middle of the time step $[M_e^\infty]_{j-\frac{1}{2}}$.

$$[R_e]_n = \sum_{j=1}^{n} [M_e^\infty]_{n-j+\frac{1}{2}} \int_{t_{j-1}}^{t_j} [\ddot{u}_e(\tau)] d\tau = \sum_{j=1}^{n} [M_e^\infty]_{n-j+\frac{1}{2}} ([\dot{u}_e]_j - [\dot{u}_e]_{j-1}) \tag{4.20}$$

The discretized form of the convolution integral is expressed in velocities.

The forecasting method makes use of the fact that the waves propagate with finite velocity. For compressible elasto-dynamics the maximum wave velocity equals that of the dilatational wave c_p. A wave originating at one location will thus reach another location with a time delay. In the forecasting method a unit impulse applied at the interior boundary will start influencing the exterior boundary after a finite time delay. The latter is evaluated as the ratio of the shortest distance between the interior and exterior boundaries to c_p. After selecting the finite-element length in the radial direction and the time step, the number of rows in the radial direction can be determined by requiring that the velocities at the exterior boundary during the first k time steps vanish (Figure 4-1b). The number of rows should also be large enough to represent adequately the wave pattern behind the wave front before the wave reaches the exterior boundary. This guarantees that the interaction forces at the interior boundary, i.e. the unit-impulse response matrix (Eq. 4.6), are accurate. Typically, k is selected as 3 or 4, resulting in 2 or 3 rows of finite elements in the radial direction for a time step which is about half of the finite-element length divided by c_p. Returning to Eq. 4.20 and substituting $[\dot{u}_e]_j = 0$ for $0 \leq j \leq k$ yields

$$[R_e]_n = \sum_{j=k+1}^{n} [M_e^\infty]_{n-j+\frac{1}{2}} ([\dot{u}_e]_j - [\dot{u}_e]_{j-1}) \quad (4.21)$$

Note that for $0 \leq j \leq k$, $[R_e]_j = 0$.

For the nth time step, $[\dot{u}_e]_n$ in Eq. 4.21 is unknown. Substituting Eq. 4.13b and using Eq. 4.14b leads to

$$[R_e]_n = \gamma \Delta t [M_e^\infty]_{\frac{1}{2}} [\ddot{u}_e]_n + [\tilde{R}_e]_n \quad (4.22)$$

with the known

$$[\tilde{R}_e]_n = (1-\gamma)\Delta t [M_e^\infty]_{\frac{1}{2}} [\ddot{u}_e]_{n-1} + \sum_{j=k+2}^{n-1} [M_e^\infty]_{n-j+\frac{1}{2}} ([\dot{u}_e]_j - [\dot{u}_e]_{j-1})$$

$$+ [M_e^\infty]_{n-k-\frac{1}{2}} [\dot{u}_e]_{k+1} \quad (4.23)$$

For $k \geq 1$ (which is always satisfied), all $[M_e^\infty]_j$ appearing in Eq. 4.23 are known as they are calculated by forecasting.

Substituting Eq. 4.22 in Eq. 4.12b and then in Eq. 4.15 using Eq. 4.10b results in

$$\begin{bmatrix} [D_{ii}] & [D_{im}] & 0 \\ [D_{mi}] & [D_{mm}] & [D_{me}] \\ 0 & [D_{em}] & [D_{ee}] + \gamma \Delta t [M_e^\infty]_{\frac{1}{2}} \end{bmatrix} \begin{bmatrix} [\ddot{u}_i]_n \\ [\ddot{u}_m]_n \\ [\ddot{u}_e]_n \end{bmatrix} = \begin{bmatrix} [P_i]_n \\ 0 \\ -[\tilde{R}_e]_n \end{bmatrix} - [K][\tilde{u}]_n \quad (4.24)$$

This equation represents the equations of motion of the unbounded medium with the structure-medium interface coinciding with the interior boundary, discretized in time and space, which will lead to the unit-impulse response matrix at the structure-medium interface.

After completing the time discretization, the forecasting algorithm can be discussed. The following sequence is processed for the nth time step. Up to and including the $(n-1)$th step, the motion, i.e. the displacements, velocities and accelerations for all degrees of freedom, and the unit-impulse response matrices at the interior boundary $[M_i^\infty]_j$ $(j = 1, \ldots, n-1)$ are known. In addition, from previous time steps $[M_e^\infty]_{\frac{1}{2}}$, $\ldots, [M_e^\infty]_{n-k-\frac{3}{2}}$ are known based on forecasting (Eq. 4.8).

4.3 TIME DISCRETIZATION AND NUMERICAL ALGORITHM

To be able to calculate $[\tilde{R}_e]_n$ (Eq. 4.23), $[M_e^\infty]_{n-k-\frac{1}{2}}$ is first determined based on forecasting

$$[M_e^\infty]_{n-k-\frac{1}{2}} = [M_e^\infty((n-k-\frac{1}{2})\Delta t)] = \left(\frac{r_e}{r_i}\right)^{s-1} [M_i^\infty(\frac{r_i}{r_e}(n-k-\frac{1}{2})\Delta t)] \quad (4.25)$$

As already discussed, k can be chosen arbitrarily. When $k \geq 1$ (which is always satisfied), $(r_i/r_e)(n-k-\frac{1}{2})\Delta t < (n-1)\Delta t$ holds. This permits the right-hand side of Eq. 4.25 to be calculated by interpolating $[M_i^\infty]_j$ ($j = 1, \ldots, n-1$) which then yields $[M_e^\infty]_{n-k-\frac{1}{2}}$.

In Eq. 4.24 $[\ddot{u}_m]_n$, $[\ddot{u}_e]_n$ and $[P_i]_n$ are unknowns which are determined by processing this equation. $[u]_n$ and $[\dot{u}]_n$ are obtained from Eqs. 4.13 and 4.14. Applying Eqs. 4.12a and 4.6 yields

$$[M_i^\infty]_n = \frac{1}{c}[P_i]_n \quad (4.26)$$

which is the unit-impulse response matrix at the structure-medium interface of the unbounded medium. This completes the nth time step.

The time delay caused by waves propagating with finite velocity is calculated rigorously only for a continuous formulation. After spatial and time discretizations, wave propagation is represented approximately. To be able to reduce the number of rows of finite elements in the radial direction and thus cut down the computational effort, the following points must be considered:

1. To describe the wave pattern behind the wave front accurately, at least one finite element is necessary. For times before the wave front travels this distance, the result is unreliable, and thus the unit-impulse response matrix at the interior boundary is disregarded.
2. To be able to eliminate the effect of the fictitious high-frequency components in the time history of the interaction forces at the interior boundary on the unit-impulse response matrix (Eq. 4.26), an averaging over m consecutive values is performed, yielding the unit-impulse response matrix at the centre of the time interval $[M_i^\infty]_{n-\frac{m}{2}}$ ($m \geq 1$, typically 3). To be able to calculate the last term on the right-hand side of Eq. 4.23 from Eq. 4.25, the following inequality

$$\frac{r_i}{r_e}\left(n-k-\frac{1}{2}\right) \leq n-m-1 \quad (4.27)$$

must be satisfied. This is achieved by choosing a sufficiently large k. Introducing the conservative assumption $r_i = r_e$, Eq. 4.27 is simplified to

$$k \geq m + \frac{1}{2} \quad (4.28)$$

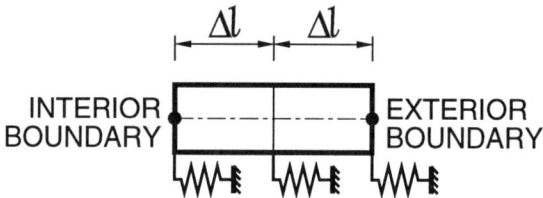

Figure 4-3 Finite-element region of semi-infinite rod on elastic foundation

4.4 ACCURACY

To illustrate the implementation of the forecasting method, a series of examples with increasing complexity is examined. This also permits the accuracy to be evaluated. All calculations are performed with the Newmark method using $\beta = 0.25$ and $\gamma = 0.5$.

4.4.1 Semi-infinite Rod on Elastic Foundation

The semi-infinite rod on an elastic foundation is addressed. This one-dimensional case with an analytical solution is described in Appendix A.2 (Figure A-3a). In this case the characteristic length r which is equal to r_0 does not vary in the longitudinal ("radial") direction. The ratio r_e/r_i is thus equal to 1.

The finite-element region with its interior boundary coinciding with the structure-medium interface consists of only 2 elements of length $\Delta l = 0.1 r_0$ (Figure 4-3). A time step $\Delta t = 0.04 r_0/c_l$ and $m = 3$ are selected, corresponding to $k = 4$. The inequality of Eq. 4.28 is satisfied. The result of the first time step being unreliable is disregarded. Excellent agreement of the acceleration unit-impulse response function $M^\infty(\bar{t})$ (non-dimensionalized with $\rho c_l A$) with the exact solution of Eq. A.2.23 (Figure A-5) results (Figure 4-4).

As an application, the interaction force $R(t)$ caused by a rounded triangular displacement pulse at the end of the rod described in Eq. A.2.24 is calculated. Analogous to Eq. 4.20

$$R_n = \sum_{j=1}^{n} M^\infty_{n-j-\frac{1}{2}} (\dot{u}_{0j} - \dot{u}_{0j-1}) \tag{4.29}$$

follows with the velocities determined from Eq. A.2.24. The analysis is performed with the time step $\Delta t = 0.12 r_0/c_l$. The interaction force $R(\bar{t})$ (non-dimensionalized with $K^\infty u_0$) calculated using the forecasting method hardly deviates from the exact solution of Figure A-6 (Figure 4-5).

4.4 ACCURACY

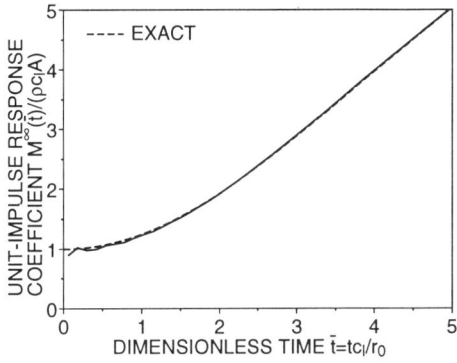

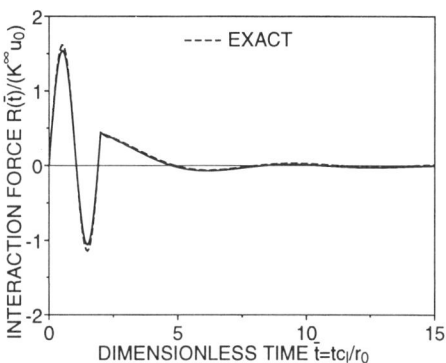

Figure 4-4 Acceleration unit-impulse response coefficient of semi-infinite rod on elastic foundation

Figure 4-5 Interaction force caused by rounded triangular displacement pulse applied to semi-infinite rod on elastic foundation

4.4.2 Out-of-plane Motion of Semi-infinite Wedge

The out-of-plane motion of the semi-infinite wedge described in Appendix A.4.1 (Figure A-11) is addressed. The analytical solution of the dynamic-stiffness coefficient (Eq. A.4.30) is plotted for the opening angle $\alpha = 30°$ in Figure A-12.

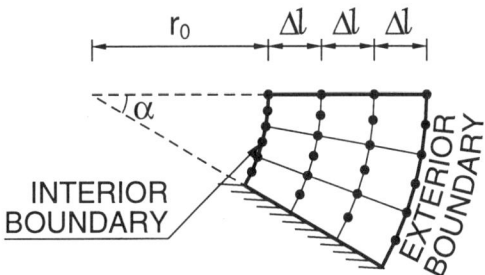

Figure 4-6 Finite-element region of semi-infinite wedge (not to scale in radial direction)

The mesh consists of 3×3 isoparametric finite elements, each element having 3 nodes in the circumferential direction and 2 nodes in the radial direction (Figure 4-6). The length in the radial direction equals $\Delta l = 0.02 r_0$. The time step $\Delta t = 0.01 r_0/c_s$ and $m = 2$ are chosen. With this mesh and time step, the wave with velocity c_s will reach the exterior boundary at the end of the 6th time step. Thus, $k = 5$ follows, which also satisfies Eq. 4.28.

The acceleration unit-impulse response matrix $[M^\infty(t)]$ of order 6×6 is determined. The enforced displacement pattern on the structure-medium interface (Eq. A.4.7) defines the $\{\phi\}$ vector in Eq. A.0.1. The corresponding coefficient $M^\infty(t)$ is calculated using Eq. A.0.1. In order to compare with the analytical solution in the frequency domain, the dynamic-stiffness coefficient $S^\infty(\omega)$ is evaluated as described in Eqs. 2.23 and 2.25. With the static-stiffness coefficient K^∞ of Eq. A.4.31 the decomposition of Eq. A.0.3 yields the dimensionless spring coefficient $k(a_0)$ and damping coefficient $c(a_0)$ which are plotted in Figure 4-7.

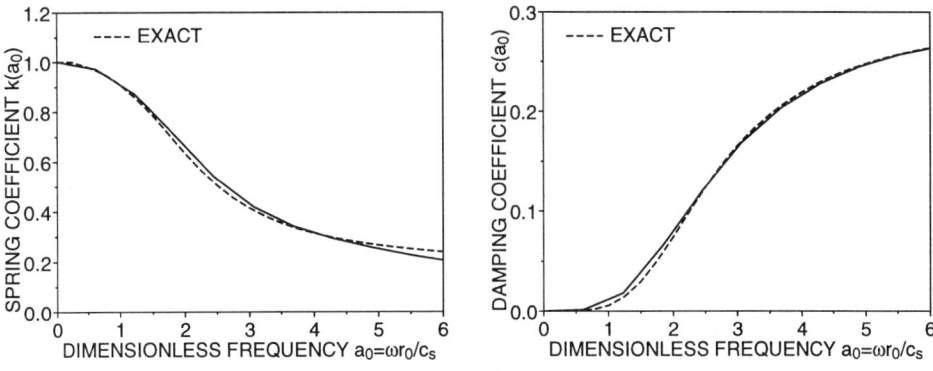

Figure 4-7 Dynamic-stiffness coefficient in frequency domain of out-of-plane motion of semi-infinite wedge

4.4.3 In-plane Motion of Semi-infinite Wedge

The in-plane motion of the semi-infinite wedge is examined, which is described in Appendix A.4.2 (Figure A-13). The homogeneous case $G_1 = G_2 = G_3 = G$ is processed with Poisson's ratio $\nu = 0.25$.

The same finite-element discretization as for the out-of-plane motion (Figure 4-6) is selected with $\Delta l = 0.025 r_0$. The time step Δt equals $0.01 r_0/c_s$ (shear-wave velocity c_s) with $m = 2$. During the time interval $m\Delta t$ the dilatational wave propagates with $c_p = \sqrt{3} c_s$ across 2 elements and the shear wave across 1.2 elements. As two wave velocities are involved, 3 rows of finite elements are necessary in contrast to the semi-infinite rod on an elastic foundation. k follows as 4 based on c_p. Eq. 4.28 is satisfied.

The acceleration unit-impulse response matrix $[M^\infty(t)]$ of order 12×12 is calculated. The enforced displacement pattern on the structure-medium interface (Figure A-13) defines the $\{\phi\}$ vector in Eq. A.0.1. The equivalent coefficient $M^\infty(t)$ (non-dimensionalized with $\rho c_s r_0$) is plotted as a function of $\bar{t} = t c_s / r_0$ in Figure 4-8, which agrees well with the result determined with an extended mesh of finite elements with the same length Δl and time step Δt. 35 rows of finite elements are necessary up to $\bar{t} = 2$. The procedure to calculate the extended mesh is explained at the beginning of Appendix A.

To further evaluate the accuracy, the results of dynamic condensation in the

4.4 ACCURACY

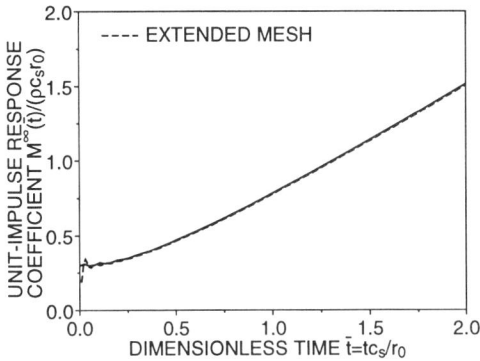

Figure 4-8 Acceleration unit-impulse response coefficient of in-plane motion of semi-infinite wedge

frequency domain are used. This procedure is outlined at the beginning of Appendix A. Starting from $M^\infty(t)$, the displacement dynamic-stiffness coefficient $S^\infty(\omega)$ is calculated as specified in Eqs. 2.23 and 2.25. The dynamic-stiffness coefficient is decomposed in a spring and dashpot coefficient (Eq. A.0.3) with $K^\infty = 0.759G$. The agreement is excellent (Figure 4-9).

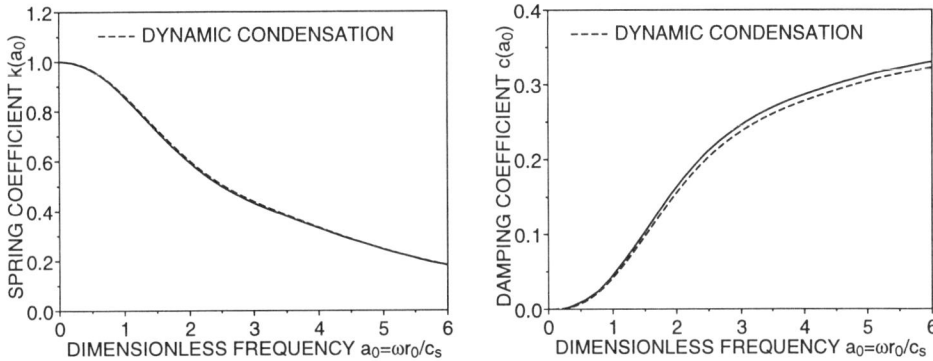

Figure 4-9 Dynamic-stiffness coefficient in frequency domain of in-plane motion of semi-infinite wedge

4.4.4 Strip Foundation with Rectangular Cross-section Embedded in Half-plane

The in-plane motion of a strip foundation with a rectangular cross-section embedded in a half-plane is addressed, which is described in Appendix A.6 (Figure A-16). First, the isotropic homogeneous case $G_1 = G_2 = G_3 = G$ is examined.

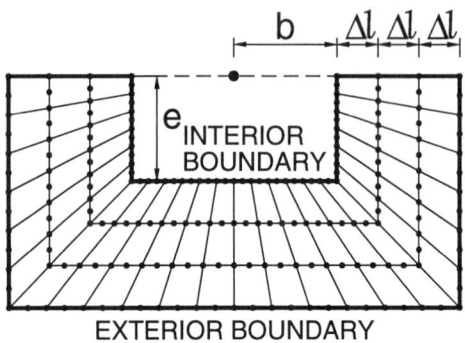

Figure 4-10 Finite-element region of half-plane used for embedded strip foundation (not to scale in radial direction)

The finite-element mesh (Figure 4-10) of the forecasting method consists of 24×3 isoparametric elements, each element having 3 nodes in the circumferential direction and 2 nodes in the radial direction The length in the radial direction equals $\Delta l = 0.025b$. The time step Δt equals $0.01b/c_s$ with $m = 3$, yielding $k = 4$ which satisfies Eq. 4.28.

As described in Section 4.4.3 for the in-plane motion of the semi-infinite wedge, the acceleration unit-impulse response matrix $[M^\infty(t)]$ of order 98×98 is determined. After enforcing the rigid-body constraints on the structure-medium interface which defines the $\{\phi\}$ vector in Eq. A.0.1, the equivalent coefficients corresponding to the horizontal, vertical and rocking degrees of freedom are calculated. To be able to perform a comparison in the frequency domain, the corresponding displacement dynamic-stiffness coefficients $S^\infty(\omega)$ are evaluated from $M^\infty(t)$ using Eqs. 2.23 and 2.25. The results of the boundary-element method [56] and of dynamic condensation (see beginning of Appendix A) are used. The dynamic-stiffness coefficients decomposed as in Eq. A.0.3, but non-dimensionalized for the translational motions with G and for the rocking motion with Gb^2, are plotted in Figure 4-11. Good agreement of the results of the forecasting method with those of dynamic condensation, which are available for the entire range of frequencies, is observed.

4.4 ACCURACY

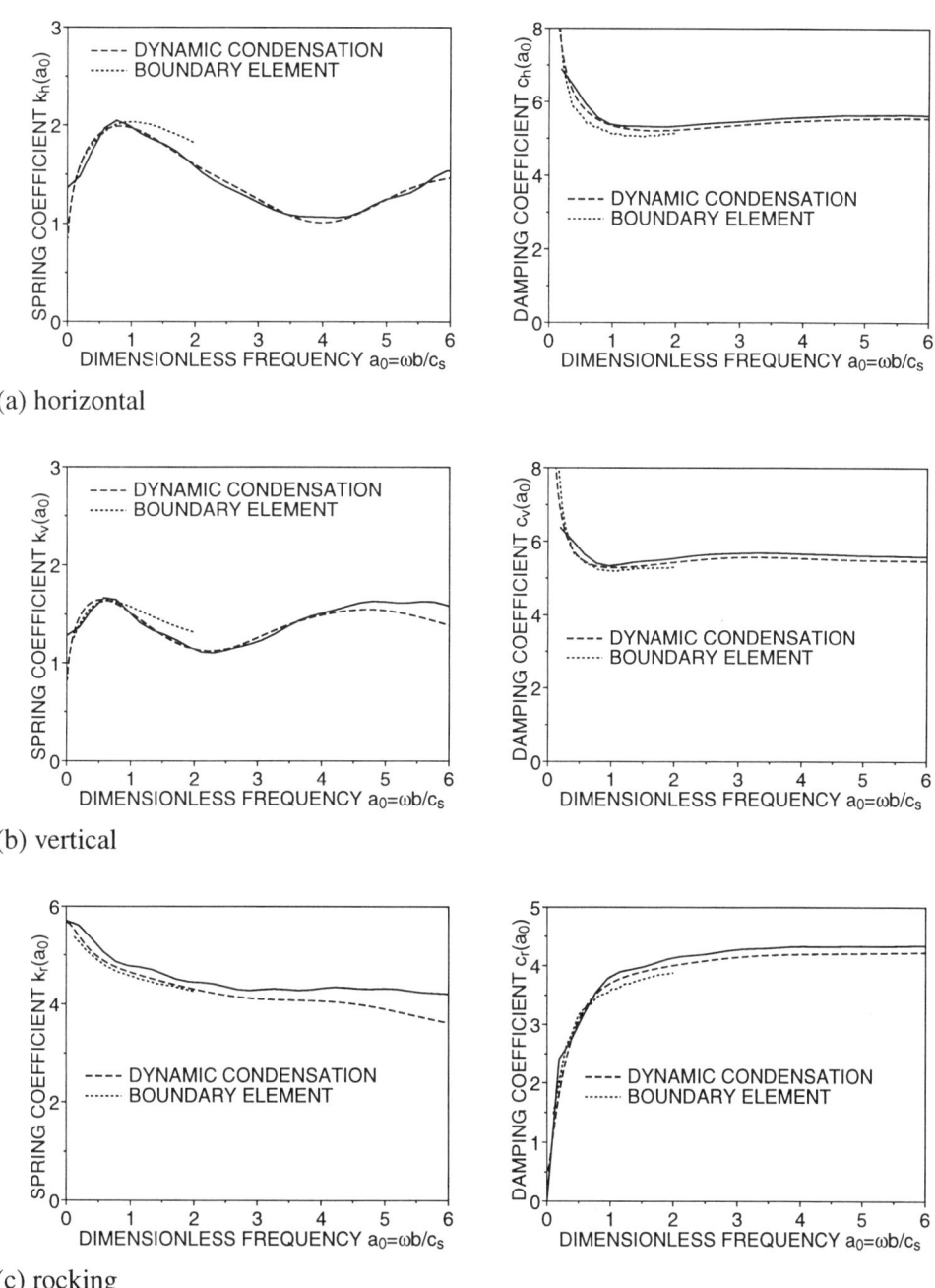

Figure 4-11 Dynamic-stiffness coefficients in frequency domain of strip foundation embedded in isotropic half-plane

5

CONSISTENT INFINITESIMAL FINITE-ELEMENT CELL METHOD FOR WAVE PROPAGATION

This chapter is based on References [50] and [70] for the two- and three-dimensional scalar wave equations and on References [68] and [51] for the two- and three-dimensional vector wave equations.

The concept and the characteristics of the consistent infinitesimal finite-element cell method are discussed in Section 5.1. The fundamental equations in continuous frequency and time are derived in Section 5.2, followed by the time discretization in Section 5.3. The accuracy of the consistent infinitesimal finite-element cell method is evaluated in Section 5.4.

5.1 CONCEPT BASED ON INFINITESIMAL CELL WIDTH

The consistent infinitesimal finite-element cell method is a concise implementation to calculate the unit-impulse response matrix of the unbounded medium based on similarity and finite elements. Only the structure-medium interface is discretized with surface finite elements resulting in a reduction of the spatial dimension by one. The radial direction towards infinity is represented exactly and the circumferential directions are modelled "exactly in the finite-element sense". In contrast to the forecasting method (Chapter 4) which only works for wave propagation in the time domain, the consistent infinitesimal finite-element cell method is in addition applicable to incompressible elasto-dynamics, diffusion and statics in both time and frequency domains. All the advantages of the similarity-based finite-element modelling of the unbounded medium discussed in Section 1.5.1 also apply to the consistent infinitesimal finite-element cell method.

Without addressing any details, the concept of the consistent infinitesimal finite-element cell method can be described. The unit-impulse response matrix or the dynamic-stiffness matrix at the structure-medium interface of the unbounded medium is to be determined. As these two matrices are related by the Fourier transformation

(Chapter 2), the unit-impulse response matrix only is addressed in describing the concept.

In the application only the structure-medium interface is discretized. In the derivation of the consistent infinitesimal finite-element cell equation a fictitious similar interface at an infinitesimal distance measured in the radial direction is introduced (Figure 5-1a) to make use of the relationship between the unit-impulse response matrices of similar interfaces of the unbounded medium. The similar

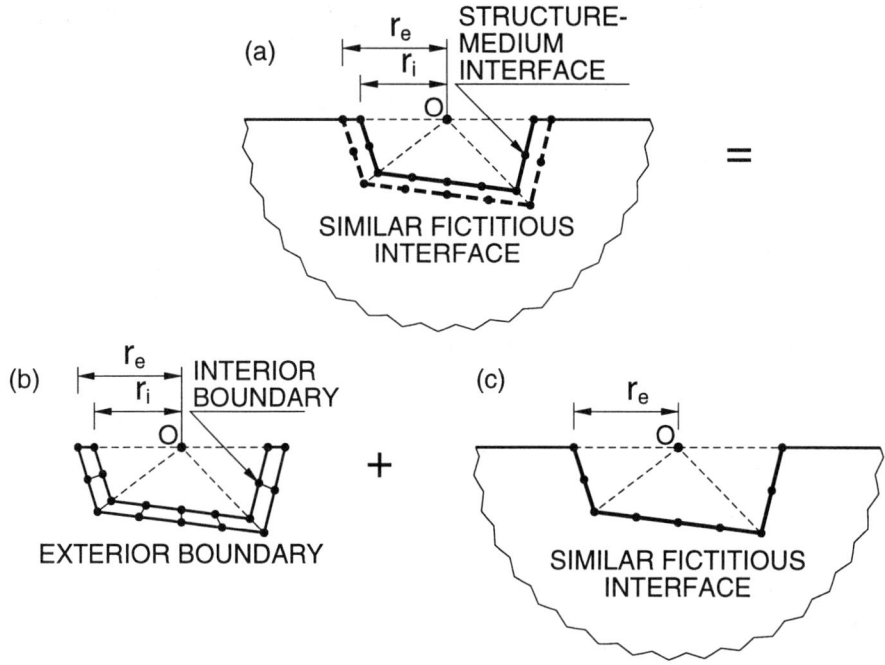

Figure 5-1 Concept of consistent infinitesimal finite-element cell method with infinitesimal cell width leading to finite-element discretization of structure-medium interface only

interfaces are defined by the characteristic length r: r_i for the structure-medium interface, r_e for the fictitious interface. r_e follows from r_i as

$$r_e = (1+w)r_i \tag{5.1}$$

with the infinitesimal dimensionless length w. The dynamic-stiffness matrices and the unit-impulse response matrices of the unbounded medium at these two interfaces are related as specified in Eq. 3.12 and Eq. 3.21, respectively.

The region between the structure-medium interface and the fictitious interface is a cell of infinitesimal width which is discretized with finite elements (Figure 5-1b). Its interior and exterior boundaries coincide with the structure-medium interface and the fictitious interface, respectively. The arrangement of the nodes on the two boundaries must satisfy similarity. w in Eq. 5.1 is thus called the *infinitesimal dimensionless cell width*. Adding the infinitesimal finite-element cell to the unbounded medium defined by the fictitious interface (Figure 5-1c) results in the unbounded medium defined by the structure-medium interface. The same applies to their unit-impulse response matrices. This *assemblage enforcing compatibility and equilibrium* results in another relationship linking the unit-impulse response matrices at the two interfaces. Substituting the relationship based on *similarity* mentioned above permits the unit-impulse response matrix at the structure-medium interface to be expressed as a function of the property matrices (static-stiffness and mass matrices) of the finite-element cell.

In the infinitesimal finite-element cell method [65], [66], the dimensionless cell width measured in the radial direction is selected as a very small number for which the computer will provide reliable results. It is appealing to perform the limit of the cell width *analytically*, which leads to the *consistent* formulation. In this no arbitrary infinitesimal cell width has to be chosen, which also eliminates the associated potential numerical problem. The mathematical formulation of this consistent approach is more concise. After performing the limit analytically, *only one interface, the structure-medium interface* which *has to be discretized* and where all unknowns are defined, is present (Figure 5-1a). Only the consistent infinitesimal finite-element cell method is described in this chapter.

5.2 FUNDAMENTAL EQUATIONS

The derivation, which is performed in the frequency domain for the sake of simplicity, proceeds as follows. To be able to perform the limit analytically, the static-stiffness and mass matrices of the infinitesimal finite-element cell are decomposed in a power series of the dimensionless cell width, which leads to the coefficient matrices. This is performed for any type of finite-element discretization of the structure-medium interface addressing the three-dimensional vector wave equation (Section 5.2.1). The coefficient matrices are then listed for the three-dimensional scalar wave equation (Section 5.2.2), the two-dimensional vector wave equation (Section 5.2.3) and the two-dimensional scalar wave equation (Section 5.2.4). Explicit expressions for the 2-node line element of the two-dimensional vector wave equation (Section 5.2.3) and scalar wave equation (Section 5.2.4) are also listed. The force-displacement relationship of the finite-element cell is established. Assembling the cell and the unbounded medium which enforces compatibility and equilibrium yields an equation for the dynamic-stiffness matrices of the unbounded medium referred to the interior and exterior boundaries of the cell (Section 5.2.7). Substituting the equation of the dynamic-stiffness matrices based on similarity and then performing the limit of the infinitesimal cell width analytically leads to the consistent infinitesimal finite-element cell equation in the frequency domain (Section 5.2.8). Applying the inverse Fourier transformation yields the consistent infinitesimal finite-element cell equation in the time domain, which permits the acceleration unit-impulse response matrix to be calculated (Section 5.2.9). As a special case the similarity can be located at infinity, as for the horizontally layered unbounded medium (Section 5.2.10). As an extension the unbounded medium with material properties varying as power functions in the radial direction is addressed (Section 5.2.11).

5.2.1 Coefficient Matrices of Finite-element Cell for Three-dimensional Vector Wave Equation

In a three-dimensional unbounded medium-structure-interaction analysis the structure-medium interface is, in general, a doubly-curved finite surface, which is discretized using a general two-dimensional mesh of isoparametric surface finite elements (shaded area in Figure 5-2a). Starting from this mesh a three-dimensional

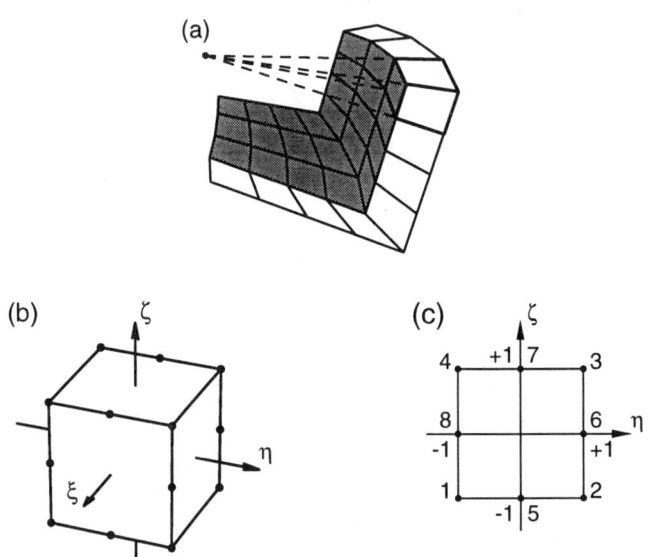

Figure 5-2 (a) Structure-medium interface with adjacent three-dimensional finite-element cell; (b) Three-dimensional parent element of finite element of cell; (c) Two-dimensional parent element of finite element on discretized structure-medium interface

finite-element cell satisfying similarity with a single element in the radial direction is constructed for the derivation of the consistent infinitesimal finite-element cell

5.2 FUNDAMENTAL EQUATIONS

equation. In the radial direction of the finite element of the cell a linear shape function is introduced. To construct the static-stiffness and mass matrices of the isoparametric three-dimensional finite element, its parent element is addressed. As an example the parent element with 8 nodes on the structure-medium interface is shown in Figure 5-2b. The ξ-axis of the parent element corresponds to the radial direction which points from the structure-medium interface towards infinity. At the interior boundary (which coincides with the structure-medium interface) and at the exterior boundary the values $\xi_i = -1$ and $\xi_e = +1$ result.

On the interior boundary $\{N(\eta,\zeta)\}$ denotes the shape functions of the two-dimensional surface finite element. The corresponding 8-node quadrilateral parent element is shown in Figure 5-2c. The following shape functions apply with the standard indicated numbering scheme

$$N_1(\eta,\zeta) = \frac{1}{4}(1-\eta)(1-\zeta) - \frac{1}{2}(N_8(\eta,\zeta) + N_5(\eta,\zeta)) \quad (5.2a)$$

$$N_2(\eta,\zeta) = \frac{1}{4}(1+\eta)(1-\zeta) - \frac{1}{2}(N_5(\eta,\zeta) + N_6(\eta,\zeta)) \quad (5.2b)$$

$$N_3(\eta,\zeta) = \frac{1}{4}(1+\eta)(1+\zeta) - \frac{1}{2}(N_6(\eta,\zeta) + N_7(\eta,\zeta)) \quad (5.2c)$$

$$N_4(\eta,\zeta) = \frac{1}{4}(1-\eta)(1+\zeta) - \frac{1}{2}(N_7(\eta,\zeta) + N_8(\eta,\zeta)) \quad (5.2d)$$

$$N_5(\eta,\zeta) = \frac{1}{2}(1-\eta^2)(1-\zeta) \quad (5.2e)$$

$$N_6(\eta,\zeta) = \frac{1}{2}(1+\eta)(1-\zeta^2) \quad (5.2f)$$

$$N_7(\eta,\zeta) = \frac{1}{2}(1-\eta^2)(1+\zeta) \quad (5.2g)$$

$$N_8(\eta,\zeta) = \frac{1}{2}(1-\eta)(1-\zeta^2) \quad (5.2h)$$

The shape functions of the three-dimensional parent element $\{\hat{N}(\xi,\eta,\zeta)\} = \{\hat{N}\}$ (the symbol circumflex ^ denotes the three-dimensional finite element) are generated from those of the two-dimensional parent element $\{N(\eta,\zeta)\} = \{N\}$. It is appropriate to decompose the shape function vector $\{\hat{N}\}$ of the three-dimensional element into two subvectors corresponding to the nodes on the interior boundary i and the exterior boundary e. For the conciseness of the derivation, a subscript, e.g. j, is introduced as an index which assumes either the value of i or e. This subscript is placed outside the brace used to denote the vector. When the subscript is assigned a fixed value i or e, it is moved inside the brace. This convention leads to a nomenclature which is consistent with that of the forecasting method (Chapter 4). The same convention applies also to coordinate vectors, stiffness and mass matrices and some intermediate quantities. After this decomposition of the shape function vector with respect to the interior and exterior boundaries

$$\{\hat{N}\} = \left\{ \begin{array}{c} \{\hat{N}_i\} \\ \{\hat{N}_e\} \end{array} \right\} \qquad (5.3)$$

the subvectors are equal to

$$\{\hat{N}\}_j = \frac{1}{2}(1+\xi_j\xi)\{N\} \qquad (j=i,e) \qquad (5.4)$$

with a typical element

$$\hat{N}_{jk} = \frac{1}{2}(1+\xi_j\xi)N_k \qquad (j=i,e;\quad k=1,2,\ldots) \qquad (5.5)$$

j denotes the interior or exterior boundary and k the node number of the two-dimensional parent element. Owing to similarity, the coordinates of the nodes on the exterior boundary $\{x_e\}$, $\{y_e\}$, $\{z_e\}$ can be expressed by those of the nodes on the interior boundary $\{x\}$, $\{y\}$, $\{z\}$ and the dimensionless cell width w as (Eq. 5.1)

$$\{x_e\} = (1+w)\{x\} \qquad (5.6a)$$
$$\{y_e\} = (1+w)\{y\} \qquad (5.6b)$$
$$\{z_e\} = (1+w)\{z\} \qquad (5.6c)$$

Note that the subscript i for the coordinates of the interior boundary, the structure-medium interface, is omitted (as in an actual application only the latter is discretized). The nodal coordinates of the three-dimensional finite element $\{\hat{x}\}$, $\{\hat{y}\}$, $\{\hat{z}\}$ are constructed by assembling those of the nodes on the interior and exterior boundaries $\{x\}$ and $\{x_e\}$, $\{y\}$ and $\{y_e\}$, $\{z\}$ and $\{z_e\}$.

Introducing the isoparametric mapping rule and with Eqs. 5.3, 5.4 and 5.6

$$\hat{x} = \{\hat{N}\}^T\{\hat{x}\} = \{\hat{N}_i\}^T\{x\} + \{\hat{N}_e\}^T\{x_e\} = \left(1+\frac{w}{2}(1+\xi)\right)\{N\}^T\{x\} \qquad (5.7a)$$

$$\hat{y} = \left(1+\frac{w}{2}(1+\xi)\right)\{N\}^T\{y\} \qquad (5.7b)$$

$$\hat{z} = \left(1+\frac{w}{2}(1+\xi)\right)\{N\}^T\{z\} \qquad (5.7c)$$

follow. The Jacobian matrix of the three-dimensional finite element equals

$$[\hat{J}] = \begin{bmatrix} \hat{x}_{,\xi} & \hat{y}_{,\xi} & \hat{z}_{,\xi} \\ \hat{x}_{,\eta} & \hat{y}_{,\eta} & \hat{z}_{,\eta} \\ \hat{x}_{,\zeta} & \hat{y}_{,\zeta} & \hat{z}_{,\zeta} \end{bmatrix} = \begin{bmatrix} \frac{w}{2} & & \\ & 1+\frac{w}{2}(1+\xi) & \\ & & 1+\frac{w}{2}(1+\xi) \end{bmatrix} [J] \qquad (5.8)$$

5.2 FUNDAMENTAL EQUATIONS

where the abbreviation

$$[J] = \begin{bmatrix} \{N\}^T\{x\} & \{N\}^T\{y\} & \{N\}^T\{z\} \\ \{N_{,\eta}\}^T\{x\} & \{N_{,\eta}\}^T\{y\} & \{N_{,\eta}\}^T\{z\} \\ \{N_{,\zeta}\}^T\{x\} & \{N_{,\zeta}\}^T\{y\} & \{N_{,\zeta}\}^T\{z\} \end{bmatrix} \quad (5.9)$$

is introduced. The determinant of the Jacobian matrix equals

$$|\hat{J}| = \frac{w}{2}\left(1 + \frac{w}{2}(1+\xi)\right)^2 |J| \quad (5.10)$$

The inverse of the Jacobian matrix follows as

$$[\hat{J}]^{-1} = [J]^{-1} \begin{bmatrix} \frac{2}{w} & & \\ & \frac{1}{1+\frac{w}{2}(1+\xi)} & \\ & & \frac{1}{1+\frac{w}{2}(1+\xi)} \end{bmatrix} \quad (5.11)$$

where $[J]^{-1}$ is denoted as

$$[J]^{-1} = \begin{bmatrix} j_{11} & j_{12} & j_{13} \\ j_{21} & j_{22} & j_{23} \\ j_{31} & j_{32} & j_{33} \end{bmatrix} \quad (5.12)$$

The derivatives of the shape functions are formulated as

$$\begin{Bmatrix} \hat{N}_{jk,\hat{x}} \\ \hat{N}_{jk,\hat{y}} \\ \hat{N}_{jk,\hat{z}} \end{Bmatrix} = [\hat{J}]^{-1} \begin{Bmatrix} \hat{N}_{jk,\xi} \\ \hat{N}_{jk,\eta} \\ \hat{N}_{jk,\zeta} \end{Bmatrix} = \frac{\xi_j}{w} \begin{Bmatrix} j_{11} \\ j_{21} \\ j_{31} \end{Bmatrix} N_k$$

$$+ \frac{1+\xi_j\xi}{2\left(1+\frac{w}{2}(1+\xi)\right)} \left(\begin{Bmatrix} j_{21} \\ j_{22} \\ j_{23} \end{Bmatrix} N_{k,\eta} + \begin{Bmatrix} j_{13} \\ j_{23} \\ j_{33} \end{Bmatrix} N_{k,\zeta} \right) \quad (5.13)$$

The strain-nodal displacement matrix with the strain vector arranged as $[\varepsilon_x, \varepsilon_y, \varepsilon_z, \gamma_{yz}, \gamma_{xz}, \gamma_{xy}]^T$ equals

$$[B]_{jk} = \begin{bmatrix} \hat{N}_{jk,\hat{x}} & 0 & 0 \\ 0 & \hat{N}_{jk,\hat{y}} & 0 \\ 0 & 0 & \hat{N}_{jk,\hat{z}} \\ 0 & \hat{N}_{jk,\hat{z}} & \hat{N}_{jk,\hat{y}} \\ \hat{N}_{jk,\hat{z}} & 0 & \hat{N}_{jk,\hat{x}} \\ \hat{N}_{jk,\hat{y}} & \hat{N}_{jk,\hat{x}} & 0 \end{bmatrix} = \frac{\xi_j}{w}[B^1]_k + \frac{1+\xi_j\xi}{2\left(1+\frac{w}{2}(1+\xi)\right)}[B^2]_k \quad (5.14)$$

where

$$[B^1]_k = \begin{bmatrix} j_{11} & 0 & 0 \\ 0 & j_{21} & 0 \\ 0 & 0 & j_{31} \\ 0 & j_{31} & j_{21} \\ j_{31} & 0 & j_{11} \\ j_{21} & j_{11} & 0 \end{bmatrix} N_k \qquad (5.15a)$$

$$[B^2]_k = \begin{bmatrix} j_{12} & 0 & 0 \\ 0 & j_{22} & 0 \\ 0 & 0 & j_{32} \\ 0 & j_{32} & j_{22} \\ j_{32} & 0 & j_{12} \\ j_{22} & j_{12} & 0 \end{bmatrix} N_{k,\eta} + \begin{bmatrix} j_{13} & 0 & 0 \\ 0 & j_{23} & 0 \\ 0 & 0 & j_{33} \\ 0 & j_{33} & j_{23} \\ j_{33} & 0 & j_{13} \\ j_{23} & j_{13} & 0 \end{bmatrix} N_{k,\zeta} \qquad (5.15b)$$

Assembling $[B^1]_k$ and $[B^2]_k$ of all nodes yields $[B^1]$ and $[B^2]$, which are not functions of ξ. $[B]$ is decomposed with respect to the interior and exterior boundaries as

$$[B] = [\,[B_i]\quad [B_e]\,] \qquad (5.16)$$

with

$$[B]_j = \frac{\xi_j}{w}[B^1] + \frac{1+\xi_j\xi}{2\left(1+\frac{w}{2}(1+\xi)\right)}[B^2] \qquad (j=i,e) \qquad (5.17)$$

The static-stiffness matrix of the three-dimensional finite element equals

$$[K] = \int_V [B]^T [D][B]\,dV \qquad (5.18)$$

with the integration over the volume V of the element. $[D]$ is the three-dimensional (in general, anisotropic) elasticity matrix of the stress-strain relationship and is positive definite. (As an example the elasticity matrix $[D]$ of the transversely isotropic material is constructed from engineering elastic moduli in Appendix A.8.) $[K]$ is decomposed into submatrices with respect to the interior and exterior boundaries as ($j = i, e; l = i, e$)

$$[K]_{jl} = \int_V [B]_j^T [D][B]_l\,dV = \int_{-1}^{+1}\int_{-1}^{+1}\int_{-1}^{+1} [B]_j^T [D][B]_l |\hat{J}|\,d\xi\,d\eta\,d\zeta \qquad (5.19)$$

Substituting Eqs. 5.17 and 5.10 in Eq. 5.19 yields a polynomial in ξ which can be integrated analytically. Decomposition with respect to the dimensionless cell width w

5.2 FUNDAMENTAL EQUATIONS

for later use in the consistent infinitesimal finite-element cell equation yields without introducing any approximation

$$[K]_{jl} = \frac{1}{w}[K^0]_{jl} + [K^1]_{jl} + w[K^2]_{jl} \tag{5.20}$$

where

$$[K^0]_{jl} = \xi_j \xi_l [E^0] \tag{5.21a}$$

$$[K^1]_{jl} = \xi_j \xi_l [E^0] + \frac{\xi_l}{2}[E^1] + \frac{\xi_j}{2}[E^1]^T \tag{5.21b}$$

$$[K^2]_{jl} = \frac{\xi_j \xi_l}{3}[E^0] + \left(\frac{\xi_l}{4} + \frac{\xi_j \xi_l}{12}\right)[E^1]$$

$$+ \left(\frac{\xi_j}{4} + \frac{\xi_j \xi_l}{12}\right)[E^1]^T + \left(\frac{1}{4} + \frac{\xi_j \xi_l}{12}\right)[E^2] \tag{5.21c}$$

with

$$[E^0] = \int_{-1}^{+1}\int_{-1}^{+1}[B^1]^T[D][B^1]|J|\,d\eta\,d\zeta \tag{5.22a}$$

$$[E^1] = \int_{-1}^{+1}\int_{-1}^{+1}[B^2]^T[D][B^1]|J|\,d\eta\,d\zeta \tag{5.22b}$$

$$[E^2] = \int_{-1}^{+1}\int_{-1}^{+1}[B^2]^T[D][B^2]|J|\,d\eta\,d\zeta \tag{5.22c}$$

The coefficient matrices specified in Eq. 5.22 are determined by numerical integration. Note that $[E^0]$, $[E^1]$ and $[E^2]$ depend on the discretization of the structure-medium interface and not on that in the radial direction. $[E^0]$ is positive definite which can be proven as follows.

For conciseness Eq. 5.15a is rewritten as

$$[B^1]_k = [b^1]N_k \tag{5.23}$$

with

$$[b^1] = \begin{bmatrix} j_{11} & 0 & 0 \\ 0 & j_{21} & 0 \\ 0 & 0 & j_{31} \\ 0 & j_{31} & j_{21} \\ j_{31} & 0 & j_{11} \\ j_{21} & j_{11} & 0 \end{bmatrix} \tag{5.24}$$

Assemblage $[B^1]_k$ of all nodes of the surface finite element yields

$$[B^1] = [b^1][N] \tag{5.25}$$

with

$$[N] = \begin{bmatrix} N_1 & & & N_2 & & \cdots \\ & N_1 & & & N_2 & \cdots \\ & & N_1 & & & N_2 & \cdots \end{bmatrix} \quad (5.26)$$

Substituting Eq. 5.25 in Eq. 5.22a leads to

$$[E^0] = \int_{-1}^{+1}\int_{-1}^{+1} [N]^T[b^1]^T[D][b^1][N]|J|\mathrm{d}\eta\mathrm{d}\zeta \quad (5.27)$$

For a properly constructed surface finite element $[J]$ is regular which results in j_{11}, j_{21}, j_{31} not vanishing identically. The rank of $[b^1]$ is thus 3. As $[D]$ is positive definite, the product $[b^1]^T[D][b^1]$ will also be positive definite. Invoking the definition of positive definiteness of a matrix, Eq. 5.27 is premultiplied and postmultiplied by the non-trivial nodal displacement vector $\{u\}$ leading to a quadratic form

$$\{u\}^T[E^0]\{u\} = \int_{-1}^{+1}\int_{-1}^{+1} \{u\}^T[N]^T[b^1]^T[D][b^1][N]\{u\}|J|\mathrm{d}\eta\mathrm{d}\zeta \quad (5.28)$$

$[N]\{u\}$ does not vanish identically within the surface finite element. Thus

$$\{u\}^T[E^0]\{u\} > 0 \quad (5.29)$$

applies, i.e. $[E^0]$ is positive definite. The positive definiteness of $[E^0]$ can also be verified based on a physical interpretation. For $w \ll 1$, the second and third terms in Eq. 5.20 can be neglected compared with the first one. The static-stiffness matrix of the cell is then formulated using Eq. 5.21a with $\xi_i = -1$, $\xi_e = +1$ as

$$[K] = \begin{bmatrix} [K_{ii}] & [K_{ie}] \\ [K_{ei}] & [K_{ee}] \end{bmatrix} \approx \frac{1}{w}\begin{bmatrix} [E^0] & -[E^0] \\ -[E^0] & [E^0] \end{bmatrix} \quad (5.30)$$

Fixing the exterior boundary and applying the displacement vector $\{u_i\}$ at the interior boundary, the virtual work equals

$$\left\{\begin{array}{c}\{u_i\}\\\{0\}\end{array}\right\}^T [K] \left\{\begin{array}{c}\{u_i\}\\\{0\}\end{array}\right\} = \frac{1}{w}\{u_i\}^T[E^0]\{u_i\} \quad (5.31)$$

where Eq. 5.30 is substituted. As the virtual work for a real physical system is always positive, $[E^0]$ must be positive definite. By inspection, $[E^2]$ is semi-positive definite (and symmetric).

The mass matrix with the mass density ρ equals

$$[M] = \int_V \rho[\hat{N}]^T[\hat{N}]\mathrm{d}V \quad (5.32)$$

5.2 FUNDAMENTAL EQUATIONS

Its submatrices are decomposed with respect to the interior and exterior boundaries as $(j = i, e; l = i, e)$

$$[M]_{jl} = \int_V \rho[\hat{N}]_j^T [\hat{N}]_l dV = \int_{-1}^{+1} \int_{-1}^{+1} \int_{-1}^{+1} \rho[\hat{N}]_j^T [\hat{N}]_l |\hat{J}| d\xi d\eta d\zeta \qquad (5.33)$$

where

$$[\hat{N}]_j = \frac{1}{2}(1 + \xi_j \xi)[N] \qquad (j = i, e) \qquad (5.34)$$

Eq. 5.10 is substituted in Eq. 5.33. Again the integration in the ξ–direction is performed analytically. As will become apparent, the terms of order in w higher than 1 can be neglected when the limit $w \to 0$ is performed analytically. Thus

$$[M]_{jl} = w[M^2]_{jl} + O(w^2) = \frac{w}{4}\left(1 + \frac{\xi_j \xi_l}{3}\right)[M^0] + O(w^2) \qquad (5.35)$$

applies with the coefficient matrix

$$[M^0] = \int_{-1}^{+1} \int_{-1}^{+1} \rho[N]^T [N]|J| d\eta d\zeta \qquad (5.36)$$

Note that $[M^0]$ is positive definite.

To derive the consistent infinitesimal finite-element cell equation, the following relationships will be used

$$[K_{ii}^0] = -[K_{ie}^0] = -[K_{ei}^0] = [K_{ee}^0] = [E^0] \qquad (5.37a)$$

$$[K_{ie}^1] + [K_{ee}^1] = -\left([K_{ii}^1] + [K_{ie}^1]\right)^T = [E^1] \qquad (5.37b)$$

$$[K_{ii}^2] + [K_{ie}^2] + [K_{ei}^2] + [K_{ee}^2] = [E^2] \qquad (5.37c)$$

$$[M_{ii}^2] + [M_{ie}^2] + [M_{ei}^2] + [M_{ee}^2] = [M^0] \qquad (5.38)$$

Eq. 5.37 follows from Eq. 5.21 with $\xi_i = -1$, $\xi_e = +1$ and Eq. 5.38 from Eq. 5.35. The right-hand sides of Eqs. 5.37 and 5.38 are defined in Eqs. 5.22 and 5.36, respectively. $[E^0]$, $[E^1]$, $[E^2]$ and $[M^0]$ are the coefficient matrices of one surface finite element.

As for the static-stiffness and mass matrices the coefficient matrices $[E^0]$, $[E^1]$, $[E^2]$ and $[M^0]$ are assembled to form those of the structure-medium interface. The latter will thus be banded. To simplify the nomenclature, the same symbols are used for the assembled coefficient, static-stiffness and mass matrices in the following. For the assembled static-stiffness and mass matrices Eqs. 5.37 and 5.38 still hold. In particular,

$$[K_{ii}^0] + [K_{ie}^0] + [K_{ei}^0] + [K_{ee}^0] = 0 \qquad (5.39)$$

applies (from Eq. 5.37a) and

$$[K_{ii}^1] + [K_{ie}^1] + [K_{ei}^1] + [K_{ee}^1] = 0 \qquad (5.40)$$

results from Eq. 5.37b and the symmetry of $[K^1]$.

5.2.2 Summary of Coefficient Matrices for Three-dimensional Scalar Wave Equation

For easy reference, the coefficient matrices of the finite-element cell for the three-dimensional scalar wave equation are listed. The derivation is restricted to those aspects which differ from Section 5.2.1.

The (isotropic) scalar wave equation equals

$$G(u_{,xx}+u_{,yy}+u_{,zz}) - \rho\ddot{u} = 0 \tag{5.41}$$

with the function $u = u(x,y,z,t)$ and the material constants G and ρ. The wave velocity is defined as

$$c = \sqrt{\frac{G}{\rho}} \tag{5.42}$$

Besides vanishing initial conditions, boundary conditions are prescribed as

$$u = \bar{u} \quad \text{on } S_u \tag{5.43a}$$
$$Gu_{,n} = \bar{q} \quad \text{on } S_q \tag{5.43b}$$

$\bar{u}$ is the prescribed function on the surface S_u and $\bar{q}$ the prescribed normal flux multiplied by the material constant G on S_q.

Many wave propagation problems are governed by this scalar wave equation. The physical significance of the material constants G, ρ will become apparent when the two-dimensional scalar wave equation of the out-of-plane motion is discussed in Section 5.2.4.

The discretization of the doubly-curved structure-medium interface and the construction of the infinitesimal finite-element cell are the same as for the vector wave equation. Thus, Eqs. 5.2 to 5.12 still apply.

The gradients of the shape functions equal

$$\{B\}_{jk} = \begin{Bmatrix} \hat{N}_{jk,\hat{x}} \\ \hat{N}_{jk,\hat{y}} \\ \hat{N}_{jk,\hat{z}} \end{Bmatrix} = [\hat{J}]^{-1} \begin{Bmatrix} \hat{N}_{jk,\xi} \\ \hat{N}_{jk,\eta} \\ \hat{N}_{jk,\zeta} \end{Bmatrix} = \frac{\xi_j}{w}\{B^1\}_k + \frac{1+\xi_j\xi}{2\left(1+\frac{w}{2}(1+\xi)\right)}\{B^2\}_k \tag{5.44}$$

where

$$\{B^1\}_k = \begin{Bmatrix} j_{11} \\ j_{21} \\ j_{31} \end{Bmatrix} N_k \tag{5.45a}$$

$$\{B^2\}_k = \begin{Bmatrix} j_{12} \\ j_{22} \\ j_{32} \end{Bmatrix} N_{k,\eta} + \begin{Bmatrix} j_{13} \\ j_{23} \\ j_{33} \end{Bmatrix} N_{k,\zeta} \tag{5.45b}$$

5.2 FUNDAMENTAL EQUATIONS

Assembling $\{B^1\}_k$ and $\{B^2\}_k$ of all nodes yields

$$[B^1] = \begin{Bmatrix} j_{11} \\ j_{21} \\ j_{31} \end{Bmatrix} \{N\}^T \qquad (5.46a)$$

$$[B^2] = \begin{Bmatrix} j_{12} \\ j_{22} \\ j_{32} \end{Bmatrix} \{N,_\eta\}^T + \begin{Bmatrix} j_{13} \\ j_{23} \\ j_{33} \end{Bmatrix} \{N,_\zeta\}^T \qquad (5.46b)$$

$[B^1]$ and $[B^2]$ are not a function of ξ. $[B]$ is decomposed with respect to the interior and exterior boundaries as specified in Eqs. 5.16 and 5.17.

The static-stiffness matrix of the three-dimensional finite element equals

$$[K] = \int_V G[B]^T [B] dV \qquad (5.47)$$

with the integration over the volume V of the element. $[K]$ is decomposed into submatrices with respect to the interior and exterior boundaries as $(j = i, e; l = i, e)$

$$[K]_{jl} = \int_V G[B]_j^T [B]_l dV = \int_{-1}^{+1} \int_{-1}^{+1} \int_{-1}^{+1} G[B]_j^T [B]_l |J| d\xi d\eta d\zeta \qquad (5.48)$$

The integration in the ξ–direction is performed analytically. Decomposition with respect to the dimensionless cell width w is formulated in Eqs. 5.20 and 5.21 with the coefficient matrices

$$[E^0] = \int_{-1}^{+1} \int_{-1}^{+1} G[B^1]^T [B^1] |J| d\eta d\zeta \qquad (5.49a)$$

$$[E^1] = \int_{-1}^{+1} \int_{-1}^{+1} G[B^2]^T [B^1] |J| d\eta d\zeta \qquad (5.49b)$$

$$[E^2] = \int_{-1}^{+1} \int_{-1}^{+1} G[B^2]^T [B^2] |J| d\eta d\zeta \qquad (5.49c)$$

The derivation of the mass matrix and the corresponding coefficient matrix is identical to the vector-wave equation (Eqs. 5.32 to 5.36). In particular

$$[M^0] = \int_{-1}^{+1} \int_{-1}^{+1} \rho [N]^T [N] |J| d\eta d\zeta \qquad (5.50)$$

applies with

$$[N] = [\ N_1 \quad N_2 \quad \ldots\ ] \qquad (5.51)$$

$[E^0]$, $[E^1]$, $[E^2]$ and $[M^0]$ are the coefficient matrices used in the consistent infinitesimal finite-element cell equation. The properties expressed in Eqs. 5.37 to 5.40 are still valid.

5.2.3 Summary of Coefficient Matrices for Two-dimensional Vector Wave Equation

The coefficient matrices of the finite-element cell for the two-dimensional vector wave equation follow straightforwardly from the three-dimensional case derived in Section 5.2.1. All terms in z and ζ are deleted. Those equations where this operation cannot be performed straightforwardly are listed in the following.

The determinant of the Jacobian matrix equals (Eq. 5.10)

$$|\hat{J}| = \frac{w}{2}\left(1 + \frac{w}{2}(1+\xi)\right)|J| \tag{5.52}$$

The strain-nodal displacement matrix equals (Eq. 5.14)

$$[B]_{jk} = \begin{bmatrix} \hat{N}_{jk,\hat{x}} & 0 \\ 0 & \hat{N}_{jk,\hat{y}} \\ \hat{N}_{jk,\hat{y}} & \hat{N}_{jk,\hat{x}} \end{bmatrix} = \frac{\xi_j}{w}[B^1]_k + \frac{1+\xi_j\xi}{2\left(1+\frac{w}{2}(1+\xi)\right)}[B^2]_k \tag{5.53}$$

where

$$[B^1]_k = \begin{bmatrix} j_{11} & 0 \\ 0 & j_{21} \\ j_{21} & j_{11} \end{bmatrix} N_k \tag{5.54a}$$

$$[B^2]_k = \begin{bmatrix} j_{12} & 0 \\ 0 & j_{22} \\ j_{22} & j_{12} \end{bmatrix} N_{k,\eta} \tag{5.54b}$$

The submatrices of the static stiffness (Eq. 5.19) with respect to the interior and exterior boundaries ($j = i, e; l = i, e$) are formulated as

$$[K]_{jl} = \int_S [B]_j^T [D][B]_l \, dS = \int_{-1}^{+1}\int_{-1}^{+1} [B]_j^T [D][B]_l |\hat{J}| \, d\xi \, d\eta \tag{5.55}$$

In the two-dimensional case the integrand in Eq. 5.55 is a ratio of 2 polynomials in ξ which can still be integrated analytically. The decomposition with respect to the cell width w neglecting the terms of order in w higher than 1 yields

$$[K]_{jl} = \frac{1}{w}[K^0]_{jl} + [K^1]_{jl} + w[K^2]_{jl} + O(w^2) \tag{5.56}$$

where

$$[K^0]_{jl} = \xi_j \xi_l [E^0] \tag{5.57a}$$

$$[K^1]_{jl} = \frac{\xi_j \xi_l}{2}[E^0] + \frac{\xi_l}{2}[E^1] + \frac{\xi_j}{2}[E^1]^T \tag{5.57b}$$

$$[K^2]_{jl} = \frac{1}{4}\left(1 + \frac{\xi_j \xi_l}{3}\right)[E^2] \tag{5.57c}$$

5.2 FUNDAMENTAL EQUATIONS

The definitions of $[E^0]$, $[E^1]$, $[E^2]$ in Eq. 5.22 and of $[M^0]$ in Eq. 5.36 still apply (deleting the integration in the ζ-direction).

For illustration, the 2-node line element with nodal coordinates (x_1, y_1), (x_2, y_2) and the following shape functions is addressed

$$N_1(\eta) = \frac{1}{2}(1-\eta) \tag{5.58a}$$

$$N_2(\eta) = \frac{1}{2}(1+\eta) \tag{5.58b}$$

For this simple element the integrations over η in the coefficient matrices $[E^0]$, $[E^1]$, $[E^2]$ and $[M^0]$ are performed analytically, yielding

$$[E^0] = \frac{2}{3}\begin{bmatrix} 2[Q^0] & [Q^0] \\ [Q^0] & 2[Q^0] \end{bmatrix} \tag{5.59a}$$

$$[E^1] = \frac{1}{3}\begin{bmatrix} -[Q^0] & [Q^0] \\ [Q^0] & -[Q^0] \end{bmatrix} + 2\begin{bmatrix} -[Q^1] & -[Q^1] \\ [Q^1] & [Q^1] \end{bmatrix} \tag{5.59b}$$

$$[E^2] = \frac{1}{3}\begin{bmatrix} [Q^0] & -[Q^0] \\ -[Q^0] & [Q^0] \end{bmatrix} + 4\begin{bmatrix} [Q^2] & -[Q^2] \\ -[Q^2] & [Q^2] \end{bmatrix} \tag{5.59c}$$

$$[M^0] = \frac{\rho a}{6}\begin{bmatrix} 2[I] & [I] \\ [I] & 2[I] \end{bmatrix} \tag{5.60}$$

The following abbreviations are introduced

$$[Q^0] = \frac{1}{4a}[C^1]^T[D][C^1] \tag{5.61a}$$

$$[Q^1] = -\frac{1}{4a}[C^2]^T[D][C^1] \tag{5.61b}$$

$$[Q^2] = \frac{1}{4a}[C^2]^T[D][C^2] \tag{5.61c}$$

where

$$[C^1] = \begin{bmatrix} \Delta_y & 0 \\ 0 & -\Delta_x \\ -\Delta_x & \Delta_y \end{bmatrix} \tag{5.62a}$$

$$[C^2] = \begin{bmatrix} \bar{y} & 0 \\ 0 & -\bar{x} \\ -\bar{x} & \bar{y} \end{bmatrix} \tag{5.62b}$$

with

$$a = x_1 y_2 - x_2 y_1 \tag{5.63a}$$

$$\Delta_x = x_2 - x_1 \tag{5.63b}$$

$$\Delta_y = y_2 - y_1 \tag{5.63c}$$

$$\bar{x} = \frac{1}{2}(x_2 + x_1) \tag{5.63d}$$

$$\bar{y} = \frac{1}{2}(y_2 + y_1) \tag{5.63e}$$

Isotropic material behaviour defined by the two Lamé constants λ and G, the shear modulus ($\lambda = 2G\nu/(1-2\nu)$ with Poisson's ratio ν), is introduced

$$[D] = \begin{bmatrix} \lambda + 2G & \lambda & 0 \\ \lambda & \lambda + 2G & 0 \\ 0 & 0 & G \end{bmatrix} \tag{5.64}$$

yielding

$$[Q^0] = \frac{1}{4a} \left(\lambda \begin{bmatrix} \Delta_y^2 & -\Delta_x \Delta_y \\ -\Delta_x \Delta_y & \Delta_x^2 \end{bmatrix} + G \begin{bmatrix} 2\Delta_y^2 + \Delta_x^2 & -\Delta_x \Delta_y \\ -\Delta_x \Delta_y & 2\Delta_x^2 + \Delta_y^2 \end{bmatrix} \right) \tag{5.65a}$$

$$[Q^1] = -\frac{1}{4a} \left(\lambda \begin{bmatrix} \Delta_y \bar{y} & -\Delta_x \bar{y} \\ -\bar{x} \Delta_y & \Delta_x \bar{x} \end{bmatrix} + G \begin{bmatrix} 2\Delta_y \bar{y} + \Delta_x \bar{x} & -\bar{x} \Delta_y \\ -\Delta_x \bar{y} & 2\Delta_x \bar{x} + \Delta_y \bar{y} \end{bmatrix} \right) \tag{5.65b}$$

$$[Q^2] = \frac{1}{4a} \left(\lambda \begin{bmatrix} \bar{y}^2 & -\bar{x}\bar{y} \\ -\bar{x}\bar{y} & \bar{x}^2 \end{bmatrix} + G \begin{bmatrix} 2\bar{y}^2 + \bar{x}^2 & -\bar{x}\bar{y} \\ -\bar{x}\bar{y} & 2\bar{x}^2 + \bar{y}^2 \end{bmatrix} \right) \tag{5.65c}$$

5.2.4 Summary of Coefficient Matrices for Two-dimensional Scalar Wave Equation

The two-dimensional scalar wave equation and the boundary conditions follow straightforwardly from Eqs. 5.41 and 5.43. The out-of-plane (anti-plane) motion is one physical interpretation in elasto-dynamics, with the displacement u, surface traction q, shear modulus G and mass density ρ.

The coefficient matrices for the two-dimensional scalar wave equation follow from those of the three-dimensional vector wave equations by performing the modifications described in Sections 5.2.2 and 5.2.3. This results in the following coefficient matrices

5.2 FUNDAMENTAL EQUATIONS

$$[E^0] = \int_{-1}^{+1} G[B^1]^T[B^1]|J|\mathrm{d}\eta \tag{5.66a}$$

$$[E^1] = \int_{-1}^{+1} G[B^2]^T[B^1]|J|\mathrm{d}\eta \tag{5.66b}$$

$$[E^2] = \int_{-1}^{+1} G[B^2]^T[B^2]|J|\mathrm{d}\eta \tag{5.66c}$$

$$[M^0] = \int_{-1}^{+1} \rho[N]^T[N]|J|\mathrm{d}\eta \tag{5.67}$$

$[B^1]$ and $[B^2]$ are defined in Eq. 5.46, deleting the coefficients associated with z and ζ. Analogously $|J|$ and $[N]$ follow from Eqs. 5.9 and 5.51.

Again, the 2-node line element with nodal coordinates (x_1, y_1), (x_2, y_2) is used for illustration. The shape function is specified in Eq. 5.58. The integrations over η in Eqs. 5.66 and 5.67 are performed analytically again, leading to Eqs. 5.59 to 5.61 with

$$[C^1] = \begin{Bmatrix} \Delta_y \\ -\Delta_x \end{Bmatrix} \tag{5.68a}$$

$$[C^2] = \begin{Bmatrix} \bar{y} \\ -\bar{x} \end{Bmatrix} \tag{5.68b}$$

Introducing the constitutive relationship of an isotropic material

$$[D] = G[I] \tag{5.69}$$

yields

$$Q^0 = \frac{G}{4a}(\Delta_x^2 + \Delta_y^2) \tag{5.70a}$$

$$Q^1 = -\frac{G}{4a}(\bar{x}\Delta_x + \bar{y}\Delta_y) \tag{5.70b}$$

$$Q^2 = \frac{G}{4a}(\bar{x}^2 + \bar{y}^2) \tag{5.70c}$$

5.2.5 Summary of Coefficient Matrices for Axisymmetric Vector Wave Equation

When the structure-medium interface is axisymmetric, the three-dimensional vector wave equation can be decomposed into two-dimensional equations using a Fourier series in the circumferential direction.

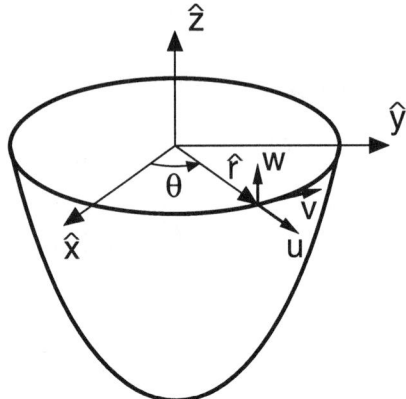

Figure 5-3 Axisymmetric structure-medium interface in cylindrical coordinates

Cylindrical coordinates $\hat{r}, \theta, \hat{z}$, are used (Figure 5-3). The displacements are written as

$$u(\hat{r},\theta,\hat{z}) = \sum_{n=0}^{\infty} u_n(\hat{r},\theta,\hat{z}) = \sum_{n=0}^{\infty} \bar{u}_n^s \cos n\theta + \sum_{n=0}^{\infty} \bar{u}_n^a \sin n\theta \qquad (5.71a)$$

$$v(\hat{r},\theta,\hat{z}) = \sum_{n=0}^{\infty} v_n(\hat{r},\theta,\hat{z}) = -\sum_{n=0}^{\infty} \bar{v}_n^s \sin n\theta + \sum_{n=0}^{\infty} \bar{v}_n^a \cos n\theta \qquad (5.71b)$$

$$w(\hat{r},\theta,\hat{z}) = \sum_{n=0}^{\infty} w_n(\hat{r},\theta,\hat{z}) = \sum_{n=0}^{\infty} \bar{w}_n^s \cos n\theta + \sum_{n=0}^{\infty} \bar{w}_n^a \sin n\theta \qquad (5.71c)$$

The amplitudes in the θ-direction are denoted by a bar. The superscripts s and a are used for the symmetric and anti-symmetric parts. As a simple example an axisymmetric foundation with a rigid structure-medium interface embedded in a half-space can be examined. The symmetric and anti-symmetric parts of $n = 0$ correspond to the vertical and torsional degrees of freedom and the symmetric part of $n = 1$ can model the horizontal and rocking degrees of freedom.

As $\cos n\theta$ and $\sin n\theta$ are orthogonal functions for the range from 0 to 2π, only one term $u_n(\hat{r},\theta,\hat{z})$, $v_n(\hat{r},\theta,\hat{z})$ and $w_n(\hat{r},\theta,\hat{z})$ is addressed. The subscript n is omitted for conciseness. The strains $\{\varepsilon\}$ are calculated as

$$\varepsilon_r = u_{,\hat{r}} \qquad (5.72a)$$

$$\varepsilon_\theta = \frac{u}{\hat{r}} + \frac{1}{\hat{r}} v_{,\theta} \qquad (5.72b)$$

$$\varepsilon_z = w_{,\hat{z}} \qquad (5.72c)$$

$$\gamma_{\theta z} = v_{,\hat{z}} + \frac{1}{\hat{r}} w_{,\theta} \qquad (5.72d)$$

5.2 FUNDAMENTAL EQUATIONS

$$\gamma_{rz} = u_{,\hat{z}} + w_{,\hat{r}} \tag{5.72e}$$

$$\gamma_{r\theta} = \frac{1}{\hat{r}} u_{,\theta} + v_{,\hat{r}} - \frac{v}{\hat{r}} \tag{5.72f}$$

Substituting Eq. 5.71 in Eq. 5.72 yields

$$\varepsilon_r = \bar{u}^s{}_{,\hat{r}} \cos n\theta + \bar{u}^a{}_{,\hat{r}} \sin n\theta \tag{5.73a}$$

$$\varepsilon_\theta = \frac{1}{\hat{r}}(\bar{u}^s - n\bar{v}^s)\cos n\theta + \frac{1}{\hat{r}}(\bar{u}^a - n\bar{v}^a)\sin n\theta \tag{5.73b}$$

$$\varepsilon_z = \bar{w}^s{}_{,\hat{z}} \cos n\theta + \bar{w}^a{}_{,\hat{z}} \sin n\theta \tag{5.73c}$$

$$\gamma_{\theta z} = \left(-\bar{v}^s{}_{,\hat{z}} - \frac{n}{\hat{r}}\bar{w}^s\right)\sin n\theta + \left(\bar{v}^a{}_{,\hat{z}} + \frac{n}{\hat{r}}\bar{w}^a\right)\cos n\theta \tag{5.73d}$$

$$\gamma_{rz} = (\bar{u}^s{}_{,\hat{z}} + \bar{w}^s{}_{,\hat{r}})\cos n\theta + (\bar{u}^a{}_{,\hat{z}} + \bar{w}^a{}_{,\hat{r}})\sin n\theta \tag{5.73e}$$

$$\gamma_{r\theta} = \left(-\frac{n}{\hat{r}}\bar{u}^s - \bar{v}^s{}_{,\hat{r}} + \frac{\bar{v}^s}{\hat{r}}\right)\sin n\theta + \left(\frac{n}{\hat{r}}\bar{u}^a + \bar{v}^a{}_{,\hat{r}} - \frac{\bar{v}^a}{\hat{r}}\right)\cos n\theta \tag{5.73f}$$

The intersection of the structure-medium interface with the plane $\theta = constant$ is discretized with line finite elements as for the two-dimensional vector wave equation in Section 5.2.3. Proceeding for one finite element as in Section 5.2.1 yields the strain-nodal displacement amplitude relationship

$$\{\varepsilon\} = [B^s]\{\bar{u}^s\} + [B^a]\{\bar{u}^a\} \tag{5.74}$$

with $\{\bar{u}\}$ containing the displacement amplitudes in the three directions in all nodes. The decomposition

$$[B^s] = [B_\alpha]\cos n\theta + [B_\beta]\sin n\theta \tag{5.75a}$$
$$[B^a] = [B_\alpha]\sin n\theta - [B_\beta]\cos n\theta \tag{5.75b}$$

applies, with the submatrices ($j = i, e;\ k = 1, 2, \ldots$)

$$[B_\alpha]_{jk} = \begin{bmatrix} \hat{N}_{jk,\hat{r}} & 0 & 0 \\ \frac{1}{\hat{r}}\hat{N}_{jk} & -\frac{n}{\hat{r}}\hat{N}_{jk} & 0 \\ 0 & 0 & \hat{N}_{jk,\hat{z}} \\ 0 & 0 & 0 \\ \hat{N}_{jk,\hat{z}} & 0 & \hat{N}_{jk,\hat{r}} \\ 0 & 0 & 0 \end{bmatrix} \tag{5.76a}$$

$$[B_\beta]_{jk} = \begin{bmatrix} 0 & 0 & 0 \\ 0 & 0 & 0 \\ 0 & 0 & 0 \\ 0 & -\hat{N}_{jk,\hat{z}} & -\dfrac{n}{\hat{r}}\hat{N}_{jk} \\ 0 & 0 & 0 \\ -\dfrac{n}{\hat{r}}\hat{N}_{jk} & \left(-\hat{N}_{jk,\hat{r}}+\dfrac{1}{\hat{r}}\hat{N}_{jk}\right) & 0 \end{bmatrix} \quad (5.76b)$$

The static-stiffness matrices of the symmetric and anti-symmetric parts of a finite element of the cell decouple, leading to

for $n = 0$

$$[K^s] = 2\pi \int_{-1}^{+1}\int_{-1}^{+1} [B_\alpha]^T [D][B_\alpha] \hat{r} |\hat{J}| d\xi d\eta \quad (5.77a)$$

$$[K^a] = 2\pi \int_{-1}^{+1}\int_{-1}^{+1} [B_\beta]^T [D][B_\beta] \hat{r} |\hat{J}| d\xi d\eta \quad (5.77b)$$

for $n \geq 1$

$$[K^s] = [K^a] = \pi \int_{-1}^{+1}\int_{-1}^{+1} [B_\alpha]^T [D][B_\alpha] \hat{r} |\hat{J}| d\xi d\eta$$
$$+ \pi \int_{-1}^{+1}\int_{-1}^{+1} [B_\beta]^T [D][B_\beta] \hat{r} |\hat{J}| d\xi d\eta \quad (5.77c)$$

Eq. 5.77 should be compared with Eq. 5.55. The determinant of the Jacobian $|\hat{J}|$ is specified in Eq. 5.52.

Proceeding as in Section 5.2.3, Eq. 5.76 is formulated similarly to Eq. 5.53

$$[B_\alpha]_{jk} = \dfrac{\xi_j}{w}[B_\alpha^1]_k + \dfrac{1+\xi_j\xi}{2\left(1+\dfrac{w}{2}(1+\xi)\right)}[B_\alpha^2]_k \quad (5.78a)$$

$$[B_\beta]_{jk} = \dfrac{\xi_j}{w}[B_\beta^1]_k + \dfrac{1+\xi_j\xi}{2\left(1+\dfrac{w}{2}(1+\xi)\right)}[B_\beta^2]_k \quad (5.78b)$$

where

$$[B_\alpha^1]_k = \begin{bmatrix} j_{11} & 0 & 0 \\ 0 & 0 & 0 \\ 0 & 0 & j_{21} \\ 0 & 0 & 0 \\ j_{21} & 0 & j_{11} \\ 0 & 0 & 0 \end{bmatrix} N_k \quad (5.79a)$$

5.2 FUNDAMENTAL EQUATIONS

$$[B_\alpha^2]_k = \begin{bmatrix} j_{12}N_{k,\eta} & 0 & 0 \\ \dfrac{N_k}{r} & -\dfrac{n}{r}N_k & 0 \\ 0 & 0 & j_{22}N_{k,\eta} \\ 0 & 0 & 0 \\ j_{22}N_{k,\eta} & 0 & j_{12}N_{k,\eta} \\ 0 & 0 & 0 \end{bmatrix} \qquad (5.79b)$$

$$[B_\beta^1]_k = \begin{bmatrix} 0 & 0 & 0 \\ 0 & 0 & 0 \\ 0 & 0 & 0 \\ 0 & -j_{21} & 0 \\ 0 & 0 & 0 \\ 0 & -j_{11} & 0 \end{bmatrix} N_k \qquad (5.79c)$$

$$[B_\beta^2]_k = \begin{bmatrix} 0 & 0 & 0 \\ 0 & 0 & 0 \\ 0 & 0 & 0 \\ 0 & -j_{22}N_{k,\eta} & -\dfrac{n}{r}N_k \\ 0 & 0 & 0 \\ -\dfrac{n}{r}N_k & (-j_{12}N_{k,\eta}+\dfrac{1}{r}N_k) & 0 \end{bmatrix} \qquad (5.79d)$$

The coefficient matrices then follow as

for the symmetric part of $n = 0$

$$[E^0] = 2\pi \int_{-1}^{+1} [B_\alpha^1]^T [D][B_\alpha^1] r |J| \mathrm{d}\eta \qquad (5.80a)$$

$$[E^1] = 2\pi \int_{-1}^{+1} [B_\alpha^2]^T [D][B_\alpha^1] r |J| \mathrm{d}\eta \qquad (5.80b)$$

$$[E^2] = 2\pi \int_{-1}^{+1} [B_\alpha^2]^T [D][B_\alpha^2] r |J| \mathrm{d}\eta \qquad (5.80c)$$

and for the anti-symmetric part of $n = 0$ Eqs. 5.80a to 5.80c still apply with the subscript β replacing α,

for the symmetric and anti-symmetric parts of $n \geq 1$

$$[E^0] = \pi \int_{-1}^{+1} [B_\alpha^1]^T [D][B_\alpha^1] r |J| \mathrm{d}\eta + \pi \int_{-1}^{+1} [B_\beta^1]^T [D][B_\beta^1] r |J| \mathrm{d}\eta \qquad (5.80d)$$

$$[E^1] = \pi \int_{-1}^{+1} [B_\alpha^2]^T [D][B_\alpha^1] r |J| d\eta + \pi \int_{-1}^{+1} [B_\beta^2]^T [D][B_\beta^1] r |J| d\eta \qquad (5.80e)$$

$$[E^2] = \pi \int_{-1}^{+1} [B_\alpha^2]^T [D][B_\alpha^2] r |J| d\eta + \pi \int_{-1}^{+1} [B_\beta^2]^T [D][B_\beta^2] r |J| d\eta \qquad (5.80f)$$

To calculate the mass matrix, the displacement - nodal displacement amplitude relationship is formulated as

$$\{u\} = [\hat{N}_\alpha] \cos n\theta \{\bar{u}^s\} - [\hat{N}_\beta] \sin n\theta \{\bar{u}^s\} + [\hat{N}_\alpha] \sin n\theta \{\bar{u}^a\} + [\hat{N}_\beta] \cos n\theta \{\bar{u}^a\} \quad (5.81)$$

with the shape functions

$$[\hat{N}_\alpha] = \begin{bmatrix} \hat{N}_1 & & \hat{N}_2 & & \cdots \\ 0 & & 0 & & \cdots \\ & \hat{N}_1 & & \hat{N}_2 & \cdots \end{bmatrix} \qquad (5.82a)$$

$$[\hat{N}_\beta] = \begin{bmatrix} 0 & & 0 & & \cdots \\ \hat{N}_1 & & \hat{N}_2 & & \cdots \\ & 0 & & 0 & \cdots \end{bmatrix} \qquad (5.82b)$$

The mass matrix follows as

for $n = 0$

$$[M^s] = 2\pi \int_{-1}^{+1} \int_{-1}^{+1} \rho [\hat{N}_\alpha]^T [\hat{N}_\alpha] \hat{r} |\hat{J}| d\xi d\eta \qquad (5.83a)$$

$$[M^a] = 2\pi \int_{-1}^{+1} \int_{-1}^{+1} \rho [\hat{N}_\beta]^T [\hat{N}_\beta] \hat{r} |\hat{J}| d\xi d\eta \qquad (5.83b)$$

for $n \geq 1$

$$[M^s] = [M^a] = \pi \int_{-1}^{+1} \int_{-1}^{+1} \rho [\hat{N}_\alpha]^T [\hat{N}_\alpha] \hat{r} |\hat{J}| d\xi d\eta$$

$$+ \pi \int_{-1}^{+1} \int_{-1}^{+1} \rho [\hat{N}_\beta]^T [\hat{N}_\beta] \hat{r} |\hat{J}| d\xi d\eta \qquad (5.83c)$$

The coefficient matrix $[M^0]$ is written as

for the symmetric part of $n = 0$

$$[M^0] = 2\pi \int_{-1}^{+1} \rho [N_\alpha]^T [N_\alpha] r |J| d\eta \qquad (5.84a)$$

and for the anti-symmetric part of $n = 0$ Eq. 5.84a still applies with the subscript β replacing α,

5.2 FUNDAMENTAL EQUATIONS

for the symmetric and anti-symmetric parts of $n \geq 1$

$$[M^0] = \pi \int_{-1}^{+1} \rho [N_\alpha]^T [N_\alpha] r |J| d\eta + \pi \int_{-1}^{+1} \rho [N_\beta]^T [N_\beta] r |J| d\eta \qquad (5.84b)$$

Of course, for later use the spatial dimension introduced in Eq. 3.12 $s = 3$ applies for the axisymmetric case.

5.2.6 Summary of Coefficient Matrices for Axisymmetric Scalar Wave Equation

For easy reference, the coefficient matrices of the finite-element cell for the three-dimensional scalar wave equation with an axisymmetric structure-medium interface are listed. Only those aspects which differ from Section 5.2.5 are addressed. The three-dimensional scalar wave equation with a structure-medium interface of arbitrary geometry is examined in Section 5.2.2.

The scalar wave equation in cylindrical coordinates equals

$$G\left(u_{,\hat{r}\hat{r}} + \frac{1}{\hat{r}} u_{,\hat{r}} + \frac{1}{\hat{r}^2} u_{,\theta\theta} + u_{,\hat{z}\hat{z}}\right) - \rho \ddot{u} = 0 \qquad (5.85)$$

with the material constants G and ρ. Proceeding as in Section 5.2.5, the function is decomposed as (Eq. 5.71)

$$u(\hat{r}, \theta, \hat{z}) = \sum_{n=0}^{\infty} \bar{u}_n^s \cos n\theta + \sum_{n=0}^{\infty} \bar{u}_n^a \sin n\theta \qquad (5.86)$$

With the gradients of the function defined as

$$\gamma_r = u_{,\hat{r}} \qquad (5.87a)$$

$$\gamma_\theta = \frac{1}{\hat{r}} u_{,\theta} \qquad (5.87b)$$

$$\gamma_z = u_{,\hat{z}} \qquad (5.87c)$$

Eqs. 5.74 and 5.75 still apply with (Eq. 5.76)

$$\{B_\alpha\}_{jk} = \left\{ \begin{array}{c} \hat{N}_{jk,\hat{r}} \\ 0 \\ \hat{N}_{jk,\hat{z}} \end{array} \right\} \qquad (5.88a)$$

$$\{B_\beta\}_{jk} = \left\{ \begin{array}{c} 0 \\ -\frac{n}{\hat{r}} \hat{N}_{jk} \\ 0 \end{array} \right\} \qquad (5.88b)$$

This leads to (Eq. 5.79)

$$\{B_\alpha^1\}_k = \begin{Bmatrix} j_{11} \\ 0 \\ j_{21} \end{Bmatrix} N_k \qquad (5.89a)$$

$$\{B_\alpha^2\}_k = \begin{Bmatrix} j_{12} \\ 0 \\ j_{22} \end{Bmatrix} N_{k,\eta} \qquad (5.89b)$$

$$\{B_\beta^1\}_k = \begin{Bmatrix} 0 \\ 0 \\ 0 \end{Bmatrix} \qquad (5.89c)$$

$$\{B_\beta^2\}_k = \begin{Bmatrix} 0 \\ -\dfrac{n}{r} \\ 0 \end{Bmatrix} N_k \qquad (5.89d)$$

The coefficient matrices follow as

for the symmetric part of $n = 0$

$$[E^0] = 2\pi \int_{-1}^{+1} G[B_\alpha^1]^T [B_\alpha^1] r|J| d\eta \qquad (5.90a)$$

$$[E^1] = 2\pi \int_{-1}^{+1} G[B_\alpha^2]^T [B_\alpha^1] r|J| d\eta \qquad (5.90b)$$

$$[E^2] = 2\pi \int_{-1}^{+1} G[B_\alpha^2]^T [B_\alpha^2] r|J| d\eta \qquad (5.90c)$$

for the symmetric and anti-symmetric parts of $n \geq 1$

$$[E^0] = \pi \int_{-1}^{+1} G[B_\alpha^1]^T [B_\alpha^1] r|J| d\eta + \pi \int_{-1}^{+1} G[B_\beta^1]^T [B_\beta^1] r|J| d\eta \qquad (5.90d)$$

$$[E^1] = \pi \int_{-1}^{+1} G[B_\alpha^2]^T [B_\alpha^1] r|J| d\eta + \pi \int_{-1}^{+1} G[B_\beta^2]^T [B_\beta^1] r|J| d\eta \qquad (5.90e)$$

$$[E^2] = \pi \int_{-1}^{+1} G[B_\alpha^2]^T [B_\alpha^2] r|J| d\eta + \pi \int_{-1}^{+1} G[B_\beta^2]^T [B_\beta^2] r|J| d\eta \qquad (5.90f)$$

To calculate the mass matrix, the function-nodal function amplitude relationship is formulated as

$$\{u\} = [\hat{N}] \cos n\theta \{\bar{u}^s\} + [\hat{N}] \sin n\theta \{\bar{u}^a\} \qquad (5.91)$$

5.2 FUNDAMENTAL EQUATIONS

with the shape functions

$$[\hat{N}] = [\ \hat{N}_1 \ \ \hat{N}_2 \ \ \ldots \ ] \tag{5.92}$$

The coefficient matrix $[M^0]$ follows as

for $n = 0$

$$[M^0] = 2\pi \int_{-1}^{+1} \rho[N]^T[N] r |J| d\eta \tag{5.93a}$$

for $n \geq 1$

$$[M^0] = \pi \int_{-1}^{+1} \rho[N]^T[N] r |J| d\eta \tag{5.93b}$$

5.2.7 Assemblage of Finite-element Cell and Unbounded Medium

The assemblage enforcing compatibility and equilibrium formulated in the frequency domain links the dynamic-stiffness matrix of the unbounded medium at the structure-medium interface coinciding with the interior boundary of the finite-element cell to that at the fictitious interface corresponding to the exterior boundary. This relationship involves the dynamic-stiffness matrix of the cell which can be expressed by its static-stiffness and mass matrices.

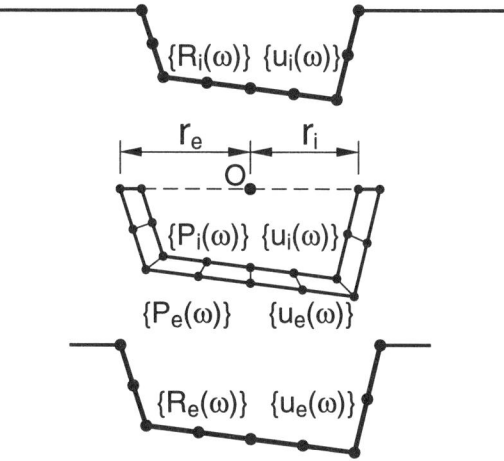

Figure 5-4 Finite-element region between interior and exterior boundaries and corresponding interfaces of unbounded medium

The force-displacement relationship of the finite-element cell located between the interior and exterior boundaries (Figure 5-4) is written as

$$[S(\omega)]\{u(\omega)\} = \{P(\omega)\} \tag{5.94}$$

with the amplitudes of the displacements $\{u(\omega)\}$ and of the nodal forces $\{P(\omega)\}$. The dynamic-stiffness matrix equals

$$[S(\omega)] = [K] - \omega^2[M] \tag{5.95}$$

with the static-stiffness matrix $[K]$ and the mass matrix $[M]$ of the finite-element cell. Partitioning Eq. 5.94 into the nodes lying on the interior and exterior boundaries yields

$$\begin{bmatrix} [S_{ii}(\omega)] & [S_{ie}(\omega)] \\ [S_{ei}(\omega)] & [S_{ee}(\omega)] \end{bmatrix} \begin{Bmatrix} \{u_i(\omega)\} \\ \{u_e(\omega)\} \end{Bmatrix} = \begin{Bmatrix} \{P_i(\omega)\} \\ \{P_e(\omega)\} \end{Bmatrix} \tag{5.96}$$

The interaction force-displacement relationship of the unbounded medium at the interfaces corresponding to the interior and exterior boundaries is formulated as (Eq. 2.1)

$$\{R_i(\omega)\} = [S_i^\infty(\omega)]\{u_i(\omega)\} \tag{5.97a}$$
$$\{R_e(\omega)\} = [S_e^\infty(\omega)]\{u_e(\omega)\} \tag{5.97b}$$

Note that by using the same displacement amplitudes at the interfaces (Eqs. 5.96 and 5.97), compatibility is enforced. Formulating equilibrium at the interior and exterior boundaries relates the interaction force amplitudes of the unbounded medium to the force amplitudes of the cell

$$\{P_i(\omega)\} = \{R_i(\omega)\} \tag{5.98a}$$
$$\{P_e(\omega)\} = -\{R_e(\omega)\} \tag{5.98b}$$

Eliminating $\{P(\omega)\}$ and $\{R(\omega)\}$ by substituting Eqs. 5.96 and 5.97 in Eq. 5.98 leads to

$$\begin{bmatrix} [S_{ii}(\omega)] & [S_{ie}(\omega)] \\ [S_{ei}(\omega)] & [S_{ee}(\omega)] \end{bmatrix} \begin{Bmatrix} \{u_i(\omega)\} \\ \{u_e(\omega)\} \end{Bmatrix} = \begin{bmatrix} [S_i^\infty(\omega)] & 0 \\ 0 & -[S_e^\infty(\omega)] \end{bmatrix} \begin{Bmatrix} \{u_i(\omega)\} \\ \{u_e(\omega)\} \end{Bmatrix} \tag{5.99}$$

The relationship linking $[S_i^\infty(\omega)]$ and $[S_e^\infty(\omega)]$ can be derived by eliminating $\{u_e(\omega)\}$ from Eq. 5.99 resulting in

$$([S_i^\infty(\omega)] - [S_{ii}(\omega)] + [S_{ie}(\omega)]([S_e^\infty(\omega)] + [S_{ee}(\omega)])^{-1}[S_{ei}(\omega)])\{u_i(\omega)\} = 0 \tag{5.100}$$

As Eq. 5.100 is satisfied for an arbitrary $\{u_i(\omega)\}$, the coefficient matrix must vanish yielding

$$[S_i^\infty(\omega)] = [S_{ii}(\omega)] - [S_{ie}(\omega)]([S_e^\infty(\omega)] + [S_{ee}(\omega)])^{-1}[S_{ei}(\omega)] \tag{5.101}$$

This Eq. 5.101 will be used in dynamic condensation (Section 7.3).

5.2 FUNDAMENTAL EQUATIONS

5.2.8 Consistent Infinitesimal Finite-element Cell Equation in Frequency Domain

The consistent infinitesimal finite-element cell equation is derived using the relationship of the dynamic-stiffness matrices based on similarity formulated in the continuous form (Eq. 3.12) and the relationship based on assemblage (Eq. 5.101), for which the limit of infinitesimal cell width must be performed.

Eq. 5.101 is reformulated as

$$([S_e^\infty(\omega)] + [S_{ee}(\omega)])[S_{ie}(\omega)]^{-1}([S_i^\infty(\omega)] - [S_{ii}(\omega)]) + [S_{ei}(\omega)] = 0 \quad (5.102)$$

The dynamic-stiffness submatrices of the cell are written as follows. For instance,

$$[S_{ii}(\omega)] = [K_{ii}] - \omega^2[M_{ii}] \quad (5.103)$$

applies with $[K_{ii}]$ specified in Eq. 5.20 and $[M_{ii}]$ in Eq. 5.35. This yields

$$[S_{ii}(\omega)] = \frac{1}{w}[K_{ii}^0] + [K_{ii}^1] + w[K_{ii}^2] - w\omega^2[M_{ii}^2] + O(w^2) \quad (5.104)$$

Using Eq. 5.37a leads to

$$[S_{ii}(\omega)] = \frac{1}{w}[E^0] + [K_{ii}^1] + w([K_{ii}^2] - \omega^2[M_{ii}^2]) + O(w^2) \quad (5.105a)$$

Analogously

$$[S_{ie}(\omega)] = -\frac{1}{w}[E^0] + [K_{ie}^1] + w([K_{ie}^2] - \omega^2[M_{ie}^2]) + O(w^2) \quad (5.105b)$$

$$[S_{ei}(\omega)] = -\frac{1}{w}[E^0] + [K_{ei}^1] + w([K_{ei}^2] - \omega^2[M_{ei}^2]) + O(w^2) \quad (5.105c)$$

$$[S_{ee}(\omega)] = \frac{1}{w}[E^0] + [K_{ee}^1] + w([K_{ee}^2] - \omega^2[M_{ee}^2]) + O(w^2) \quad (5.105d)$$

follow. The inverse $[S_{ie}(\omega)]^{-1}$ appearing in Eq. 5.102 is expressed as a polynomial in w with the unknown coefficient matrices $[A]$ and $[B]$

$$[S_{ie}(\omega)]^{-1} = -w[E^0]^{-1} + w^2[A] + w^3[B] + O(w^4) \quad (5.106)$$

The coefficient matrix of w is equal to the inverse of that of $1/w$ in Eq. 5.105b. $[A]$ and $[B]$ follow from

$$[I] = [S_{ie}(\omega)][S_{ie}(\omega)]^{-1} = [I] - w([K_{ie}^1][E^0]^{-1} + [E^0][A])$$
$$- w^2(([K_{ie}^2] - \omega^2[M_{ie}^2])[E_0]^{-1} - [K_{ie}^1][A] + [E^0][B]) + O(w^3) \quad (5.107)$$

Setting the coefficient matrices of w and w^2 equal to zero yields

$$[A] = -[E^0]^{-1}[K^1_{ie}][E^0]^{-1} \tag{5.108a}$$

$$[B] = -[E^0]^{-1}([K^1_{ie}][E^0]^{-1}[K^1_{ie}] - ([K^2_{ie}] - \omega^2[M^2_{ie}]))[E^0]^{-1} \tag{5.108b}$$

Substituting Eqs. 5.105 and 5.106 with Eq. 5.108 in Eq. 5.102 leads to

$$\underbrace{[K^1_{ii}] + [K^1_{ie}] + [K^1_{ei}] + [K^1_{ee}]}_{1}$$

$$- w\underbrace{([S^\infty_e(\omega)] + [K^1_{ie}] + [K^1_{ee}])[E^0]^{-1}([S^\infty_i(\omega)] - [K^1_{ii}] - [K^1_{ie}])}_{2}$$

$$+ \underbrace{[S^\infty_e(\omega)] - [S^\infty_i(\omega)]}_{3} + w\underbrace{([K^2_{ii}] + [K^2_{ie}] + [K^2_{ei}] + [K^2_{ee}])}_{4}$$

$$- \omega^2\underbrace{([M^2_{ii}] + [M^2_{ie}] + [M^2_{ei}] + [M^2_{ee}]))}_{5} = O(w^2) \tag{5.109}$$

The sum identified by 1 in Eq. 5.109 vanishes according to Eq. 5.40. Eq. 5.37b is substituted in the term identified by 2. The sums identified by 4 and 5 are transformed using Eqs. 5.37c and 5.38, respectively. Dividing Eq. 5.109 by w results in

$$([S^\infty_e(\omega)] + [E^1])[E^0]^{-1}([S^\infty_i(\omega)] + [E^1]^T) - \frac{[S^\infty_e(\omega)] - [S^\infty_i(\omega)]}{w}$$

$$- [E^2] + \omega^2[M^0] = O(w) \tag{5.110}$$

The limit of $w \to 0$ can now be performed. With w defined in Eq. 5.1

$$\lim_{w \to 0} \frac{[S^\infty_e(\omega)] - [S^\infty_i(\omega)]}{w} = \lim_{r_e \to r_i} r_i \frac{[S^\infty_e(\omega)] - [S^\infty_i(\omega)]}{r_e - r_i} = r[S^\infty(\omega)]_{,r} \tag{5.111}$$

results with $[S^\infty(\omega)] = [S^\infty_i(\omega)]$ and $r = r_i$. With $[S^\infty_e(\omega)] = [S^\infty_i(\omega)] + O(w)$, the limit of Eq. 5.110 equals

$$([S^\infty(\omega)] + [E^1])[E^0]^{-1}([S^\infty(\omega)] + [E^1]^T) - r[S^\infty(\omega)]_{,r} - [E^2] + \omega^2[M^0] = O(w) \tag{5.112}$$

Substituting $r[S^\infty(\omega)]_{,r}$ of Eq. 3.12 based on similarity in Eq. 5.112 leads to

$$([S^\infty(\omega)] + [E^1])[E^0]^{-1}([S^\infty(\omega)] + [E^1]^T) - (s-2)[S^\infty(\omega)]$$

$$- \omega[S^\infty(\omega)]_{,\omega} - [E^2] + \omega^2[M^0] = 0 \tag{5.113}$$

In an actual application a specific structure-medium interface is addressed which fixes r. The dynamic-stiffness matrix thus becomes a function of ω only. The partial derivative $[S^\infty(r, \omega)]_{,\omega}$ in Eq. 3.12 is replaced by $[S^\infty(\omega)]_{,\omega}$.

This represents the *consistent infinitesimal finite-element cell equation formulated in the frequency domain*. It is a system of nonlinear ordinary differential equations of first order in the independent variable which is the frequency ω. Its solution in the frequency domain is discussed in Chapter 7.

5.2 FUNDAMENTAL EQUATIONS

5.2.9 Consistent Infinitesimal Finite-element Cell Equation in Time Domain

As discussed in Section 2.2, the interaction force-acceleration relationship of the unbounded medium involving the acceleration unit-impulse response matrix is used in the formulation of the consistent infinitesimal finite-element cell method. The displacement dynamic-stiffness matrix $[S^\infty(\omega)]$ appearing in Eq. 5.113 is thus transformed to the acceleration dynamic-stiffness matrix $[M^\infty(\omega)]$ based on Eq. 2.14. Dividing Eq. 5.113 by $(i\omega)^4$ and substituting Eq. 2.14 yield

$$[M^\infty(\omega)][E^0]^{-1}[M^\infty(\omega)] + [E^1][E^0]^{-1}\frac{[M^\infty(\omega)]}{(i\omega)^2} + \frac{[M^\infty(\omega)]}{(i\omega)^2}[E^0]^{-1}[E^1]^T -$$
$$s\frac{[M^\infty(\omega)]}{(i\omega)^2} + \frac{1}{\omega}[M^\infty(\omega)]_{,\omega} - \frac{1}{(i\omega)^4}([E^2] - [E^1][E^0]^{-1}[E^1]^T) - \frac{1}{(i\omega)^2}[M^0] = 0$$
(5.114)

Applying the inverse Fourier transformation to Eq. 5.114 results in

$$\int_0^t [M^\infty(t-\tau)][E^0]^{-1}[M^\infty(\tau)]d\tau + \left([E^1][E^0]^{-1} - \frac{s+1}{2}\right)\int_0^t \int_0^\tau [M^\infty(\tau')]d\tau'd\tau$$
$$+ \int_0^t \int_0^\tau [M^\infty(\tau')]d\tau'd\tau \left([E^0]^{-1}[E^1]^T - \frac{s+1}{2}\right) + t\int_0^t [M^\infty(\tau)]d\tau$$
$$- \frac{t^3}{6}([E^2] - [E^1][E^0]^{-1}[E^1]^T)H(t) - t[M^0]H(t) = 0 \quad (5.115)$$

where

$$\mathcal{F}^{-1}\left\langle \frac{1}{\omega}[M^\infty(\omega)]_{,\omega} \right\rangle = \int_0^t \tau[M^\infty(\tau)]d\tau = t\int_0^t [M^\infty(\tau)]d\tau - \int_0^t \int_0^\tau [M^\infty(\tau')]d\tau'd\tau$$
(5.116)

$\mathcal{F}^{-1}\langle\,\rangle$ denotes the inverse Fourier transform of $\langle\,\rangle$. Note that in the integral equation (Eq. 5.115) a convolution occurs.

To achieve a concise formulation, the positive definite coefficient matrix $[E^0]$ is decomposed by Cholesky's method as

$$[E^0] = [U]^T[U] \quad (5.117)$$

where $[U]$ is an upper-triangular matrix. Substituting Eq. 5.117 in Eq. 5.115, which is premultiplied by $([U]^{-1})^T$ and postmultiplied by $[U]^{-1}$, yields the *consistent infinitesimal finite-element cell equation in the time domain*

$$\int_0^t [m^\infty(t-\tau)][m^\infty(\tau)]d\tau + [e^1]\int_0^t \int_0^\tau [m^\infty(\tau')]d\tau'd\tau + \int_0^t \int_0^\tau [m^\infty(\tau')]d\tau'd\tau[e^1]^T$$
$$+ t\int_0^t [m^\infty(\tau)]d\tau - \frac{t^3}{6}[e^2]H(t) - t[m^0]H(t) = 0 \quad (5.118)$$

where

$$[m^\infty(t)] = ([U]^{-1})^T [M^\infty(t)][U]^{-1} \tag{5.119}$$

and with the coefficient matrices

$$[e^1] = ([U]^{-1})^T [E^1][U]^{-1} - \frac{s+1}{2}[I] \tag{5.120a}$$

$$[e^2] = ([U]^{-1})^T ([E^2] - [E^1][E^0]^{-1}[E^1]^T)[U]^{-1} \tag{5.120b}$$

$$[m^0] = ([U]^{-1})^T [M^0][U]^{-1} \tag{5.120c}$$

Note that $[e^2]$ is symmetric and $[m^0]$ is positive definite.

After determining $[m^\infty(t)]$ from Eq. 5.118, the acceleration unit-impulse response matrix follows from Eq. 5.119 as

$$[M^\infty(t)] = [U]^T [m^\infty(t)][U] \tag{5.121}$$

Eq. 5.118 is based on the finite-element assemblage enforcing equilibrium and compatibility of the infinitesimal finite-element cell and the unbounded medium as well as on similarity. It is applicable to both two- and three-dimensional scalar and vector wave equations with the corresponding coefficient matrices specified in Sections 5.2.1 to 5.2.4.

The asymptotic expansion at early time of the acceleration unit-impulse response matrix $[M^\infty(t)]$ is discussed at the end of Section 7.2, together with the corresponding asymptotic expansion at high frequency of the dynamic-stiffness matrix $[S^\infty(\omega)]$.

5.2.10 Two-dimensional Layered Unbounded Medium as a Special Case

A layered unbounded medium is often encountered in practice (Figure 5-5a). The unbounded medium is enclosed by two parallel boundaries extending to infinity and

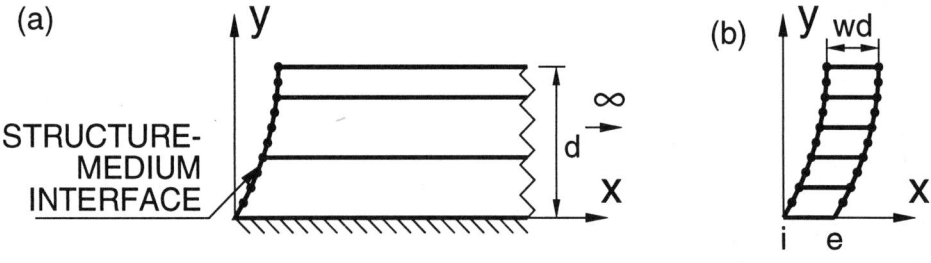

Figure 5-5 Two-dimensional layered unbounded medium: (a) structure-medium interface discretized with line finite elements; (b) two-dimensional finite-element cell adjacent to structure-medium interface used in derivation

5.2 FUNDAMENTAL EQUATIONS

the structure-medium interface which can be curved. Often the material properties vary in the direction perpendicular to the boundaries extending to infinity.

For the two-dimensional case, the similarity centre is located at infinity ($x \to -\infty$). As an approximation the similarity centre can be chosen at a large distance which permits the consistent infinitesimal finite-element cell method to be applied without any modification. Typically, the distance to the similarity centre equals $100d$ with the depth d of the layer.

In the rigorous formulation, as the similarity centre is located at infinity, the characteristic lengths of similar structure-medium interfaces of the unbounded medium do not change. The characteristic length can thus be chosen as the (constant) d. The dimensionless frequency

$$a_0 = \frac{\omega d}{c_s} \tag{5.122}$$

is also constant. This fact permits a streamlined formulation.

The structure-medium interface is discretized by line finite elements (Figure 5-5a). Any inhomogeneity in the y-direction, if present, is taken into consideration. The derivation of the coefficient matrices follows closely Section 5.2.1, which the reader should be familiar with. A cell of constant width wd measured in the x-direction with the dimensionless cell width w is introduced (Figure 5-5b). The interior boundary of the cell coincides with the structure-medium interface whose nodes have the coordinates $\{x\}$, $\{y\}$ and the shape functions $\{N\}$ (Eq. 5.2). The coordinates of the exterior nodes $\{x_e\}$, $\{y_e\}$ are expressed as (Eq. 5.6)

$$\{x_e\} = \{x\} + wd \tag{5.123a}$$

$$\{y_e\} = \{y\} \tag{5.123b}$$

The isoparametric mapping for a finite element of the cell equals (Eq. 5.7)

$$\{\hat{x}\} = \{N\}^T\{x\} + \frac{wd}{2}(1+\xi) \tag{5.124a}$$

$$\{\hat{y}\} = \{N\}^T\{y\} \tag{5.124b}$$

with the shape functions $\{N\}^T$ (Eq. 5.2). The Jacobian matrix equals (Eq. 5.8)

$$[\hat{J}] = \begin{bmatrix} \{\hat{x}\}_{,\xi} & \{\hat{y}\}_{,\xi} \\ \{\hat{x}\}_{,\eta} & \{\hat{y}\}_{,\eta} \end{bmatrix} = \begin{bmatrix} \frac{w}{2} & \\ & 1 \end{bmatrix}[J] \tag{5.125}$$

with (Eq. 5.9)

$$[J] = \begin{bmatrix} d & 0 \\ \{N_{,\eta}\}^T\{x\} & \{N_{,\eta}\}^T\{y\} \end{bmatrix} \tag{5.126}$$

Its inverse is written as (Eq. 5.11)

$$[\hat{J}]^{-1} = [J]^{-1} \begin{bmatrix} \frac{2}{w} & \\ & 1 \end{bmatrix} \quad (5.127)$$

with (Eq. 5.12)

$$[J]^{-1} = \frac{1}{d\{N_{,\eta}\}^T\{y\}} \begin{bmatrix} \{N_{,\eta}\}^T\{y\} & 0 \\ -\{N_{,\eta}\}^T\{x\} & d \end{bmatrix} \quad (5.128)$$

The derivatives of the shape functions are formulated as (Eq. 5.13)

$$\begin{Bmatrix} \hat{N}_{jk,\hat{x}} \\ \hat{N}_{jk,\hat{y}} \end{Bmatrix} = [\hat{J}]^{-1} \begin{Bmatrix} \hat{N}_{jk,\xi} \\ \hat{N}_{jk,\eta} \end{Bmatrix}$$

$$= \frac{\xi_j}{wd\{N_{,\eta}\}^T\{y\}} \begin{Bmatrix} \{N_{,\eta}\}^T\{y\} \\ -\{N_{,\eta}\}^T\{x\} \end{Bmatrix} N_k + \frac{1+\xi_j\xi}{2\{N_{,\eta}\}^T\{y\}} \begin{Bmatrix} 0 \\ 1 \end{Bmatrix} N_{k,\eta} \quad (5.129)$$

For the two-dimensional vector wave equation (Eq. 5.53)

$$[B]_{jk} = \begin{bmatrix} \hat{N}_{jk,\hat{x}} & 0 \\ 0 & \hat{N}_{jk,\hat{y}} \\ \hat{N}_{jk,\hat{y}} & \hat{N}_{jk,\hat{x}} \end{bmatrix} = \frac{\xi_j}{w}[B^1]_k + \frac{1+\xi_j\xi}{2}[B^2]_k \quad (5.130)$$

applies with (Eq. 5.54)

$$[B^1]_k = \frac{1}{d\{N_{,\eta}\}^T\{y\}} \begin{bmatrix} \{N_{,\eta}\}^T\{y\} & 0 \\ 0 & -\{N_{,\eta}\}^T\{x\} \\ -\{N_{,\eta}\}^T\{x\} & \{N_{,\eta}\}^T\{y\} \end{bmatrix} N_k \quad (5.131a)$$

$$[B^2]_k = \frac{1}{\{N_{,\eta}\}^T\{y\}} \begin{bmatrix} 0 & 0 \\ 0 & 1 \\ 1 & 0 \end{bmatrix} N_{k,\eta} \quad (5.131b)$$

For the two-dimensional scalar wave equation (Eq. 5.44)

$$\{B\}_{jk} = \begin{Bmatrix} \hat{N}_{jk,\hat{x}} \\ \hat{N}_{jk,\hat{y}} \end{Bmatrix} = \frac{\xi_j}{w}\{B^1\}_k + \frac{1+\xi_j\xi}{2}\{B^2\}_k \quad (5.132)$$

5.2 FUNDAMENTAL EQUATIONS

holds with (Eq. 5.45)

$$\{B^1\}_k = \frac{1}{d\{N_{,\eta}\}^T\{y\}} \begin{Bmatrix} \{N_{,\eta}\}^T\{y\} \\ -\{N_{,\eta}\}^T\{x\} \end{Bmatrix} N_k \quad (5.133a)$$

$$\{B^2\}_k = \frac{1}{\{N_{,\eta}\}^T\{y\}} \begin{Bmatrix} 0 \\ 1 \end{Bmatrix} N_{k,\eta} \quad (5.133b)$$

The submatrices of the static stiffness with respect to the interior and exterior boundaries ($j = i, e; l = i, e$) are formulated as (Eq. 5.19)

$$[K]_{jl} = \int_S [B]_j^T [D][B]_l \, dS \quad (5.134)$$

with the matrix $[B]$ corresponding to the vector wave equation (Eq. 5.130) or the scalar wave equation (Eq. 5.132) and the associated elasticity matrix $[D]$. Integrating analytically in the ξ-direction yields (Eq. 5.20)

$$[K]_{jl} = \frac{1}{w}[K^0]_{jl} + [K^1]_{jl} + w[K^2]_{jl} \quad (5.135)$$

where (Eq. 5.21)

$$[K^0]_{jl} = \xi_j \xi_l [E^0] \quad (5.136a)$$

$$[K^1]_{jl} = \frac{\xi_l}{2}[E^1] + \frac{\xi_j}{2}[E^1]^T \quad (5.136b)$$

$$[K^2]_{jl} = \left(\frac{1}{4} + \frac{\xi_j \xi_l}{12}\right)[E^2] \quad (5.136c)$$

with (Eq. 5.22)

$$[E^0] = \int_{-1}^{+1} [B^1]^T [D][B^1] |J| \, d\eta \quad (5.137a)$$

$$[E^1] = \int_{-1}^{+1} [B^2]^T [D][B^1] |J| \, d\eta \quad (5.137b)$$

$$[E^2] = \int_{-1}^{+1} [B^2]^T [D][B^2] |J| \, d\eta \quad (5.137c)$$

The submatrices of the mass matrix with respect to the interior and exterior boundaries ($j = i, e; l = i, e$) are written as (Eq. 5.33)

$$[M]_{jl} = \int_S \rho [\hat{N}]_j^T [\hat{N}]_l \, dS \quad (5.138)$$

Integrating analytically in the ξ-direction results in (Eq. 5.35)

$$[M]_{jl} = w[M^2]_{jl} + O(w^2) = \frac{w}{4}\left(1 + \frac{\xi_j \xi_l}{3}\right)[M^0] + O(w^2) \quad (5.139)$$

with (Eq. 5.36)
$$[M^0] = \int_{-1}^{+1} \rho[N]^T[N]|J|d\eta \qquad (5.140)$$

The assemblage of the finite-element cell and the medium described in Section 5.2.7 still applies, yielding (Eq. 5.101)
$$[S_i^\infty(\omega)] = [S_{ii}(\omega)] - [S_{ie}(\omega)]([S_e^\infty(\omega)] + [S_{ee}(\omega)])^{-1}[S_{ei}(\omega)] \qquad (5.141)$$

When the characteristic lengths of the structure-medium interfaces do no change in the x-direction, the corresponding dynamic-stiffness matrices of the unbounded medium are identical (Eq. 3.6)
$$[S_i^\infty(\omega)] = [S_e^\infty(\omega)] = [S^\infty(\omega)] \qquad (5.142)$$

Substituting Eq. 5.142 in Eq. 5.141 and performing analytically the limit of $w \to 0$ as in Section 5.2.8 leads to (Eq. 5.113)
$$([S^\infty(\omega)] + [E^1])[E^0]^{-1}([S^\infty(\omega)] + [E^1]^T) - [E^2] + \omega^2[M^0] = 0 \qquad (5.143)$$

This is the consistent infinitesimal finite-element cell equation in the frequency domain for the layered unbounded medium. Compared with Eq. 5.113, the term with the derivative with respect to ω, $[S^\infty(\omega)]_{,\omega}$, vanishes. This permits $[S^\infty(\omega)]$ for distinct frequencies to be determined directly. Eq. 5.143 is solved in the consistent boundary formulation (References [30], [33], [55], [23], [24], [8])

The consistent infinitesimal finite-element cell equation in the time domain is derived as in Section 5.2.9. After performing the Cholesky's decomposition (Eq. 5.117) which defines the upper-triangular matrix $[U]$, the coefficient matrices are defined as (Eq. 5.120)

$$[e^1] = ([U]^{-1})^T [E^1][U]^{-1} \qquad (5.144a)$$
$$[e^2] = ([U]^{-1})^T ([E^2] - [E^1][E^0]^{-1}[E^1]^T)[U]^{-1} \qquad (5.144b)$$
$$[m^0] = ([U]^{-1})^T [M^0][U]^{-1} \qquad (5.144c)$$

The consistent infinitesimal finite-element cell equation in the time domain for the layered unbounded medium (Eq. 5.118) is written as

$$\int_0^t [m^\infty(t-\tau)][m^\infty(\tau)]d\tau + [e^1]\int_0^t \int_0^\tau [m^\infty(\tau')]d\tau'd\tau$$
$$+ \int_0^t \int_0^\tau [m^\infty(\tau')]d\tau'd\tau[e^1]^T - \frac{t^3}{6}[e^2]H(t) - t[m^0]H(t) = 0 \qquad (5.145)$$

The acceleration unit-impulse response matrix equals (Eq. 5.121)
$$[M^\infty(t)] = [U]^T[m^\infty(t)][U] \qquad (5.146)$$

5.2 FUNDAMENTAL EQUATIONS

5.2.11 Variation of Material Properties in Radial Direction

In all derivations up to now the unbounded medium is assumed to be homogeneous in the radial direction towards infinity, i.e. the elasticity matrix $[D]$ and the mass density ρ are constant in the radial direction r. As an extension, $[D]$ and ρ are assumed to be power functions of r. The derivation is discussed for the isotropic case where the shear modulus and mass density vary as

$$G(r) = G_0 \left(\frac{r}{r_0}\right)^g \tag{5.147}$$

$$\rho(r) = \rho_0 \left(\frac{r}{r_0}\right)^m \tag{5.148}$$

G_0 and ρ_0 are the shear modulus and mass density at the structure-medium interface with the characteristic length r_0. The powers g and m are real numbers which can be selected as positive or negative. The unbounded medium whose material properties are defined by Eqs. 5.147 and 5.148 is denoted as inhomogeneous in this section.

The relationship for the dynamic-stiffness matrices at similar structure-medium interfaces of the unbounded medium which is homogeneous in the radial direction is specified in Eq. 3.6 with the dimensionless frequency defined in Eq. 3.5. As will be demonstrated further on, for the inhomogeneous medium, $G(r)$ defined in Eq. 5.147 replaces G in Eq. 3.6, and c_s in Eq. 3.5 is calculated with $G(r)$ and $\rho(r)$. This yields

$$[S^\infty(r,\omega)] = G_0 \frac{r^{g+s-2}}{r_0^g} [\bar{S}^\infty(a_0)] \tag{5.149}$$

where the dimensionless frequency equals

$$a_0 = \frac{\omega}{c_{s0}} \frac{r^{1-\frac{g}{2}+\frac{m}{2}}}{r_0^{-\frac{g}{2}+\frac{m}{2}}} \tag{5.150}$$

with $c_{s0} = \sqrt{G_0/\rho_0}$. Note that a_0 varies with r.

Eqs. 5.149 and 5.150 are verified as follows. Conceptually, the unbounded medium is discretized with similar finite-element cells up to infinity; i.e. $\gamma = r_j/r_{j-1}$ is a constant (Figure 5-6). The static-stiffness and mass matrices of the jth cell are determined by scaling those of the first cell $G_0 r_0^{s-2}[\bar{K}^1]$ and $\rho_0 r_0^s[\bar{M}^1]$, where $[\bar{K}^1]$ and $[\bar{M}^1]$ are dimensionless. The dynamic-stiffness matrix of the jth cell is formulated as

$$[S^j(\omega)] = G_0 r_0^{s-2} \gamma^{(j-1)(g+s-2)} [\bar{K}^1] - \omega^2 \rho_0 r_0^s \gamma^{(j-1)(m+s)} [\bar{M}^1] \tag{5.151}$$

Assembling $[S^j(\omega)]$ of all cells ($j = 1,\ldots,\infty$) yields a formal representation of the dynamic property of the unbounded medium with the structure-medium interface characterized by r_0

$$[S(r_0,\omega)] = [K(r_0)] - \omega^2[M(r_0)] \tag{5.152}$$

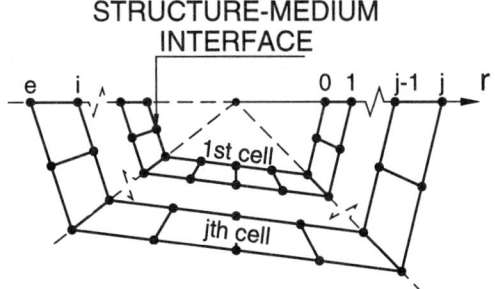

Figure 5-6 Finite-element cells of unbounded medium with material properties varying in the radial direction

where the matrices of infinite order equal

$$[K(r_0)] = G_0 r_0^{s-2} \begin{bmatrix} [\bar{K}_{ii}^1] & [\bar{K}_{ie}^1] & & \\ [\bar{K}_{ei}^1] & [\bar{K}_{ee}^1] + \gamma^{g+s-2}[\bar{K}_{ii}^1] & \gamma^{g+s-2}[\bar{K}_{ie}^1] & \\ & \gamma^{g+s-2}[\bar{K}_{ei}^1] & \gamma^{g+s-2}[\bar{K}_{ee}^1] + \gamma^{2(g+s-2)}[\bar{K}_{ii}^1] & \ddots \\ & & \ddots & \ddots \end{bmatrix}$$

$$= G_0 r_0^{s-2} [\bar{K}] \tag{5.153}$$

and

$$[M(r_0)] = \rho_0 r_0^{s} \begin{bmatrix} [\bar{M}_{ii}^1] & [\bar{M}_{ie}^1] & & \\ [\bar{M}_{ei}^1] & [\bar{M}_{ee}^1] + \gamma^{m+s}[\bar{M}_{ii}^1] & \gamma^{m+s}[\bar{M}_{ie}^1] & \\ & \gamma^{m+s}[\bar{M}_{ei}^1] & \gamma^{m+s}[\bar{M}_{ee}^1] + \gamma^{2(m+s)}[\bar{M}_{ii}^1] & \ddots \\ & & \ddots & \ddots \end{bmatrix}$$

$$= \rho_0 r_0^{s} [\bar{M}] \tag{5.154}$$

The subscripts i and e refer to the interior and exterior boundaries of the cells.

Analogously, assembling $[S^j(\omega)]$ of all but the first cell ($j = 2, \ldots, \infty$) yields

$$[S(r_1, \omega)] = [K(r_1)] - \omega^2 [M(r_1)] \tag{5.155}$$

where

$$[K(r_1)] = G_0 \frac{r_1^{g+s-2}}{r_0^g} [\bar{K}] \tag{5.156}$$

$$[M(r_1)] = \rho_0 \frac{r_1^{m+s}}{r_0^m} [\bar{M}] \tag{5.157}$$

5.2 FUNDAMENTAL EQUATIONS

Substituting Eqs. 5.156 and Eq. 5.157 in Eq. 5.155 leads to

$$[S(r_1, \omega)] = G_0 \frac{r_1^{g+s-2}}{r_0^g} \left([\bar{K}] - a_0^2 [\bar{M}] \right) \tag{5.158}$$

where the dimensionless frequency is defined as

$$a_0 = \frac{\omega}{c_{s0}} \frac{r_1^{1-\frac{g}{2}+\frac{m}{2}}}{r_0^{-\frac{g}{2}+\frac{m}{2}}} \tag{5.159}$$

This procedure can be generalized to any structure-medium interface with the characteristic length r by assembling $[S^j(\omega)]$ of all cells located between this structure-medium interface and infinity resulting in

$$[S(r, \omega)] = G_0 \frac{r^{g+s-2}}{r_0^g} \left([\bar{K}] - a_0^2 [\bar{M}] \right) \tag{5.160}$$

with

$$a_0 = \frac{\omega}{c_{s0}} \frac{r^{1-\frac{g}{2}+\frac{m}{2}}}{r_0^{-\frac{g}{2}+\frac{m}{2}}} \tag{5.161}$$

Conceptually, eliminating all degrees of freedom with the exception of those on the structure-medium interface with r in Eq. 5.160 yields the corresponding dynamic-stiffness matrix of the unbounded medium $[S^\infty(r, \omega)]$. The latter is proportional to $G_0 r^{g+s-2}/r_0^g$ and is a function of a_0 ($= [\bar{S}^\infty(a_0)]$). This is written as Eq. 5.149 with a_0 as in Eq. 5.150.

Proceeding as in Section 3.1 leads to the relationship based on similarity in continuous form

$$r[S^\infty(r, \omega)]_{,r} = (g+s-2)[S^\infty(r, \omega)] + \left(1 - \frac{g}{2} + \frac{m}{2}\right) \omega [S^\infty(r, \omega)]_{,\omega} \tag{5.162}$$

This equation is in the same form as that for the homogeneous medium (Eq. 3.12) with changes occurring only in the scalar coefficients on the right-hand side.

The relationship derived from the assemblage of the finite-element cell and the unbounded medium described in Section 5.2.7 still applies for the inhomogeneous unbounded medium.

The limit of the infinitesimal cell width performed analytically as discussed in Section 5.2.8 leads to Eq. 5.109. Straightforward algebraic manipulations yield the vanishing of the sum identified by 1 and the same expressions for the items identified by 2, 4 and 5 as for the homogeneous case. The coefficient matrices $[E^0]$, $[E^1]$, $[E^2]$ (Eq. 5.22) and $[M^0]$ (Eq. 5.36) are evaluated with the material properties at the structure-medium interface. Eq. 5.112 still applies. Substituting Eq. 5.162 in Eq. 5.112 results in

$$([S^\infty(\omega)] + [E^1])[E^0]^{-1}([S^\infty(\omega)] + [E^1]^T) - (g+s-2)[S^\infty(\omega)]$$
$$- \left(1 - \frac{g}{2} + \frac{m}{2}\right) \omega[S^\infty(\omega)]_{,\omega} - [E^2] + \omega^2[M^0] = 0 \quad (5.163)$$

This represents the *consistent infinitesimal finite-element cell equation in the frequency domain for varying material properties in the radial direction*. It is in the same form as that for the medium with constant material properties in the radial direction (Eq. 5.113).

As the two equations (Eqs. 5.163 and Eq. 5.113) only differ in two scalar coefficients, the two cases can be treated identically. The consistent infinitesimal finite-element cell equation in the time domain for the inhomogeneous medium equals (Eq. 5.118)

$$\int_0^t [m^\infty(t-\tau)][m^\infty(\tau)]d\tau + [e^1]\int_0^t\int_0^\tau [m^\infty(\tau')]d\tau'd\tau + \int_0^t\int_0^\tau [m^\infty(\tau')]d\tau'd\tau[e^1]^T$$
$$+ \left(1 - \frac{g}{2} + \frac{m}{2}\right) t \int_0^t [m^\infty(\tau)]d\tau - \frac{t^3}{6}[e^2]H(t) - t[m^0]H(t) = 0 \quad (5.164)$$

where

$$[e^1] = ([U]^{-1})^T [E^1][U]^{-1} - \frac{1}{2}\left(1 - \frac{g}{2} + \frac{3m}{2} + s\right)[I] \quad (5.165)$$

with $[e^2]$ and $[m^0]$ defined in Eqs. 5.120b and 5.120c. All procedures for solving the consistent infinitesimal finite-element cell equation in the time and frequency domains to be discussed in later sections still apply. The application to the diffusion equation addressed in Chapter 9 is also straightforward for the inhomogeneous unbounded medium.

5.2.12 Insight on Radiation Damping

The discretization of an unbounded medium with similar finite-element cells described in Section 5.2.11 can also be used to determine when radiation damping actually occurs in an unbounded medium whose material properties vary as power functions defined in Eqs. 5.147 and 5.148. As the relative contributions of the elastic restoring force and the inertial force are the same for the unbounded medium expressed as matrices of infinite order (Eqs. 5.153 and 5.154) and for the individual cell (Eq. 5.151), it is sufficient to address one cell to examine if radiation damping occurs. The dynamic-stiffness matrix of a cell with the characteristic length r at the interior boundary specified in Eq. 5.151 is written as (with superscript c for cell)

$$[S^c(\omega)] = [K^c] - \omega^2[M^c] \quad (5.166)$$

5.2 FUNDAMENTAL EQUATIONS

with the static-stiffness and mass matrices of the cell

$$[K^c] = G_0 r_0^{s-2} \left(\frac{r}{r_0}\right)^{g+s-2} [\bar{K}^1] \tag{5.167}$$

$$[M^c] = \rho_0 r_0^s \left(\frac{r}{r_0}\right)^{m+s} [\bar{M}^1] \tag{5.168}$$

$[K^c]$ multiplied by the displacement amplitudes $\{u(\omega)\}$ represents the amplitudes of the elastic restoring force and $-\omega^2[M^c]\{u(\omega)\}$ the amplitudes of the inertial force of the cell. $[S^c(\omega)]$ can be reformulated as

$$[S^c(\omega)] = G_0 r_0^{s-2} \left(\frac{r}{r_0}\right)^{g+s-2} ([\bar{K}^1] - a_0^2[\bar{M}^1]) \tag{5.169}$$

with the dimensionless frequency a_0 (Eq. 5.150) repeated for convenience

$$a_0 = \frac{\omega}{c_{s0}} \frac{r^{1-\frac{g}{2}+\frac{m}{2}}}{r_0^{-\frac{g}{2}+\frac{m}{2}}} \tag{5.170}$$

a_0 is a dimensionless variable describing the relative contributions of the elastic restoring part and the inertial part to the dynamic-stiffness matrix as a function of r and ω.

For $r \to \infty$ and $\omega \neq 0$, three cases exist, which are determined by the power of r. For

$$1 - \frac{g}{2} + \frac{m}{2} > 0 \tag{5.171}$$

$a_0 \to \infty$ for $r \to \infty$ applies. The inertial force will always dominate over the elastic restoring force for any ω. Waves propagating towards infinity exist, and thus radiation damping will always be present.
For

$$1 - \frac{g}{2} + \frac{m}{2} < 0 \tag{5.172}$$

$a_0 \to 0$ for $r \to \infty$ holds. The elastic restoring force will always dominate for any ω. In this case, only evanescent waves which do not propagate towards infinity arise, and radiation damping will never exist. This can be achieved in an unbounded medium whose shear modulus increases sufficiently in the radial direction and/or whose mass density decreases sufficiently, so that Eq. 5.172 is satisfied.

For the intermediate case

$$1 - \frac{g}{2} + \frac{m}{2} = 0 \tag{5.173}$$

a_0 is independent of r, but still depends on ω. For sufficiently small and large frequencies the elastic restoring force and the inertial force dominate, respectively. Thus, a cut-off frequency exists below which no radiation damping occurs.

The criterion (Eq. 5.173) derived on physical grounds can be verified from the consistent infinitesimal finite-element cell equation (Eq. 5.163). Substituting Eq. 5.173 in Eq. 5.163 eliminates the derivative with respect to ω. A quadratic matrix equation in $[S^\infty(\omega)]$ results

$$([S^\infty(\omega)] + [E^1])[E^0]^{-1}([S^\infty(\omega)] + [E^1]^T) - (g+s-2)[S^\infty(\omega)] \\ - [E^2] + \omega^2[M^0] = 0 \quad (5.174)$$

For $\omega = 0$, $[S^\infty(\omega = 0)] = [K^\infty]$ is real, and for $\omega \to \infty$, $[S^\infty(\omega \to \infty)] = i\omega[C_\infty]$ applies (see Section 7.2). A cut-off frequency thus exists. The cut-off frequency is calculated from Eq. 5.174 as the smallest ω for which $[S^\infty(\omega)]$ is complex.

The criterion for the existence of a cut-off frequency (Eq. 5.173) is satisfied for the rod on an elastic foundation described in Appendix A.2. The a_0 specified in Eq. A.2.8 does not depend on the radial coordinate. The presence of a cut-off frequency is confirmed from the dynamic-stiffness coefficient (Eq. A.2.14). The out-of-plane motion of a circular cavity embedded in a full-plane with a variation of the shear modulus in the radial direction is used to illustrate the three possible cases (Eqs. 5.171, 5.172 and 5.173) in Section 5.4.3.

5.3 TIME DISCRETIZATION

After discretization with respect to time, Eq. 5.118 yields an equation for the acceleration unit-impulse response matrix at each time station n. It is assumed that the response matrix is piece-wise constant over each time step, i.e. $[m^\infty]_n$ applies at time $t = (n - 1/2)\Delta t$ ($n \geq 1$). When calculating the interaction forces using Eq. 2.11, this assumption for $[M^\infty(t)]$ is the same as that for the acceleration in the constant-acceleration Newmark method. The integral terms in Eq. 5.118 are discretized as

$$[I]_n = \int_0^{n\Delta t} [m^\infty(\tau)]d\tau = [I]_{n-1} + \Delta t [m^\infty]_n \quad (5.175a)$$

$$[J]_n = \int_0^{n\Delta t} \int_0^\tau [m^\infty(\tau')]d\tau' d\tau = [J]_{n-1} + \Delta t [I]_{n-1} + \frac{\Delta t^2}{2}[m^\infty]_n \quad (5.175b)$$

$$\int_0^t [m^\infty(t-\tau)][m^\infty(\tau)]d\tau = \Delta t \sum_{j=1}^n [m^\infty]_{n-j+1}[m^\infty]_j \quad (5.176)$$

5.3 TIME DISCRETIZATION

5.3.1 First Time Step

As in Eq. 5.118 the convolution-integral term leads to a *quadratic* equation in the unknown matrix for the first time step and a special treatment is necessary. Substituting Eqs. 5.175 and 5.176 with $n = 1$ in Eq. 5.118 yields

$$[m^\infty]_1^2 + \frac{\Delta t}{2}([e^1] + [I])[m^\infty]_1 + [m^\infty]_1 \frac{\Delta t}{2}([e^1]^T + [I]) - \frac{\Delta t^2}{6}[e^2] - [m^0] = 0 \quad (5.177)$$

Eq. 5.177 is the algebraic Riccati equation. It has a unique symmetric semi-positive definite solution [72]. This corresponds to the unbounded medium acting as an energy sink and not an energy source (Section 1.2).

An efficient procedure applies the Schur factorization. A brief description follows which is based on Reference [26]. The proof is also discussed in Reference [39]. Eq. 5.177 is solved introducing the matrix

$$[Z] = \begin{bmatrix} -\frac{\Delta t}{2}([e^1]^T + [I]) & -[I] \\ -\frac{\Delta t^2}{6}[e^2] - [m^0] & \frac{\Delta t}{2}([e^1] + [I]) \end{bmatrix} \quad (5.178)$$

A real orthogonal transformation $[V]$ is applied to $[Z]$ which yields the real Schur form with the matrix $[S]$ in the quasi-upper triangular form consisting of 2×2 blocks or 1×1 blocks on the diagonal

$$[V]^T[Z][V] = [S] = \begin{bmatrix} [S_{11}] & [S_{12}] \\ 0 & [S_{22}] \end{bmatrix} \quad (5.179)$$

$[S]$ is arranged in such a way that the real parts of the eigenvalues of $[S_{11}]$ are negative and those of $[S_{22}]$ are positive. $[V]$ is partitioned conformably as

$$[V] = \begin{bmatrix} [V_{11}] & [V_{12}] \\ [V_{21}] & [V_{22}] \end{bmatrix} \quad (5.180)$$

The solution of Eq. 5.177 is formulated as

$$[m^\infty]_1 = [V_{21}][V_{11}]^{-1} \quad (5.181)$$

The Schur factorization of Eq. 5.179 is widely applied in solving eigenvalue problems. Efficient and accurate programmes in the public domain exist (for instance LAPACK [2]).

The acceleration unit-impulse response matrix at the first time station follows from equation Eq. 5.121 as

$$[M^\infty]_1 = [U]^T[m^\infty]_1[U] \quad (5.182)$$

It is obvious that $[M^\infty]_1$ is symmetric semi-positive definite.

It is worth mentioning that in the limit $\Delta t \to 0$, Eq. 5.177 results in

$$[m^\infty]_1^2 = [m^0] \tag{5.183}$$

Eq. 5.183 is solved determining the positive eigenvalues of the positive definite matrix $[m^0]$. Substituting $[m^\infty]_1$ in Eq. 5.182 yields $[M^\infty]_1$. In this limit $[M^\infty]_1$ is the value at $t = 0$ $[M^\infty(t=0)]$, which is equal to the dashpot coefficient matrix $[C_\infty]$ (Eq. 2.19a). For a detailed discussion of the early-time response, the reader should consult Section 7.2.

5.3.2 nth Time Step

For $n \geq 2$ the convolution-integral term in Eq. 5.118 results in *linear* terms in the unknown $[m^\infty]_n$. Discretizing Eq. 5.118 leads to the equation for the transformed acceleration unit-impulse response matrix $[m^\infty]_n$

$$\left([m^\infty]_1 + \frac{\Delta t}{2}[e^1]\right)[m^\infty]_n + [m^\infty]_n \left([m^\infty]_1 + \frac{\Delta t}{2}[e^1]^T\right) + t[m^\infty]_n =$$

$$-\sum_{j=2}^{n-1}[m^\infty]_{n-j+1}[m^\infty]_j - [e^1]\left(\frac{[J]_{n-1}}{\Delta t} + [I]_{n-1}\right)$$

$$-\left(\frac{[J]_{n-1}}{\Delta t} + [I]_{n-1}\right)[e^1]^T + \frac{t^3}{6\Delta t}[e^2] + \frac{t}{\Delta t}([m^0] - [I]_{n-1}) \tag{5.184}$$

Eq. 5.184 is the Lyapunov equation in the form

$$[A][X] + [X][A]^T + t[X] = [C] \tag{5.185}$$

with $[X] = [m^\infty]_n$ and the other matrices defined straightforwardly.

A summary of the solution procedure for $[A][X] + [X][A]^T = [C]$ described in Reference [5] follows. A Schur factorization is applied to $[A]$

$$[V]^T[A][V] = [S] \tag{5.186}$$

with the orthogonal transformation matrix $[V]$, which results in the quasi-upper triangular matrix $[S]$ consisting of 2×2 blocks or 1×1 blocks on the diagonal. Eq. 5.185 is premultiplied by $[V]^T$ and postmultiplied by $[V]$ yielding

$$[S][Y] + [Y][S]^T + t[Y] = [V]^T[C][V] \tag{5.187}$$

where

$$[Y] = [V]^T[X][V] \tag{5.188}$$

The orthogonal matrix $[V]$ satisfies $[V][V]^T = [V]^T[V] = [I]$. Eq. 5.187 is rewritten as

$$\left([S] + \frac{t}{2}[I]\right)[Y] + [Y]\left([S]^T + \frac{t}{2}[I]\right) = [V]^T[C][V] \qquad (5.189)$$

The matrix $[S] + 0.5t[I]$ presents the real Schur form. This permits Eq. 5.189 to be solved (for each time station n) successively by addressing the 2×2 blocks or 1×1 blocks, leading to $[Y]$. $[X]$ then follows as (Eq. 5.188)

$$[X] = [V][Y][V]^T \qquad (5.190)$$

Note that $[A]$ is independent of the time station. The Schur factorization is thus performed only once. The term $t[X]$ does not increase the computational effort.

To prove that $[m^\infty]_n$ is symmetric, the transpose of Eq. 5.184 is subtracted from Eq. 5.184 with the symmetric matrices $[e^2]$, $[m^0]$ and $[m^\infty]_j$, $(j = 1, \ldots, n-1)$ leading to

$$\left([m^\infty]_1 + \frac{\Delta t}{2}[e^1]\right)\left([m^\infty]_n - [m^\infty]_n^T\right) + \left([m^\infty]_n - [m^\infty]_n^T\right)\left([m^\infty]_1 + \frac{\Delta t}{2}[e^1]^T\right)$$
$$+ t\left([m^\infty]_n - [m^\infty]_n^T\right) = 0 \quad (5.191)$$

This linear equation is only satisfied for $[m^\infty]_n = [m^\infty]_n^T$.

After determining $[m^\infty]_n$, the acceleration unit-impulse response matrix is calculated from Eq. 5.121 as

$$[M^\infty]_n = [U]^T[m^\infty]_n[U] \qquad (5.192)$$

$[M^\infty]_n$ is also symmetric.

5.4 ACCURACY

5.4.1 Spherical Cavity Embedded in Full-space

The spherical cavity embedded in a full-space with a uniform normal displacement prescribed on the structure-medium interface described in Appendix A.1 (Figure A-1) is addressed. The analytical solution of the acceleration unit-impulse response coefficient is specified in Eq. A.1.48.

First, the consistent infinitesimal finite-element cell method is applied to this one-dimensional wave propagation problem. The coefficient matrices are established and this permits the consistent infinitesimal finite-element cell equation in the time domain to be formulated. It is then verified that the analytical solution of the acceleration unit-impulse response coefficient satisfies this equation.

The structure-medium interface coincides with the wall of the spherical cavity at r_0. The finite-element cell consists of a hollow sphere with the displacement u_i at the interior radius $r_i = r_0$ and the displacement u_e at the exterior radius $r_e = (1+w)r_i$. The linear shape function of this 2-node finite element equals (Eq. 5.3)

$$\{\hat{N}\} = \left\{ \begin{array}{c} \{\hat{N}_i\} \\ \{\hat{N}_e\} \end{array} \right\} = \left\{ \begin{array}{c} \frac{1}{2}(1-\xi)N \\ \frac{1}{2}(1+\xi)N \end{array} \right\} \tag{5.193}$$

with

$$N = 1 \tag{5.194}$$

corresponding to uniform motion on the structure-medium interface. The isoparametric mapping rule is formulated for the coordinate as (Eq. 5.7)

$$\hat{r} = \frac{1}{2}(1-\xi)r_i + \frac{1}{2}(1+\xi)r_e = \left(1 + \frac{w}{2}(1+\xi)\right)r_0 \tag{5.195}$$

and for the displacement

$$u = \hat{N}_i u_i + \hat{N}_e u_e \tag{5.196}$$

The Jacobian matrix of the one-dimensional element equals

$$\hat{J} = \frac{w}{2}J \tag{5.197}$$

The coefficient J (Eq. 5.9) is calculated as

$$J = r_0 \tag{5.198}$$

The strain-nodal displacement matrix $[B]$

$$\left\{ \begin{array}{c} \varepsilon_r \\ \varepsilon_\theta \\ \varepsilon_\phi \end{array} \right\} = [B] \left\{ \begin{array}{c} u_i \\ u_e \end{array} \right\} \tag{5.199}$$

follows from substituting Eqs. 5.195 and 5.196 in Eq. A.1.2 yielding

$$[B] = [\{B_i\} \ \{B_e\}] \tag{5.200}$$

where

$$\{B\}_j = \left\{ \begin{array}{c} \hat{N}_{j,\hat{r}} \\ \frac{\hat{N}_j}{\hat{r}} \\ \frac{\hat{N}_j}{\hat{r}} \end{array} \right\} = \frac{\xi_j}{w}\{B^1\} + \frac{1+\xi_j\xi}{2\left(1+\frac{w}{2}(1+\xi)\right)}\{B^2\} \qquad (j=i,e) \tag{5.201}$$

5.4 ACCURACY

with

$$\{B^1\} = \frac{1}{r_0} \left\{ \begin{array}{c} 1 \\ 0 \\ 0 \end{array} \right\} \tag{5.202a}$$

$$\{B^2\} = \frac{1}{r_0} \left\{ \begin{array}{c} 0 \\ 1 \\ 1 \end{array} \right\} \tag{5.202b}$$

Eqs. 5.200 and 5.201 should be compared with Eqs. 5.16 and 5.17.
The cell's static-stiffness coefficient (Eq. 5.18) is written as

$$[K] = \int_V [B]^T [D][B] dV = \int_{-1}^{+1} [B]^T [D][B] 4\pi \hat{r}^2 \hat{J} d\xi \tag{5.203}$$

with the elasticity matrix $[D]$ specified in Eq. A.1.3. The coefficients associated with $[K]$ follow from Eq. 5.22

$$E^0 = \{B^1\}^T [D]\{B^1\} 4\pi r_0^2 J = 8\pi \frac{1-\nu}{1-2\nu} G r_0 \tag{5.204a}$$

$$E^1 = \{B^2\}^T [D]\{B^1\} 4\pi r_0^2 J = 16\pi \frac{\nu}{1-2\nu} G r_0 \tag{5.204b}$$

$$E^2 = \{B^2\}^T [D]\{B^2\} 4\pi r_0^2 J = 16\pi \frac{1}{1-2\nu} G r_0 \tag{5.204c}$$

The coefficient M^0 associated with the mass matrix (Eq. 5.36) equals

$$M^0 = 4\pi \rho r_0^3 \tag{5.205}$$

For this scalar case, U in Eq. 5.117 equals $\sqrt{E^0}$ resulting in (Eq. 5.120)

$$e^1 = -\frac{2(1-2\nu)}{1-\nu} \tag{5.206a}$$

$$e^2 = \frac{2(1-2\nu)(1+\nu)}{(1-\nu)^2} \tag{5.206b}$$

$$m^0 = \frac{r_0^2}{c_p^2} \tag{5.206c}$$

The consistent infinitesimal finite-element cell equation is written as (Eq. 5.118)

$$\int_0^t m^\infty(t-\tau) m^\infty(\tau) d\tau - \frac{4(1-2\nu)}{1-\nu} \int_0^t \int_0^\tau m^\infty(\tau') d\tau' d\tau + t \int_0^t m^\infty(\tau) d\tau$$

$$- \frac{2(1-2\nu)(1+\nu)}{(1-\nu)^2} \frac{t^3}{6} H(t) - \frac{r_0^2}{c_p^2} t H(t) = 0 \tag{5.207}$$

The analytical solution of $m^\infty(t)$ is calculated from $M^\infty(t)$ in Eq. A.1.48 as (Eq. 5.121)

$$m^\infty(t) = \frac{M^\infty(t)}{E^0} = \left(\frac{2(1-2\nu)}{1-\nu}t + \frac{r_0}{c_p}e^{-\frac{c_p}{r_0}t}\right)H(t) \qquad (5.208)$$

After substitution, it is straightforwardly verified that $m^\infty(t)$ satisfies Eq. 5.207.

Second, the consistent infinitesimal finite-element cell method is applied to the three-dimensional wave propagation problem of the same spherical cavity. The uniform displacement pattern is enforced at the end of the analysis, which results in a one-dimensional problem.

The structure-medium interface is discretized with two-dimensional surface finite elements. Owing to symmetry, only one octant of the structure-medium interface is discretized (Figure 5-7). Three 8-node isoparametric surface finite elements are

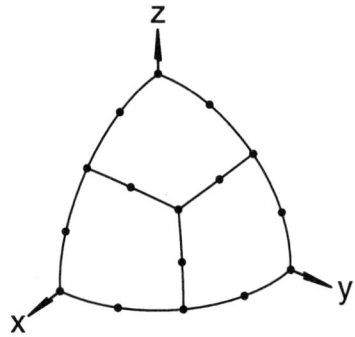

Figure 5-7 Finite-element mesh of one octant of structure-medium interface of spherical cavity

used yielding a dynamic system with 33 degrees of freedom. After determining the unit-impulse response matrix $[M^\infty(t)]$ of order 33×33, the uniform radial nodal accelerations are enforced. This defines the spatial motion pattern $\{\phi\}$. The unit-impulse response coefficient $M^\infty(t)$ follows from Eq. A.0.1, which corresponds to summing all radial nodal forces for a unit radial acceleration. The analysis is performed with $\nu = 0.25$ and with different time steps $\Delta t = 0.0692 r_0/c_p$ and $= 0.346 r_0/c_p$. The coefficient $M^\infty(t)$ non-dimensionalized with $K^\infty r_0/c_p$ is plotted as a function of the dimensionless time $\bar{t} = t c_p / r_0$ (Figure 5-8). The agreement with the exact result (Eq. A.1.49, Figure A-2c) is excellent.

As an application using the acceleration unit-impulse response coefficient calculated above with the consistent infinitesimal finite-element cell method, the displacement $u_0(t)$ for an applied pressure is determined. Discretizing the interaction

5.4 ACCURACY

force-acceleration relationship Eq. 2.11 developed in Chapter 2 at time station n yields

$$R_n = \sum_{j=1}^{n} M^{\infty}_{n-j+1} \int_{t_{j-1}}^{t_j} \dot{u}_0(\tau) d\tau = \sum_{j=1}^{n} M^{\infty}_{n-j+1}(\dot{u}_{0j} - \dot{u}_{0j-1}) \tag{5.209}$$

The velocity $\dot{u}_{0n}$ follows from Eq. 5.209 and the displacement u_{0n} is calculated as

$$u_{0n} = u_{0n-1} + \frac{\Delta t}{2}(\dot{u}_{0n} + \dot{u}_{0n-1}) \tag{5.210}$$

To be able to calculate the exact result, the analytical solution of the response coefficient $F^{\infty}(t)$ to a unit-impulse of the interaction force is derived. In the frequency domain the corresponding coefficient $F^{\infty}(\omega)$ is the inverse of the displacement dynamic-stiffness coefficient $S^{\infty}(\omega)$ (Eq. A.1.16)

$$F^{\infty}(\omega) = \frac{1}{16\pi G r_0} \frac{2(1-2\nu)c_p(c_p + i\omega r_0)}{2(1-2\nu)c_p(c_p + i\omega r_0) + (1-\nu)(i\omega r_0)^2} \tag{5.211}$$

The inverse Fourier transform equals

$$F^{\infty}(t) = \frac{1}{16\pi G r_0}\left(\alpha \cos\left(\frac{\beta t}{2}\right) - \frac{\nu\alpha^2}{(1-2\nu)\beta}\sin\left(\frac{\beta t}{2}\right)\right) e^{-\frac{\alpha t}{2}} H(t) \tag{5.212}$$

with the constants

$$\alpha = \frac{4G}{\rho c_p r_0} \tag{5.213a}$$

$$\beta = \frac{\alpha}{\sqrt{1-2\nu}} \tag{5.213b}$$

The displacement $u_0(t)$ follows as the convolution integral

$$u_0(t) = \int_0^t F^{\infty}(t-\tau)R(\tau)d\tau \tag{5.214}$$

For a spherical cavity initially at rest ($u_0 = \dot{u}_0 = 0$) the following rounded triangular pressure pulse is applied, selecting $t_0 = 3.46 r_0/c_p$

$$p(t) = \begin{cases} \dfrac{p_0}{2}\left(1 - \cos\dfrac{2\pi t}{t_0}\right) & 0 \leq t \leq t_0 \\ 0 & t > t_0 \end{cases} \tag{5.215}$$

$R(t)$ equals $4\pi r_0^2 p(t)$. The displacement u_0 determined from Eq. 5.210 with $\Delta t = 0.0692 r_0/c_p$ agrees well with that of Eq. 5.214 (Figure 5-9).

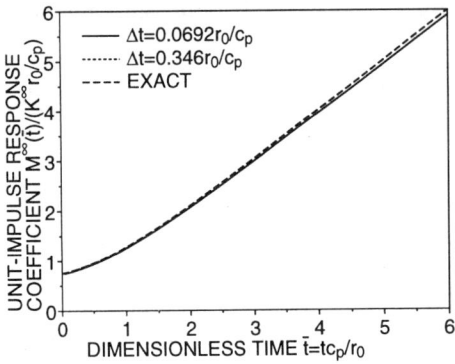
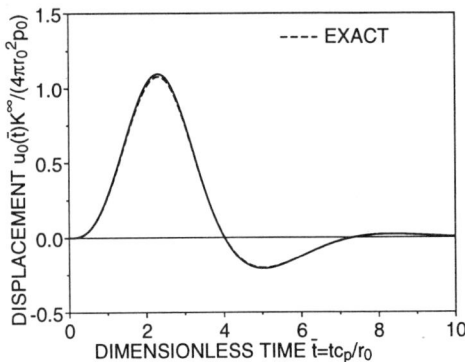

Figure 5-8 Acceleration unit-impulse response coefficient for spherical cavity

Figure 5-9 Displacement response to rounded triangular pressure pulse applied to wall of spherical cavity

5.4.2 Spherical Cavity Embedded in Full-space with Varying Material Properties in Radial Direction

As an extension to the spherical cavity embedded in a full-space addressed in Section 5.4.1, the material properties are assumed to vary in the radial direction. The consistent infinitesimal finite-element cell method for this inhomogeneous case is discussed in Section 5.2.11.

The shear modulus $G(r)$ and the mass density $\rho(r)$ vary as

$$G(r) = G_0 \left(\frac{r}{r_0}\right)^g \tag{5.216}$$

$$\rho(r) = \rho_0 \left(\frac{r}{r_0}\right)^m \tag{5.217}$$

The derivation of the equation of motion is analogous to that described in Appendix A.1. This leads to (Eq. A.1.10)

$$u(\omega)_{,rr} + \frac{g+2}{r}u(\omega)_{,r} - \frac{2(1-(g+1)\nu)}{(1-\nu)r^2}u(\omega) + \frac{\omega^2 r^{m-g}}{c_{p0}^2 r_0^{m-g}}u(\omega) = 0 \tag{5.218}$$

with $c_{p0} = \sqrt{G_0/\rho_0}\sqrt{2(1-\nu)/(1-2\nu)}$.

Eq. 5.218 is solved for two special cases. In the first, the static case, the analytical solution is derived setting $\omega = 0$

$$u_{,rr} + \frac{g+2}{r}u_{,r} - \frac{2(1-(g+1)\nu)}{(1-\nu)r^2}u = 0 \tag{5.219}$$

5.4 ACCURACY

Its general solution equals

$$u = c_1 r^{-\alpha_1} + c_2 r^{-\alpha_2} \tag{5.220}$$

with

$$\alpha_1 = \frac{1}{2}\left((g+1) - \sqrt{g^2 + 2g\frac{1-5v}{1-v} + 9}\right) \tag{5.221a}$$

$$\alpha_2 = \frac{1}{2}\left((g+1) + \sqrt{g^2 + 2g\frac{1-5v}{1-v} + 9}\right) \tag{5.221b}$$

Selecting the integration constants as in the homogeneous case ($g = 0$), $c_1 = 0$ yields

$$u = c r^{-\alpha} \tag{5.222}$$

where the subscripts 2 have been omitted. Proceeding as in Appendix A.1 results in the analytical expression for the static-stiffness coefficient

$$K^\infty = 4\pi \frac{1-v}{1-2v} G_0 r_0 \left(\frac{1-5v}{1-v} + g + \sqrt{\left(\frac{1-5v}{1-v} + g\right)^2 + \frac{8(1+v)(1-2v)}{(1-v)^2}}\right) \tag{5.223}$$

If $c_2 = 0$ were assumed, the corresponding static-stiffness coefficient would be negative for all values of g and v.

Turning to the consistent infinitesimal finite-element cell method, the coefficients E^0, E^1 and E^2 are specified in Eq. 5.204. The corresponding equation follows from Eq. 5.163 with $\omega = 0$

$$\frac{1}{8\pi}\frac{1-2v}{1-v}\frac{1}{G_0 r_0}(K^\infty)^2 - \left(\frac{1-5v}{1-v} + g\right) K^\infty - 16\pi \frac{1+v}{1-v} G_0 r_0 = 0 \tag{5.224}$$

Its positive solution is equal to Eq. 5.223.

In the second case, only the mass density varies for dynamics. The following constants apply: $g = 0$, $m = -2$. Eq. 5.218 is then written as

$$u(\omega)_{,rr} + \frac{2}{r} u(\omega)_{,r} - \frac{2}{r^2} u(\omega) + \frac{a_0^2}{r^2} u(\omega) = 0 \tag{5.225}$$

with

$$a_0 = \frac{\omega r_0}{c_{p0}} \tag{5.226}$$

Its general solution is formulated as

$$u(\omega) = c_1 r^{-\alpha_1} + c_2 r^{-\alpha_2} \tag{5.227}$$

with

$$\alpha_1 = \frac{1}{2}\left(1 - \sqrt{9 - 4a_0^2}\right) \tag{5.228a}$$

$$\alpha_2 = \frac{1}{2}\left(1 + \sqrt{9 - 4a_0^2}\right) \tag{5.228b}$$

Addressing the static case ($a_0 = 0$) α_1 yields a displacement u which tends to infinity for $r \to \infty$. Thus $c_1 = 0$. The displacement amplitude is specified as

$$u(\omega) = cr^{-\alpha} \tag{5.229}$$

where the subscripts 2 have been omitted. Proceeding as in Appendix A.1, the dynamic-stiffness coefficient equals

$$S^\infty(a_0) = 16\pi G_0 r_0 \left(1 + \frac{1-\nu}{4(1-2\nu)}\left(\sqrt{9 - 4a_0^2} - 3\right)\right) \tag{5.230}$$

As a_0 in Eq. 5.226 is independent of r, Eq. 5.173 is satisfied. A cut-off frequency thus exist which turns out to be equal to $a_0 = 1.5$.

For the consistent infinitesimal finite-element cell method the coefficient M^0 is specified in Eq. 5.205. The corresponding equation equals (Eq. 5.163)

$$\frac{1}{8\pi}\frac{1-2\nu}{1-\nu}\frac{1}{G_0 r_0}(S^\infty(\omega))^2 - \frac{1-5\nu}{1-\nu}S^\infty(\omega) - 16\pi\frac{1+\nu}{1-\nu}G_0 r_0 + \omega^2 4\pi\rho_0 r_0^3 = 0 \tag{5.231}$$

Note that the coefficient $1 - g/2 + m/2$ of the term with the derivative with respect to ω vanishes in the present case. The consistent infinitesimal finite-element cell equation in the frequency domain thus becomes a quadratic equation. Its solution coincides with the analytical expression of Eq. 5.230.

5.4.3 Out-of-plane Motion of Circular Cavity Embedded in Full-plane with Varying Shear Modulus in Radial Direction

A simple one-dimensional wave propagation problem where an analytical solution exists for the shear modulus varying in the radial direction is used to investigate the three cases of wave behaviour addressed in Section 5.2.12. The out-of-plane motion of a circular cavity of radius r_0 embedded in a full-plane with a uniform displacement u_0 enforced on its wall, the structure-medium interface, is discussed (Figure 5-10). Symmetric waves exist. The shear modulus G varies in the radial direction as

$$G(r) = G_0 \left(\frac{r}{r_0}\right)^g \tag{5.232}$$

5.4 ACCURACY

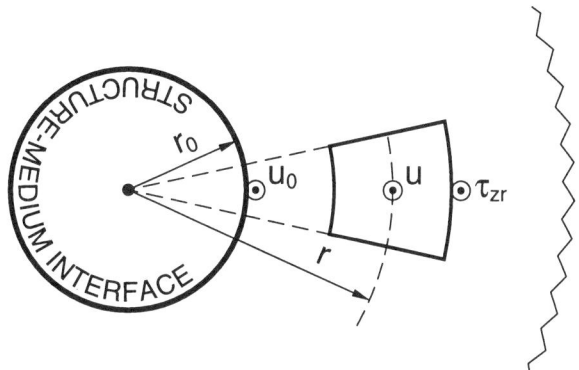

Figure 5-10 Out-of-plane motion of circular cavity embedded in full-plane with symmetric waves

and the mass density ρ is constant.

Substituting the stress-displacement relationship

$$\tau_{zr} = G(r) u_{,r} \tag{5.233}$$

into the equilibrium equation

$$\tau_{zr,r} + \frac{1}{r}\tau_{zr} - \rho \ddot{u} = 0 \tag{5.234}$$

leads to the equation of motion

$$G(r)u_{,rr} + \left(G(r)_{,r} + \frac{1}{r}G(r)\right) u_{,r} - \rho \ddot{u} = 0 \tag{5.235}$$

A solution is derived in the frequency domain. Substituting Eq. 5.232 in Eq. 5.235 formulated in the frequency domain yields

$$r^2 u(\omega)_{,rr} + (g+1) r u(\omega)_{,r} + \frac{\omega^2 r^{2-g}}{c_{s0}^2 r_0^{-g}} u(\omega) = 0 \tag{5.236}$$

with the shear-wave velocity at the structure-medium interface

$$c_{s0} = \sqrt{\frac{G_0}{\rho}} \tag{5.237}$$

To derive a solution for Eq. 5.236, the following two transformations of variables are performed.

$$u(\omega) = \bar{u}(a_0) \left(\frac{r}{r_0}\right)^{-\frac{g}{2}} \tag{5.238}$$

$$a_0 = \frac{\omega}{c_{s0}} \frac{r^{1-\frac{g}{2}}}{r_0^{-\frac{g}{2}}} \tag{5.239}$$

For the independent variable r the dimensionless frequency a_0 is used. Substituting in Eq. 5.236 leads to

$$a_0^2 \bar{u}(a_0)_{,a_0 a_0} + a_0 \bar{u}(a_0)_{,a_0} + ((\lambda a_0)^2 - (\lambda - 1)^2)\bar{u}(a_0) = 0 \tag{5.240}$$

with

$$\lambda = \frac{2}{2-g} \tag{5.241}$$

This is the Bessel differential equation. Note that Eq. 5.239 is the same as the definition of a_0 in Eq. 5.170 with $m = 0$.

The behaviour of Eq. 5.240 is governed by the boundary condition specified for a_0 as $r \to \infty$ and thus the power of r in Eq. 5.239. The same three cases as in Section 5.2.12 appear.

For $1 - 0.5g > 0$ (Eq. 5.171 with $m = 0$), i.e. $g < 2$, $\lambda a_0 \to \infty$ applies for $r \to \infty$ as verified from Eq. 5.239. The solution for the displacement follows from Eqs. 5.240 and 5.238 as

$$u(\omega) = \left(\frac{r}{r_0}\right)^{-\frac{g}{2}} \left(c_1 H^{(1)}_{|\lambda-1|}(\lambda a_0) + c_2 H^{(2)}_{|\lambda-1|}(\lambda a_0)\right) \tag{5.242}$$

$H^{(1)}_{|\lambda-1|}$ and $H^{(2)}_{|\lambda-1|}$ are the first and second kind Hankel functions of order $|\lambda - 1|$, and c_1, c_2 are the integration constants. The corresponding asymptotic behaviour for $\lambda a_0 \to \infty$ equals

$$H^{(1)}_{|\lambda-1|}(\lambda a_0) \approx \sqrt{\frac{2}{\pi \lambda a_0}} \, e^{+i\left(\lambda a_0 - \frac{|\lambda-1|\pi}{2} - \frac{\pi}{4}\right)} \tag{5.243a}$$

$$H^{(2)}_{|\lambda-1|}(\lambda a_0) \approx \sqrt{\frac{2}{\pi \lambda a_0}} \, e^{-i\left(\lambda a_0 - \frac{|\lambda-1|\pi}{2} - \frac{\pi}{4}\right)} \tag{5.243b}$$

Following the argument addressing outgoing waves in connection with Eq. A.1.14, $c_1 = 0$ is obtained. This results in

$$u(\omega) = c \left(\frac{r}{r_0}\right)^{-\frac{g}{2}} H^{(2)}_{|\lambda-1|}(\lambda a_0) \tag{5.244}$$

5.4 ACCURACY

where the subscript 2 in the integration constant is omitted.

The dynamic-stiffness coefficient $S^\infty(\omega)$ relates the displacement amplitude $u_0(\omega)$ to the interaction force amplitude

$$R(\omega) = -2\pi r_0 G_0\, u(\omega),_r \big|_{r=r_0} \quad (5.245)$$

as

$$R(\omega) = S^\infty(\omega) u_0(\omega) \quad (5.246)$$

Performing the derivative $u(\omega),_r$ in Eq. 5.244 using

$$\lambda a_0 H^{(2)}_{|\lambda-1|}(\lambda a_0),_{\lambda a_0} = |\lambda - 1| H^{(2)}_{|\lambda-1|}(\lambda a_0) - \lambda a_0 H^{(2)}_{|\lambda-1|+1}(\lambda a_0) \quad (5.247)$$

yields

$$S^\infty(\omega) = 2\pi G_0 \left(\frac{\lambda - 1 - |\lambda - 1|}{\lambda} + \frac{|\lambda|}{\lambda} \frac{\omega r_0}{c_{s0}} \frac{H^{(2)}_{|\lambda-1|+1}\left(\lambda \frac{\omega r_0}{c_{s0}}\right)}{H^{(2)}_{|\lambda-1|}\left(\lambda \frac{\omega r_0}{c_{s0}}\right)} \right) \quad (5.248)$$

The static-stiffness coefficient follows for the limit $\omega \to 0$ as

$$K^\infty = \begin{cases} 2\pi G_0 g & g > 0 \\ 0 & g \leq 0 \end{cases} \quad (5.249)$$

The dynamic-stiffness coefficient is non-dimensionalized with $2\pi G_0$ and decomposed as specified in Eq. A.0.3. The dimensionless spring and damping coefficients are plotted for $g = 0$ (homogeneous case) and $g = 1$ (linear increase of G) in Figure 5-11. As expected, radiation damping reflected by $c(\omega r_0/c_{s0}) > 0$ occurs for all non-zero frequencies.

For $1 - 0.5g < 0$ (Eq. 5.172 with $m = 0$), i.e. $g > 2$, $-\lambda a_0 \to 0$ applies for $r \to \infty$ as verified from Eq. 5.239. The solution for the displacement follows from Eqs. 5.240 and 5.238 as

$$u(\omega) = \left(\frac{r}{r_0}\right)^{-\frac{g}{2}} (c_1 Y_{1-\lambda}(-\lambda a_0) + c_2 J_{1-\lambda}(-\lambda a_0)) \quad (5.250)$$

$Y_{1-\lambda}$ and $J_{1-\lambda}$ are the second and first kind Bessel functions of order $1 - \lambda$, and c_1, c_2 are the integration constants. As $|Y_{1-\lambda}(-\lambda a_0 \to 0)| \to \infty$, $c_1 = 0$ follows. This results in

$$u(\omega) = c \left(\frac{r}{r_0}\right)^{-\frac{g}{2}} J_{1-\lambda}(-\lambda a_0) \quad (5.251)$$

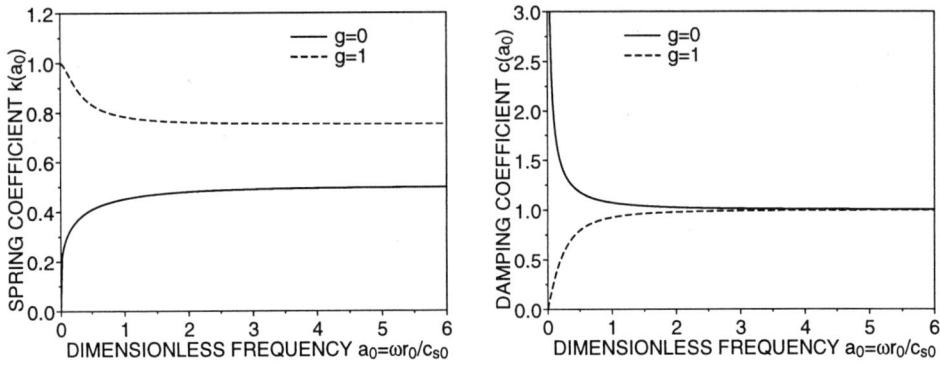

Figure 5-11 Dynamic-stiffness coefficient of out-of-plane motion of circular cavity

The dynamic-stiffness coefficient defined in Eqs. 5.245 and 5.246 equals

$$S^{\infty}(\omega) = 2\pi G_0 \left(g - \frac{\omega r_0}{c_{s0}} \frac{J_{2-\lambda}\left(-\lambda \frac{\omega r_0}{c_{s0}}\right)}{J_{1-\lambda}\left(-\lambda \frac{\omega r_0}{c_{s0}}\right)} \right) \quad (5.252)$$

$S^{\infty}(\omega)$ is real for all frequencies. The system exhibits resonances. This unbounded medium thus behaves similarly as a bounded medium. No radiation damping occurs. The dynamic-stiffness coefficient non-dimensionalized with $2\pi G_0$ is plotted for $g = 3$ in Figure 5-12.

For $1 - 0.5g = 0$ (Eq. 5.173 with $m = 0$), i.e. $g = 2$, $a_0 = \omega r_0/c_{s0}$ is independent of r. The transformation of the independent variable r to a_0 defined in Eq. 5.239 breaks

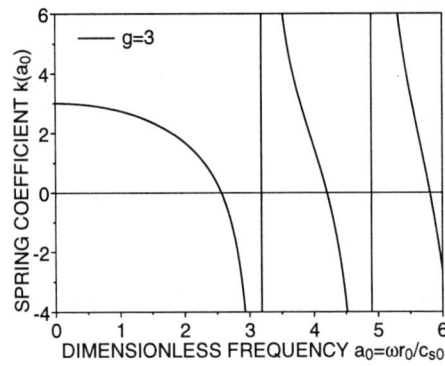

Figure 5-12 Dynamic-stiffness coefficient of out-of-plane motion of circular cavity for strong increase of shear modulus in radial direction

5.4 ACCURACY

down. Eq. 5.236 is written as

$$r^2 u(\omega)_{,rr} + 3r u(\omega)_{,r} + \left(\frac{\omega r_0}{c_{s0}}\right)^2 u(\omega) = 0 \tag{5.253}$$

The solution of Eq. 5.253 equals

$$u(\omega) = c_1 r^{-1+\sqrt{1-\left(\frac{\omega r_0}{c_{s0}}\right)^2}} + c_2 r^{-1-\sqrt{1-\left(\frac{\omega r_0}{c_{s0}}\right)^2}} \tag{5.254}$$

Addressing the static case ($\omega = 0$), the first term corresponds to a constant displacement. Thus, $c_1 = 0$ is enforced. The solution equals

$$u(\omega) = c r^{-1-\sqrt{1-\left(\frac{\omega r_0}{c_{s0}}\right)^2}} \tag{5.255}$$

The dynamic-stiffness coefficient $S^\infty(\omega)$ follows from Eqs. 5.245 and 5.246 as

$$S^\infty(\omega) = 2\pi G_0 \left(1 + \sqrt{1-\left(\frac{\omega r_0}{c_{s0}}\right)^2}\right) \tag{5.256}$$

Below the cut-off frequency $\omega = c_{s0}/r_0$ no radiation damping occurs. $S^\infty(\omega)$ is non-dimensionalized with $2\pi G_0$. The dimensionless spring and damping coefficients are plotted in Figure 5-13.

The consistent infinitesimal finite-element cell method is addressed. For this one-dimensional problem, the coefficients can be derived exactly as at the beginning of Section 5.4.1. For the uniform displacement on the structure-medium interface the shape function

$$N = 1 \tag{5.257}$$

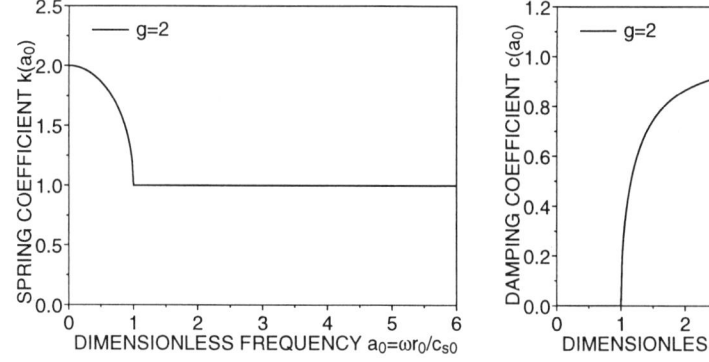

Figure 5-13 Dynamic-stiffness coefficient of out-of-plane motion of circular cavity for parabolic increase of shear modulus in radial direction

applies, and the coefficient J equals

$$J = r_0 \tag{5.258}$$

With the strain $\gamma_{zr} = u_{,r}$

$$B^1 = \frac{1}{r_0} \tag{5.259a}$$

$$B^2 = 0 \tag{5.259b}$$

are calculated. From Eqs. 5.22 and 5.36 integrated over the structure-medium interface

$$E^0 = 2\pi G_0 \tag{5.260a}$$

$$E^1 = E^2 = 0 \tag{5.260b}$$

$$M^0 = 2\pi\rho r_0^2 \tag{5.261}$$

are determined. The consistent infinitesimal finite-element cell equation in the frequency domain for varying shear modulus in the radial direction (Eq. 5.163) is formulated with $s = 2$ and $m = 0$ as

$$\frac{(S^\infty(\omega))^2}{2\pi G_0} - g S^\infty(\omega) - \left(1 - \frac{g}{2}\right) \omega \, S^\infty(\omega)_{,\omega} + \omega^2 2\pi\rho r_0^2 = 0 \tag{5.262}$$

The analytical dynamic-stiffness coefficients for the three cases (Eqs. 5.248, 5.252 and 5.256) satisfy Eq. 5.262.

5.4.4 Out-of-plane Motion of Semi-infinite Layer of Constant Depth

The out-of-plane motion of the semi-infinite layer of constant depth described in Appendix A.3.1 (Figure A-7) is addressed. The analytical expression for the acceleration unit-impulse response matrix for a quadratic finite element on the structure-medium interface is specified in Eq. A.3.37 with Eqs. A.3.36 and Eq. A.3.32.

The structure-medium interface is discretized with 8 3-node line elements. The acceleration unit-impulse response matrix $[M^\infty(t)]$ of order 16×16 is calculated with a time step $\Delta t = 0.05 d/c_s$. The shape function specified in Eq. A.3.2 defines the spatial motion patterns $\{\phi\}$ which permit the coefficients to be calculated using Eq. A.0.1. Excellent agreement with the analytical solution of Eq. A.3.37 presented in Figure A-9 results in Figure 5-14.

5.4 ACCURACY

5.4.5 Out-of-plane Motion of Semi-infinite Wedge

The out-of-plane motion of the semi-infinite wedge with an opening angle $\alpha = 30°$ described in Appendix A.4.1 (Figure A-11) is examined. The analytical expression for the dynamic-stiffness coefficient in the frequency domain is specified in Eq. A.4.30.

The structure-medium interface consisting of an arc is discretized with 3 3-node line elements of equal length. The acceleration unit-impulse response matrix $[M^\infty(t)]$ with respect to the 6 nodes is calculated. The time step is selected as $\Delta t = 0.03 r_0/c_s$. In the analytical solution a linear variation of the displacement on the arc is enforced (Eq. A.4.7) which determines the vector $\{\phi\}$ in Eq. A.0.1. The corresponding unit-impulse response coefficient $M^\infty(t)$ follows from Eq. A.0.1. To be able to perform a comparison with the analytical solution of the dynamic-stiffness coefficient (Eq. A.4.30), $S^\infty(\omega)$ is calculated from $M^\infty(t)$ using the procedure described in Eqs. 2.23 and 2.25. With the static-stiffness coefficient K^∞ of Eq. A.4.31 the decomposition of Eq. A.0.3 yields the dimensionless spring coefficient $k(a_0)$ and damping coefficient $c(a_0)$. They are plotted in Figure 5-15 and coincide with the analytical solution in Figure A-12.

5.4.6 In-plane Motion of Semi-infinite Wedge

The in-plane motion of the semi-infinite wedge defined in Appendix A.4.2 (Figure A-13) is addressed. The opening angle $\alpha = 30°$ and Poisson's ratio $\nu = 0.25$ are selected. Three regions of different shear moduli with a soft inner part compatible with similarity are present ($G_1/G_2 = G_3/G_2 = 10$). On the structure-medium interface, 6 3-node line elements of equal length are chosen. The acceleration unit-impulse response matrix $[M^\infty(t)]$ with respect to the 12 nodes is calculated with a time step selected as $\Delta t = 0.01 r_0/c_{s2}$ ($c_{s2} = \sqrt{G_2/\rho}$). The linear function in the circumferential direction of the horizontal displacement and zero vertical displacement on the structure-medium interface determine $\{\phi\}$ which leads to the acceleration unit-impulse response coefficient $M^\infty(t)$ (Eq. A.0.1). $M^\infty(\bar{t})$ calculated with the consistent infinitesimal finite-element cell method agrees well with the result determined with an extended mesh of finite elements with the element length in the radial direction $= 0.025 r_0$ and the same time step Δt (Figure 5-16 where $\bar{t} = t c_{s2}/r_0$ is the dimensionless time). 230 rows of finite elements are necessary up to $\bar{t} = 2$. The procedure for calculating the extended mesh is explained at the beginning of Appendix A.

5.4.7 Circular Cavity Embedded in Full-plane

The in-plane motion of a circular cavity embedded in a full-plane described in Appendix A.5 (Figure A-14) is addressed. The analytical expression for the dynamic-stiffness coefficient in the frequency domain is specified in Eq. A.5.22 with

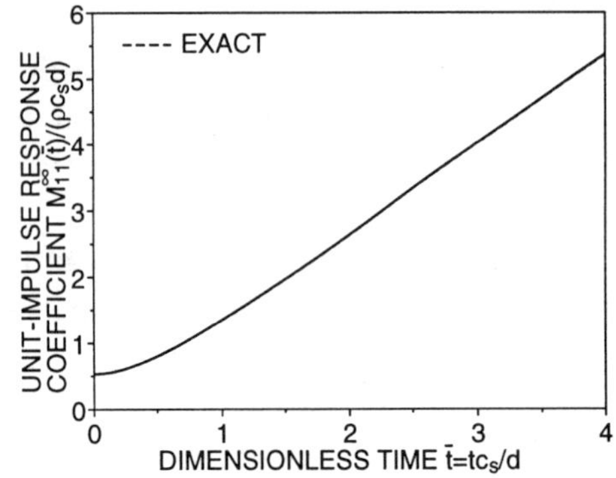

122 CONSISTENT INFINITESIMAL FINITE-ELEMENT CELL METHOD

(a) diagonal coefficient of mid-side node

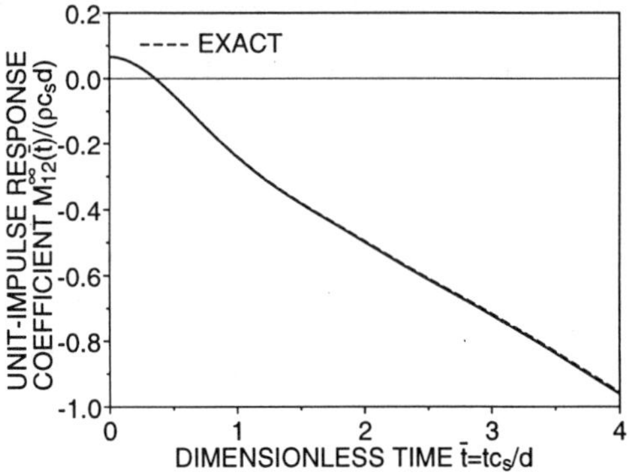

(b) off-diagonal coefficient

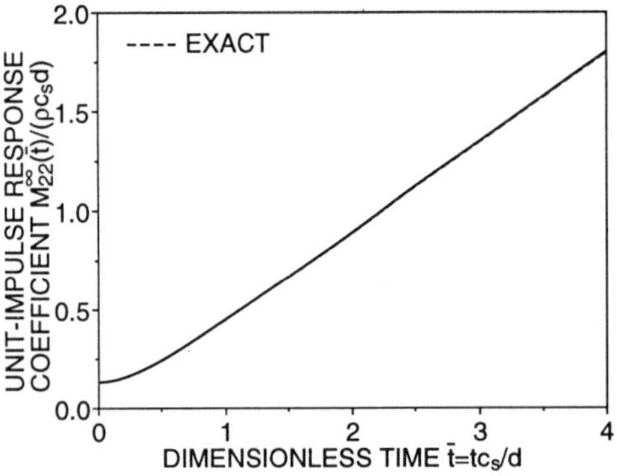

(c) diagonal coefficient of corner node

Figure 5-14 Acceleration unit-impulse response matrix of out-of-plane motion of semi-infinite layer of constant depth discretized with quadratic finite element

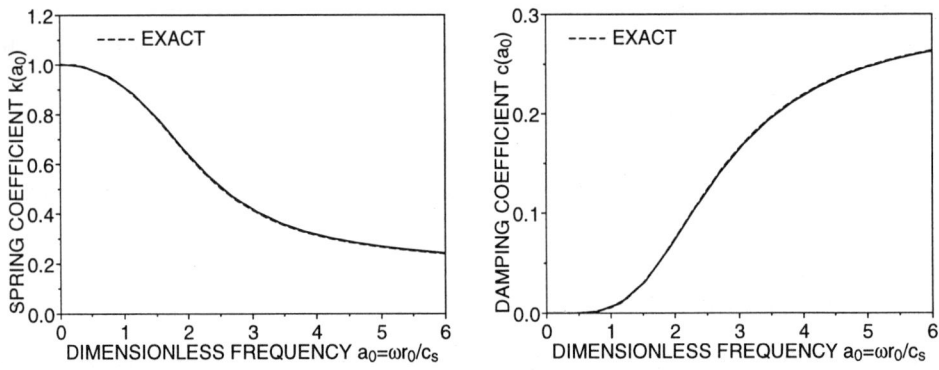

Figure 5-15 Dynamic-stiffness coefficient in frequency domain of out-of-plane motion of semi-infinite wedge

5.4 ACCURACY

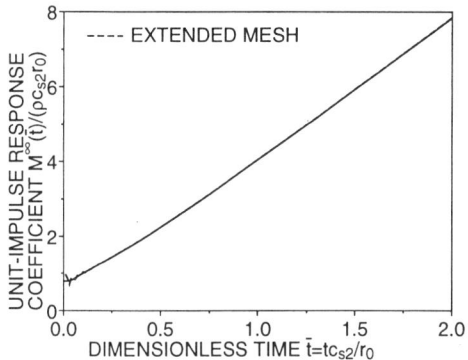

Figure 5-16 Acceleration unit-impulse response coefficient of in-plane motion of inhomogeneous semi-infinite wedge

Eq. A.5.19.

Owing to symmetry only one quarter of the structure-medium interface is discretized with 4 3-node line elements of equal length. The acceleration unit-impulse response matrix $[M^\infty(t)]$ is calculated with $\Delta t = 0.04 r_0/c_s$. The coefficient $M^\infty(t)$ follows from Eq. A.0.1 with the vector $\{\phi\}$ determined from the translational motion of the rigid structure-medium interface (Eq. A.5.5).

For a comparison with the analytical solution of the dynamic-stiffness coefficient, $S^\infty(\omega)$ is determined from $M^\infty(t)$ based on the method described in Eqs. 2.23 and 2.25. For $\nu = 1/3$, $S^\infty(\omega)$, non-dimensionalized with the shear modulus G, is decomposed in $k(a_0)$ and $c(a_0)$. Good agreement with the analytical solution of Eq. A.5.22 results (Figure 5-17).

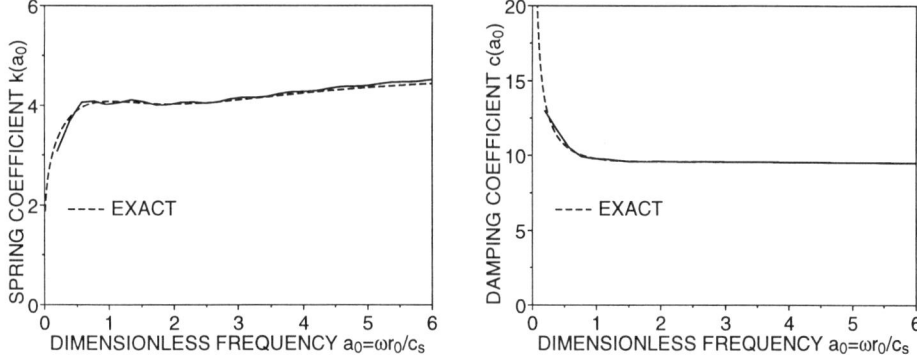

Figure 5-17 Dynamic-stiffness coefficient in frequency domain of circular cavity embedded in full-plane

5.4.8 Out-of-plane Motion of Strip Foundation with Rectangular Cross-section Embedded in Half-plane

The out-of-plane (anti-plane) motion of a strip foundation with a rectangular cross-section of width $2b$ embedded with depth e in an inhomogeneous half-plane is examined (Figure 5-18). The inhomogeneity compatible with similarity defined by the three shear moduli G_1, G_2, G_3 is specified in the figure. The mass density ρ is constant. The embedment ratio $e/b = 1$ and $G_2/G_1 = G_2/G_3 = 4$ are selected. On the structure-medium interface 24 3-node line elements of equal length are chosen. The time step equals $\Delta t = 0.03b/c_{s1}$ ($c_{s1} = \sqrt{G_1/\rho}$). The corresponding unit-impulse response matrix $[M^\infty(t)]$ of order 49×49 is determined. The rigid-body constraint is introduced to calculate the equivalent coefficient corresponding to the twisting motion around the vertical axis (Figure 5-18), which defines $\{\phi\}$ in Eq. A.0.1. The equivalent coefficient $M^\infty(\bar{t})$ normalized by $\rho c_{s1} b^3$ coincides from a practical point of view with the corresponding value determined with an extended mesh (Figure 5-19).

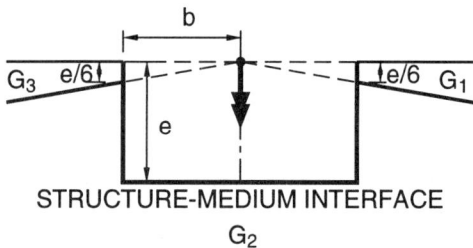

Figure 5-18 Out-of-plane motion of strip foundation with rectangular cross-section embedded in inhomogeneous half-plane

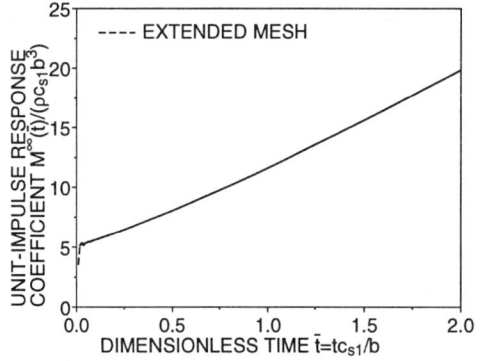

Figure 5-19 Acceleration unit-impulse response coefficient of out-of-plane motion of rigid strip foundation embedded in inhomogeneous half-plane

5.4 ACCURACY

5.4.9 In-plane Motion of Strip Foundation with Rectangular Cross-section Embedded in Half-plane

The in-plane motion of the strip foundation with a rectangular cross-section embedded in a half-plane illustrated in Appendix A.6 (Figure A-16) is calculated. A series of different analyses are performed. For the homogeneous half-plane, isotropic and transversely isotropic behaviour is examined, and a comparison with results in the literature follows. For the inhomogeneous half-plane isotropic behaviour is postulated and the results are compared with those of an analysis using an extended mesh.

For all analyses the embedment ratio $e/b = 1$ is chosen. 24 3-node line elements of equal length are introduced in the discretization of the structure-medium interface. The acceleration unit-impulse response matrix $[M^\infty(t)]$ of order 98×98 is determined. A rigid interface is assumed. $\{\phi_h\}$, $\{\phi_v\}$ and $\{\phi_r\}$ correspond to the motion patterns of the nodes on the structure-medium interface associated with the horizontal, vertical and rocking motions, respectively. The corresponding acceleration unit-impulse response coefficients $M_h^\infty(t)$, $M_v^\infty(t)$ and $M_r^\infty(t)$ then follow from Eq. A.0.1. As the results in the literature for the homogeneous half-plane are specified in the frequency domain, the acceleration unit-impulse response coefficients are transformed to the dynamic-stiffness coefficients based on Eqs. 2.23 and 2.25. After non-dimensionalization (which depends on the calculated case) a decomposition in the dimensionless spring coefficients and damping coefficients is performed.

First, the isotropic homogeneous half-plane with a common shear modulus G and Poisson's ratio $\nu = 0.25$ for the 3 zones of Figure A-16 is calculated. The time step is selected as $\Delta t = 0.05 b/c_s$. The resulting dynamic-stiffness coefficients in the frequency domain non-dimensionalized for the translational motions with G and for the rocking motion with Gb^2 agree well, as shown in Figure 5-20, with the results of dynamic condensation and of the boundary-element method in the frequency domain [56].

Second, the transversely isotropic homogeneous half-plane with common material constants for the 3 zones is addressed. The constants defining the material behaviour (Appendix A.8) are as follows: $E_{hh} = 3.864 G_{hv}$, $E_{hv} = 2.863 G_{hv}$, $\nu_{hh} = 0.301$, $\nu_{hv} = 0.185$. The elasticity matrix $[D]$ follows from Eq. A.8.3. The time step is chosen as $\Delta t = 0.03 b/c_s$ with $c_s = \sqrt{G_{hv}/\rho}$. A comparison of the dynamic-stiffness coefficients in the frequency domain (non-dimensionalized as for the isotropic case using G_{hv} instead of G) with the result of the boundary-element method in the frequency domain [56] is presented in Figure 5-21.

Third, the isotropic inhomogeneous half-plane with the shear moduli $G_2 = 4G_1 = 4G_3$ in the three zones (Figure A-16) and a common Poisson's ratio $\nu = 0.25$ is processed. The time step equals $\Delta t = 0.02 b/c_{s1}$ with $c_{s1} = \sqrt{G_1/\rho}$. The acceleration unit-impulse response coefficients non-dimensionalized with $\rho c_{s1} b$ for the two translational motions and with $\rho c_{s1} b^3$ for the rocking motion coincide from a practical point of view with the results based on an extended mesh analysis (Figure 5-22).

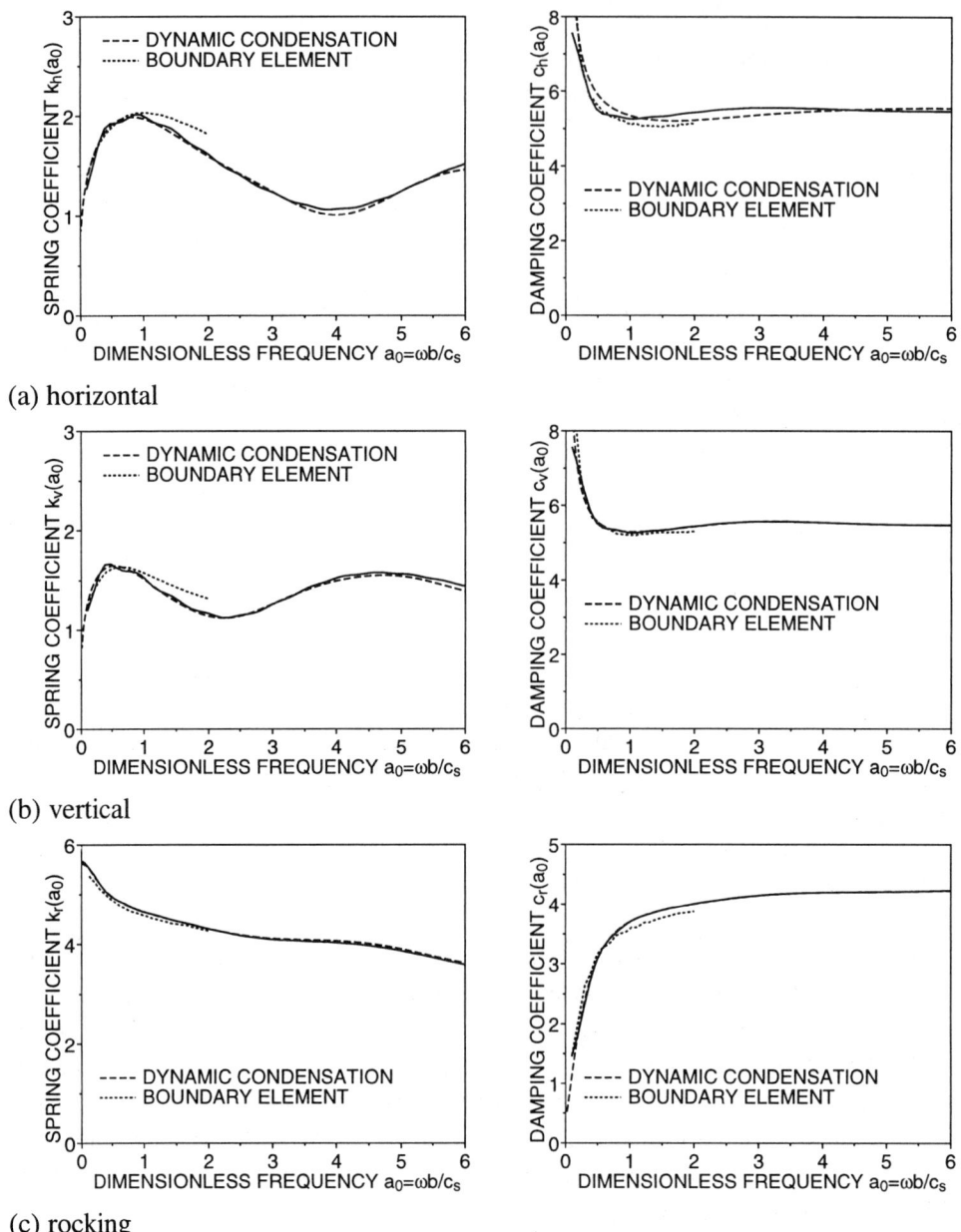

(a) horizontal

(b) vertical

(c) rocking

Figure 5-20 Dynamic-stiffness coefficients in frequency domain of rigid strip foundation embedded in isotropic homogeneous half-plane

5.4 ACCURACY

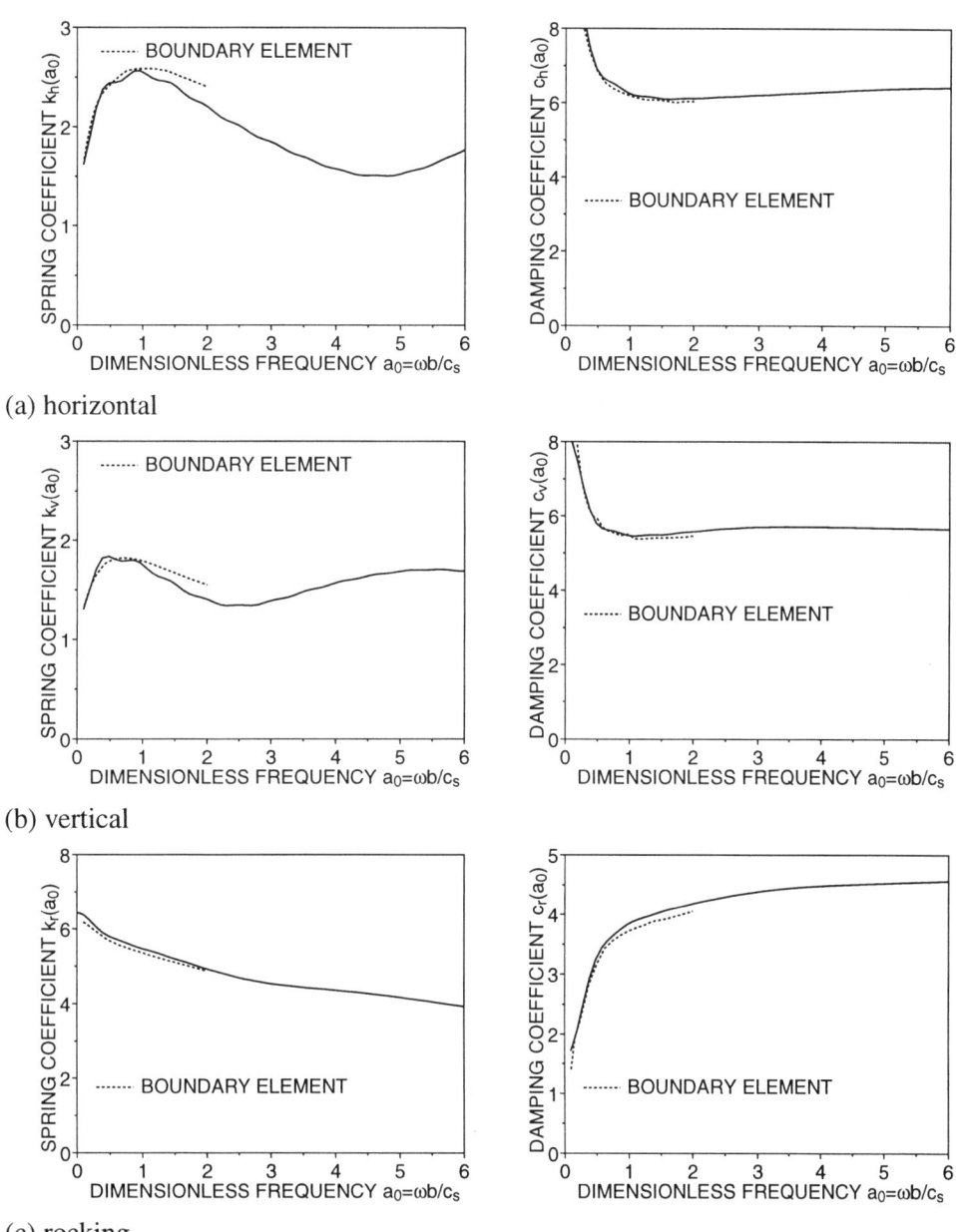

(a) horizontal

(b) vertical

(c) rocking

Figure 5-21 Dynamic-stiffness coefficients in frequency domain of rigid strip foundation embedded in transversely isotropic homogeneous half-plane

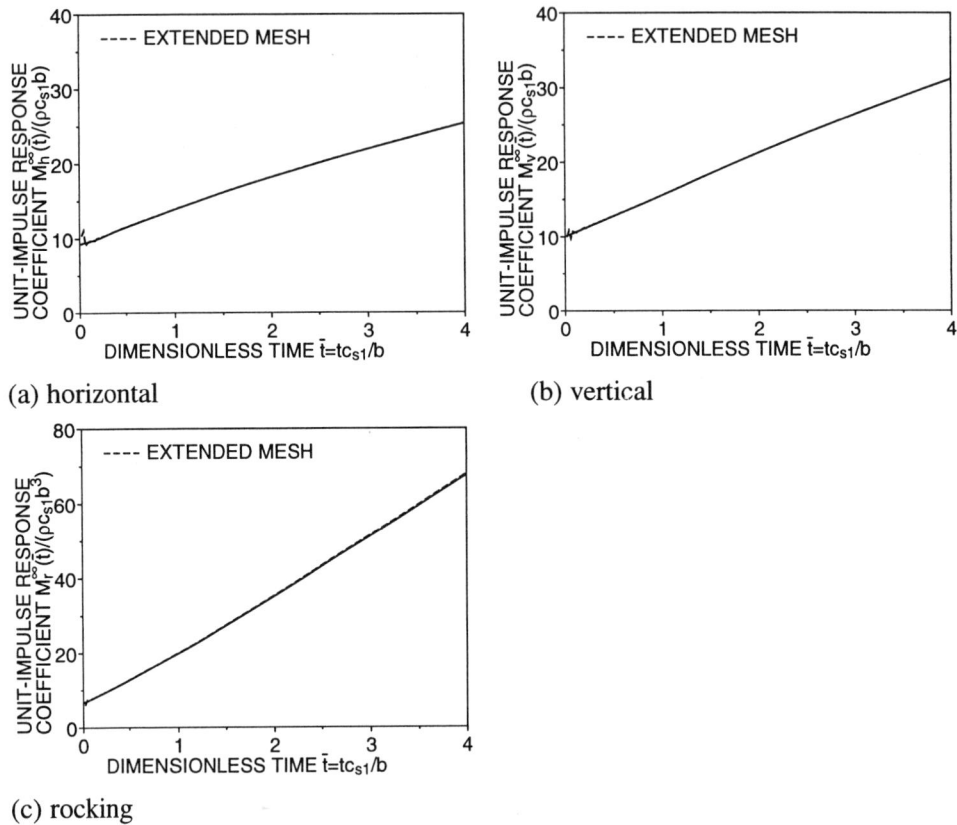

Figure 5-22 Acceleration unit-impulse response coefficients of rigid strip foundation embedded in isotropic inhomogeneous half-plane

5.4.10 Cylinder Embedded in Half-space for Vector Wave Equation

As an axisymmetric problem, a cylinder with radius r_0 embedded with depth e in a half-space with Poisson's ratio $\nu = 0.25$ for the vector wave equation is addressed (Figure 5-23). The embedment ratio $e/r_0 = 1$ is selected. The structure-medium interface is discretized with 12 3-node line elements of equal length.

The unit-impulse response matrices $[M^\infty(t)]$ for the symmetric and anti-symmetric parts of $n = 0$ and for the symmetric parts of $n = 1$ are calculated with $\Delta t = 0.02 r_0/c_s$. A rigid interface is assumed. $\{\phi_v\}$, $\{\phi_t\}$, $\{\phi_h\}$ and $\{\phi_r\}$ correspond to the amplitudes of the motion patterns of the nodes on the structure-medium interface associated with the vertical, torsional, horizontal and rocking motions, respectively. The corresponding acceleration unit-impulse response coefficients $M_v^\infty(t)$, $M_t^\infty(t)$, $M_h^\infty(t)$ and $M_r^\infty(t)$ then follow from Eq. A.0.1. For comparison, the acceleration unit-impulse response coefficients are transformed to the dynamic-stiffness coefficients in

5.4 ACCURACY

the frequency domain based on Eqs. 2.23 and 2.25. After normalization with the static-stiffness coefficients, a decomposition in the dimensionless spring coefficients and damping coefficients is performed (Eq. A.0.3). Excellent agreement with the results of the boundary-element method [3] is observed in Figure 5-24.

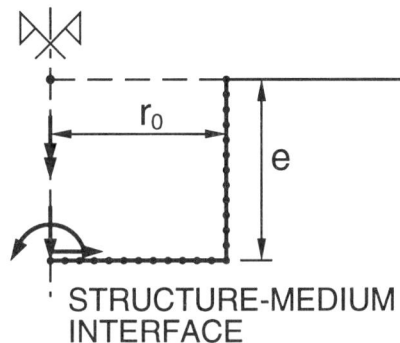

Figure 5-23 Cylinder embedded in half-space

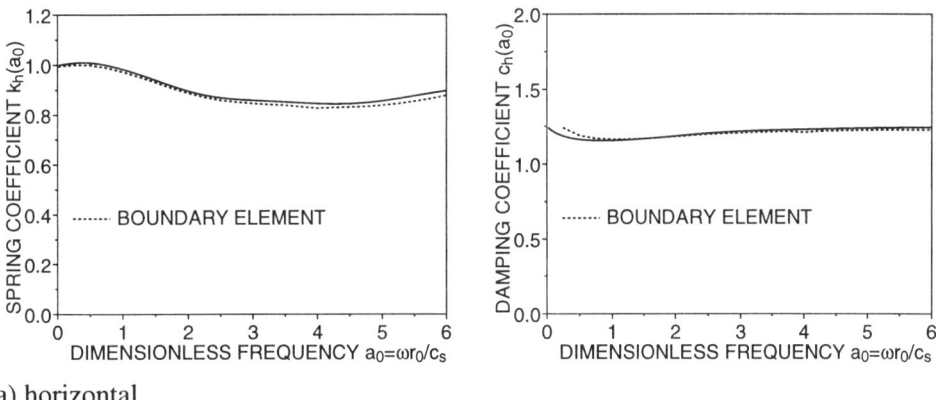

(a) horizontal

Figure 5-24 Dynamic-stiffness coefficients in frequency domain of rigid cylinder embedded in half-space

5.4.11 Prism Embedded in Half-space for Scalar Wave Equation

As a truly three-dimensional problem, a square prism of length $2b$ embedded with depth e in a half-space for the scalar wave equation is addressed (Figure 5-25). The embedment ratio $e/b = 2/3$ is selected. An inhomogeneous half-space compatible with similarity with the material constant G_1 of the unbounded medium adjacent to the

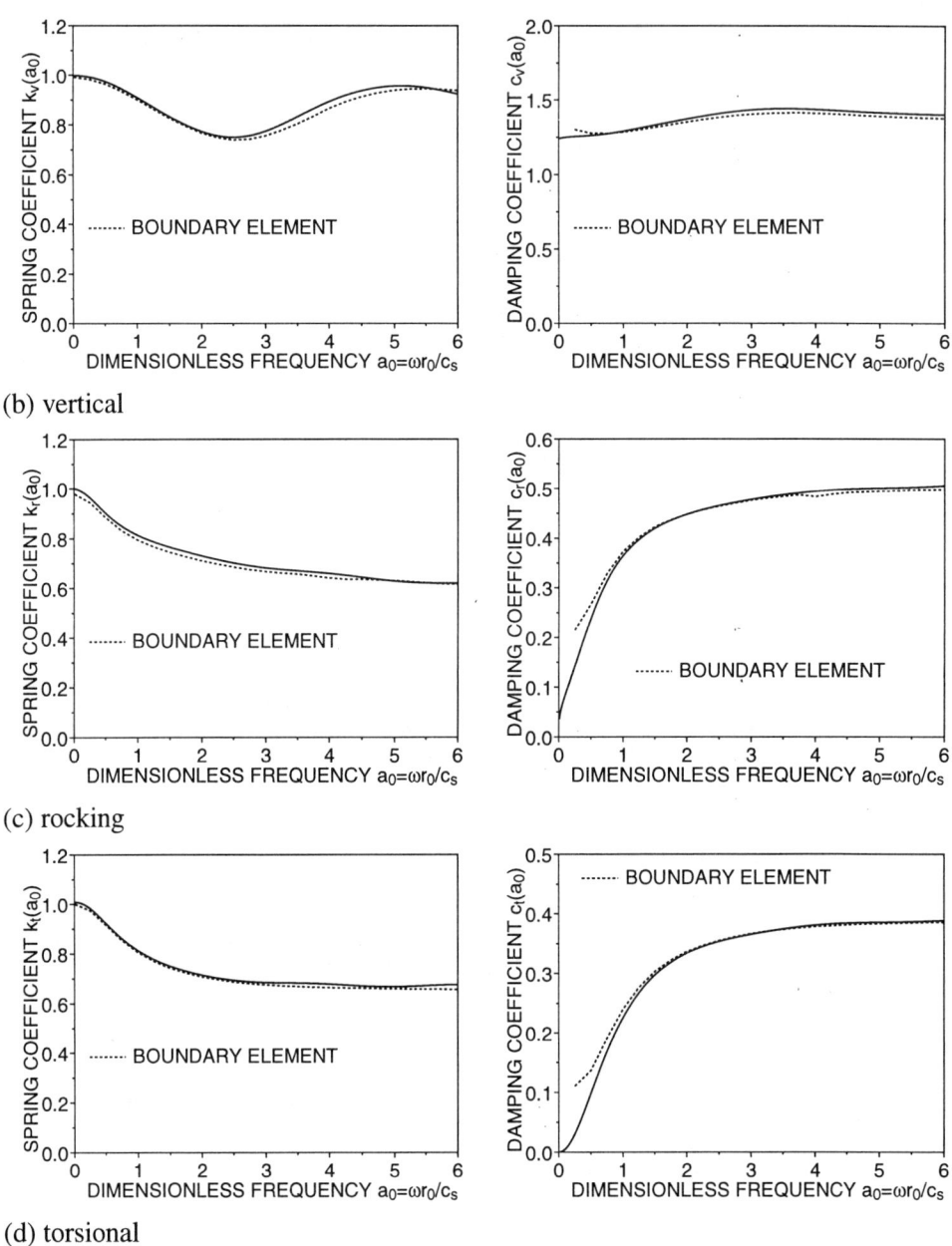

Figure 5-24 Dynamic-stiffness coefficients in frequency domain of rigid cylinder embedded in half-space

5.4 ACCURACY

walls and $G_2 = 4G_1$ adjacent to the base is calculated. The other material constant ρ does not vary. The structure-medium interface is discretized by finite elements. Owing to symmetry, only a quarter of the prism is analysed. The finite-element discretization of the three faces is shown in Figure 5-26. The corresponding unit-impulse response matrix $[M^\infty(t)]$ of order 100×100 is then established with $\Delta t = 0.04b/c_1$ ($c_1 = \sqrt{G_1/\rho}$). To ease comparison of the results, a uniform variation of the function over the structure-medium interface is enforced which defines $\{\phi\}$ in Eq. A.0.1. The equivalent coefficient $M^\infty(t)$ non-dimensionalized by $\rho c_1 b^2$ agrees very well with the corresponding value determined with an extended mesh (Figure 5-27).

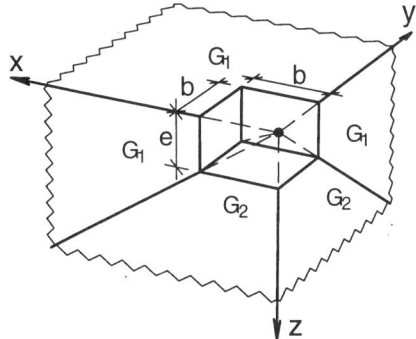

Figure 5-25 One quarter of square prism embedded in inhomogeneous half-space for scalar wave equation

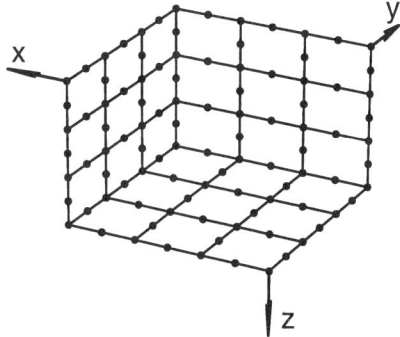

Figure 5-26 Finite-element mesh of one quarter of structure-medium interface of square prism

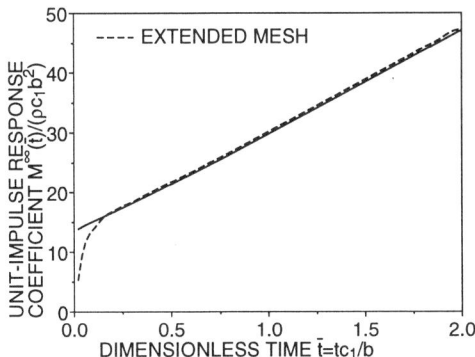

Figure 5-27 Acceleration unit-impulse response coefficient of prism embedded in inhomogeneous half-space for scalar wave equation

5.4.12 Prism Embedded in Half-space for Vector Wave Equation

The three-dimensional motion of a prism embedded in a half-space for the vector wave equation addressed in Appendix A.7 (Figure A-17) is calculated. Various analyses are performed. For all calculations the embedment ratio $e/b = 2/3$ is selected. The finite-element discretization of one quarter of the structure-medium interface is shown in Figure 5-26. The acceleration unit-impulse response matrix is calculated. A rigid interface is introduced. $\{\phi_h\}$, $\{\phi_v\}$, $\{\phi_r\}$ and $\{\phi_t\}$ correspond to the motion patterns of the nodes on the structure-medium interface associated with the horizontal, vertical, rocking and torsional motions, respectively. The acceleration unit-impulse response coefficients $M_h^\infty(t)$, $M_v^\infty(t)$, $M_r^\infty(t)$ and $M_t^\infty(t)$ are then calculated from Eq. A.0.1. As the results in the literature for the isotropic homogeneous half-space are specified in the frequency domain, the acceleration unit-impulse response coefficients are transformed to the dynamic-stiffness coefficients based on Eqs. 2.23 and 2.25. After normalization with the static-stiffness coefficients, a decomposition in the dimensionless spring coefficients and damping coefficients is performed (Eq. A.0.3).

First, the isotropic homogeneous half-space with a common shear modulus G and Poisson's ratio $\nu = 1/3$ for the 2 zones of Figure A-17 is analysed. The time step $\Delta t = 0.04b/c_s$ is used. The resulting dynamic-stiffness coefficients in the frequency domain plotted in Figure 5-28 are compared with the results of the boundary-element method for the horizontal, vertical and rocking degrees of freedom in the frequency domain [15] and with those of the hybrid method for the torsional degree of freedom [35].

Second, the transversely isotropic homogeneous half-space with common material constants for the 2 zones is addressed. The constants defining the material behaviour (Appendix A.8) are as follows: $E_{hh} = 3.864 G_{hv}$, $E_{hv} = 2.863 G_{hv}$, $\nu_{hh} = 0.301$, $\nu_{hv} = 0.185$. The elasticity matrix $[D]$ follows from Eq. A.8.3. The time step is chosen as $\Delta t = 0.04b/c_s$ with $c_s = \sqrt{G_{hv}/\rho}$. The acceleration unit-impulse response coefficients for the four degrees of freedom hardly deviate from the results determined with an extended mesh (Figure 5-29). To allow comparison with the results of the isotropic case shown in Figure 5-28, the dynamic-stiffness coefficients in the frequency domain are also determined for the transversely isotropic case shown in Figure 5-30.

Third, the isotropic inhomogeneous half-space with the shear moduli $G_2 = 4G_1$ in the 2 zones (Figure A-17) and a common Poisson's ratio $\nu = 1/3$ is calculated. The time step is selected as $\Delta t = 0.04b/c_{s1}$ with $c_{s1} = \sqrt{G_1/\rho}$. Good agreement of the acceleration unit-impulse response coefficients with the results of the extended mesh is obtained (Figure 5-31).

5.4 ACCURACY

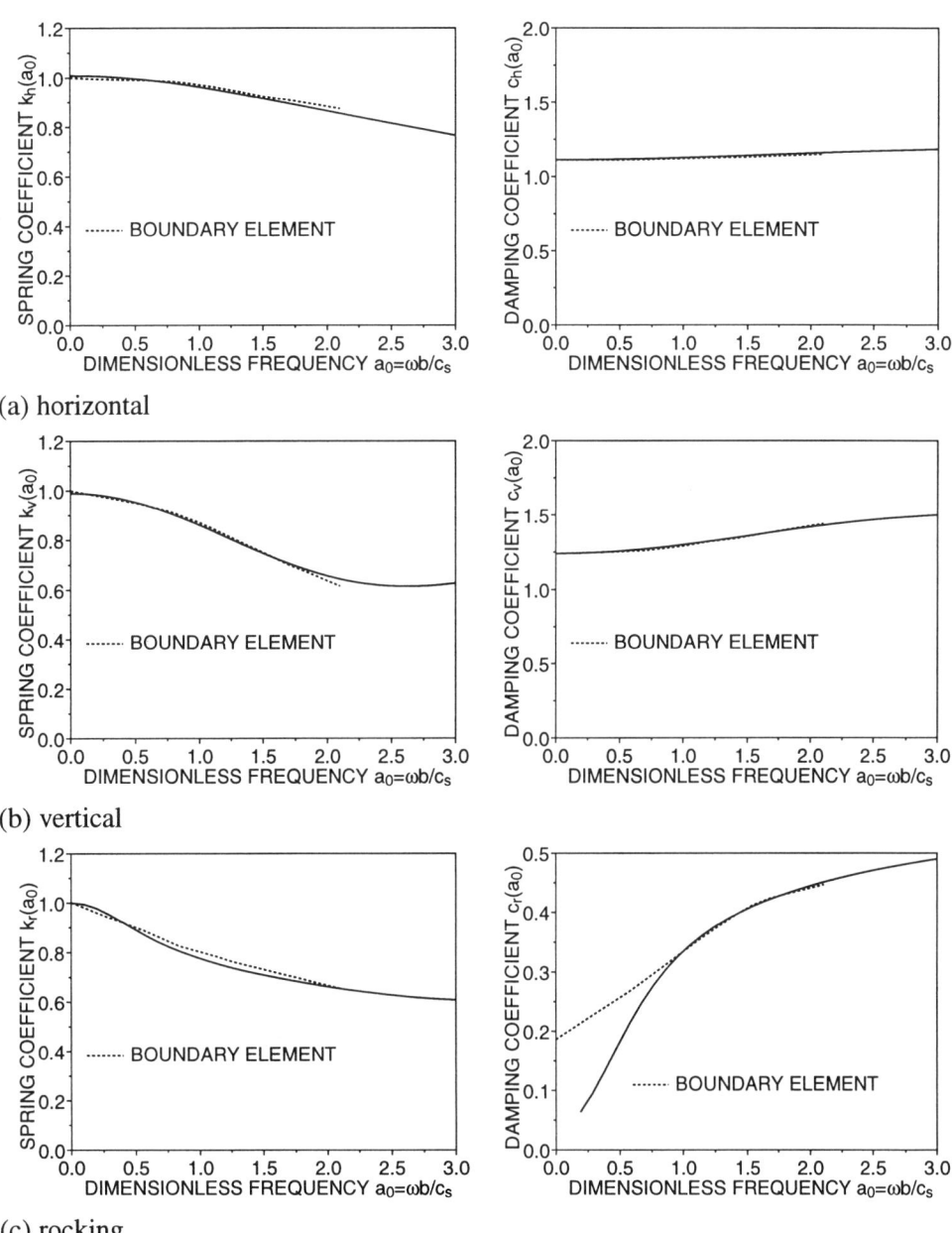

(a) horizontal

(b) vertical

(c) rocking

Figure 5-28 Dynamic-stiffness coefficients in frequency domain of rigid prism embedded in isotropic homogeneous half-space

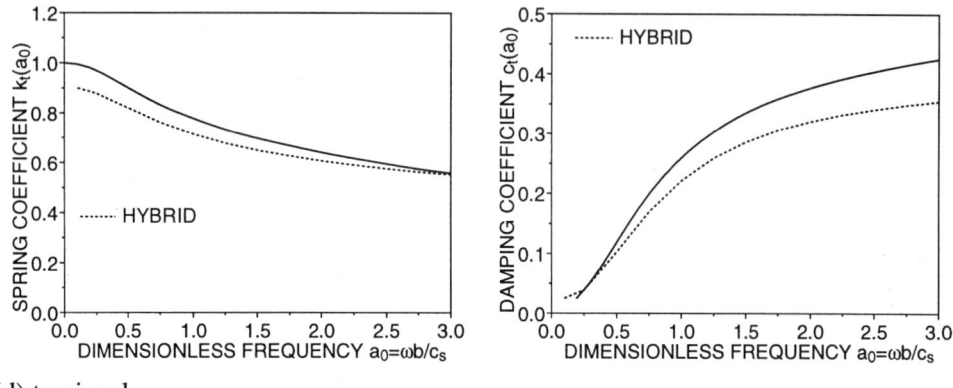

(d) torsional

Figure 5-28 Dynamic-stiffness coefficients in frequency domain of rigid prism embedded in isotropic homogeneous half-space

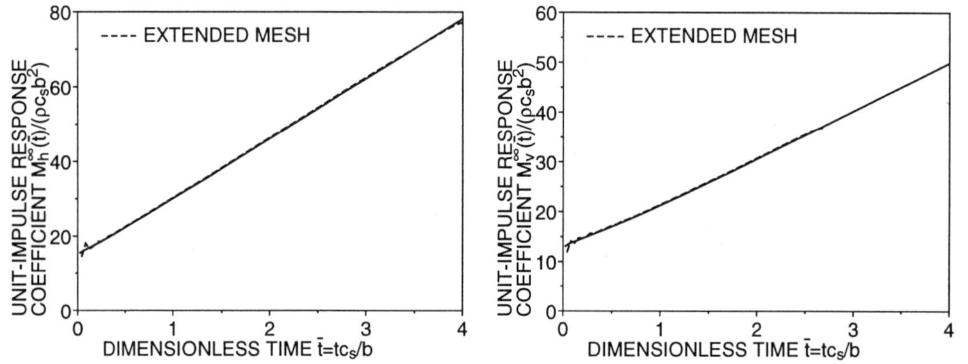

(a) horizontal

(b) vertical

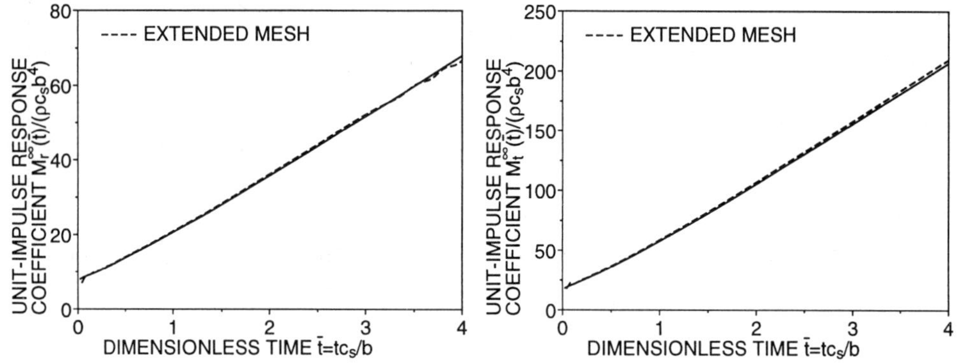

(c) rocking

(d) torsional

Figure 5-29 Acceleration unit-impulse response coefficients of rigid prism embedded in transversely isotropic half-space for vector wave equation

5.4 ACCURACY

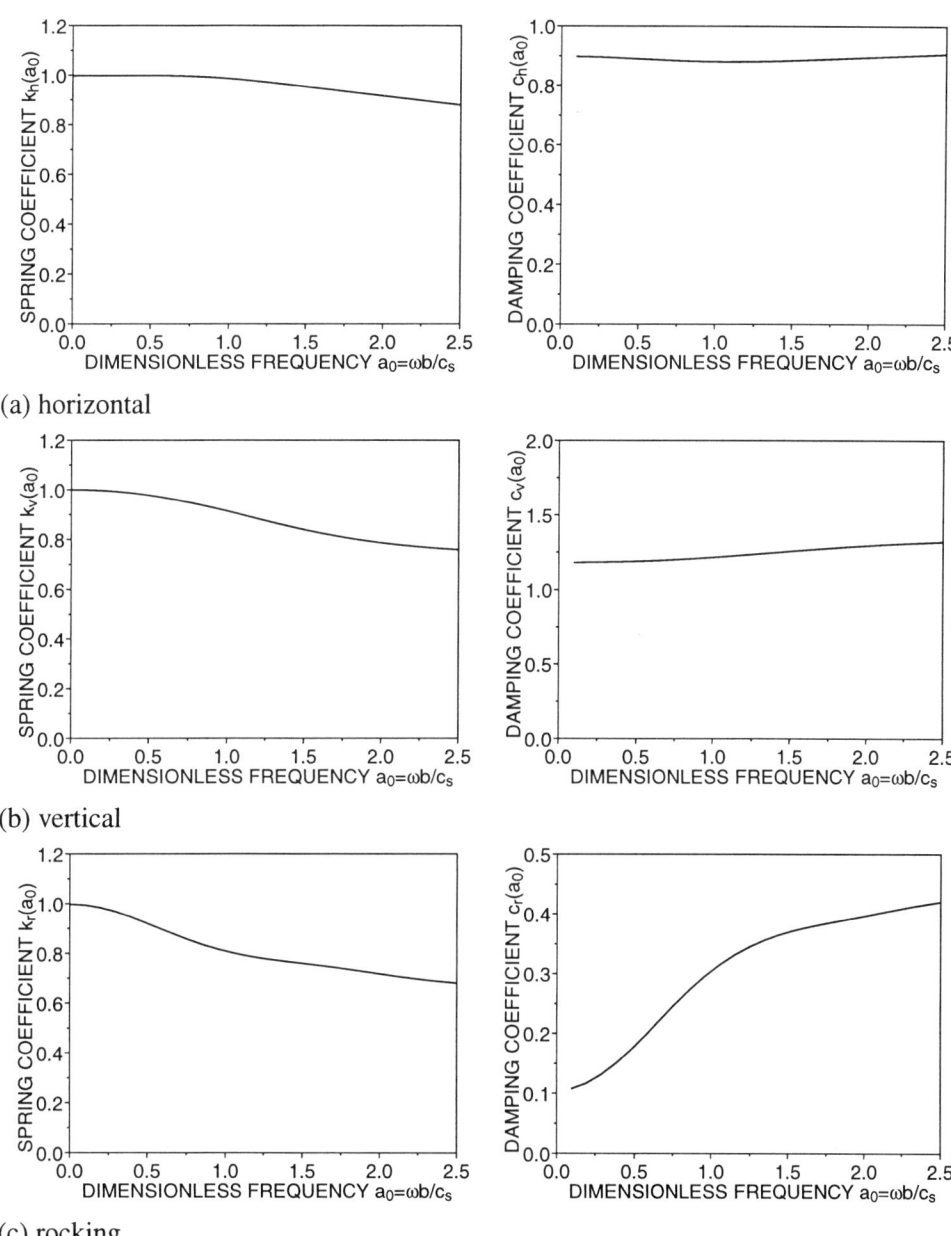

(a) horizontal

(b) vertical

(c) rocking

Figure 5-30 Dynamic-stiffness coefficients in frequency domain of rigid prism embedded in transversely isotropic homogeneous half-space for vector wave equation

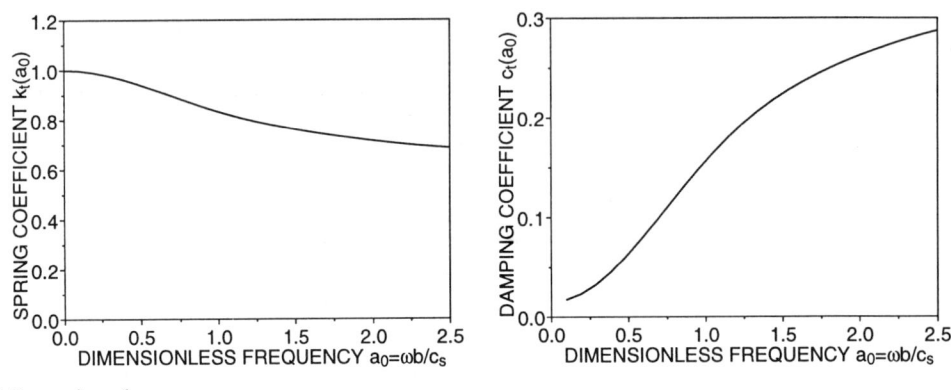

(d) torsional

Figure 5-30 Dynamic-stiffness coefficients in frequency domain of rigid prism embedded in transversely isotropic homogeneous half-space for vector wave equation

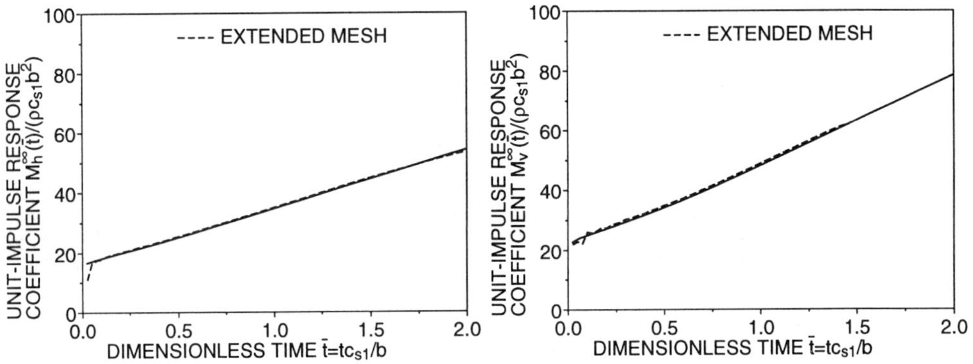

(a) horizontal

(b) vertical

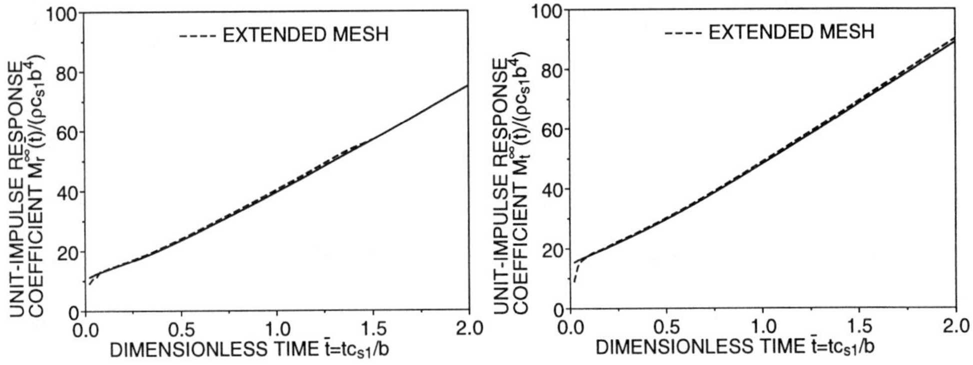

(c) rocking

(d) torsional

Figure 5-31 Acceleration unit-impulse response coefficients of rigid prism embedded in isotropic inhomogeneous half-space for vector wave equation

6

CONSISTENT INFINITESIMAL FINITE-ELEMENT CELL METHOD FOR INCOMPRESSIBLE ELASTICITY

Many important physical problems involve incompressible materials, i.e. their volumes are preserved locally during deformations. Rubber is often modelled as an incompressible elastic material. In civil engineering the saturated unbounded soil when undrained behaves as a nearly incompressible material in a one-phase formulation of a soil-structure-interaction analysis.

It is well known that the case of incompressible elasticity requires a special approach in the displacement-based finite-element method. The elasticity matrix is decomposed in the shear and volumetric parts. Selective reduced integration is used with a Poisson's ratio very close to 0.5 which introduces a certain arbitrariness. The same procedure can also be applied to the consistent infinitesimal finite-element cell method. As a further step, the limit of Poisson's ratio equal to 0.5 can be enforced *analytically* which eliminates the necessity to choose Poisson's ratio. It is the goal of this chapter to derive this streamlined formulation of the consistent infinitesimal finite-element cell method for the isotropic incompressible unbounded medium.

To avoid unnecessary duplication with the consistent infinitesimal finite-element cell method for the vector wave equation in compressible elasticity discussed in Chapter 5, only the modifications due to incompressibility are addressed. This chapter is based on Reference [53]. The fundamental equations for incompressible elasticity in continuous frequency and time are derived in Section 6.1 starting from those of compressible elasticity. The time discretization is not affected by incompressibility. The accuracy of the consistent infinitesimal finite-element cell method is evaluated in Section 6.2.

6.1 FUNDAMENTAL EQUATIONS

6.1.1 Coefficient Matrices of Finite-element Cell

The derivation of the coefficient matrices for incompressible elasticity follows that presented in Section 5.2.1 with the following modifications. These become necessary as the bulk modulus $= 2(1+\nu)/(3(1-2\nu))G$ of incompressible elasticity tends to infinity for $\nu \to 0.5$. Incompressibility corresponds to the limit of Poisson's ratio $\nu = 0.5$.

For the three-dimensional isotropic incompressible medium, the elasticity matrix is decomposed into the shear and volumetric parts as

$$[D] = G \begin{bmatrix} 2 & 0 & 0 & 0 & 0 & 0 \\ 0 & 2 & 0 & 0 & 0 & 0 \\ 0 & 0 & 2 & 0 & 0 & 0 \\ 0 & 0 & 0 & 1 & 0 & 0 \\ 0 & 0 & 0 & 0 & 1 & 0 \\ 0 & 0 & 0 & 0 & 0 & 1 \end{bmatrix} + \frac{2\nu}{1-2\nu}G \begin{bmatrix} 1 & 1 & 1 & 0 & 0 & 0 \\ 1 & 1 & 1 & 0 & 0 & 0 \\ 1 & 1 & 1 & 0 & 0 & 0 \\ 0 & 0 & 0 & 0 & 0 & 0 \\ 0 & 0 & 0 & 0 & 0 & 0 \\ 0 & 0 & 0 & 0 & 0 & 0 \end{bmatrix} = [\overline{D}] + \alpha[\overline{\overline{D}}]$$

(6.1)

with $\alpha = 2\nu/(1-2\nu)$ and $[\overline{D}]$ and $[\overline{\overline{D}}]$ defined accordingly. For $\nu \to 0.5$, $\alpha \to \infty$ results.

The (doubly-curved) structure-medium interface is discretized with any type of surface finite element. The coefficient matrices associated with the static-stiffness matrix $[E^0]$, $[E^1]$, $[E^2]$ (Eq. 5.22) of one doubly-curved surface finite element on the structure-medium interface are addressed. Substituting Eq. 6.1 in Eq. 5.22 yields the matrices

$$[E^0] = [\overline{E}^0] + \alpha[\overline{\overline{E}}^0] \quad (6.2a)$$

$$[E^1] = [\overline{E}^1] + \alpha[\overline{\overline{E}}^1] \quad (6.2b)$$

$$[E^2] = [\overline{E}^2] + \alpha[\overline{\overline{E}}^2] \quad (6.2c)$$

with

$$[\overline{E}^0] = \int_{-1}^{+1}\int_{-1}^{+1} [B^1]^T [\overline{D}][B^1]|J|\,d\eta\,d\zeta \quad (6.3a)$$

$$[\overline{\overline{E}}^0] = \int_{-1}^{+1}\int_{-1}^{+1} [B^1]^T [\overline{\overline{D}}][B^1]|J|\,d\eta\,d\zeta \quad (6.3b)$$

$$[\overline{E}^1] = \int_{-1}^{+1}\int_{-1}^{+1} [B^2]^T [\overline{D}][B^1]|J|\,d\eta\,d\zeta \quad (6.3c)$$

6.1 FUNDAMENTAL EQUATIONS

$$[\overline{\overline{E}}^1] = \int_{-1}^{+1}\int_{-1}^{+1}[B^2]^T[\overline{\overline{D}}][B^1]|J|d\eta d\zeta \qquad (6.3d)$$

$$[\overline{\overline{E}}^2] = \int_{-1}^{+1}\int_{-1}^{+1}[B^2]^T[\overline{\overline{D}}][B^2]|J|d\eta d\zeta \qquad (6.3e)$$

$$[\overline{\overline{E}}^2] = \int_{-1}^{+1}\int_{-1}^{+1}[B^2]^T[\overline{\overline{D}}][B^2]|J|d\eta d\zeta \qquad (6.3f)$$

$[B^1]$ and $[B^2]$ follow from Eq. 5.15, and $[J]$ is defined in Eq. 5.9. The coefficient matrices of the volumetric part $[\overline{\overline{E}}^1]$, $[\overline{\overline{E}}^1]$ and $[\overline{\overline{E}}^2]$ are calculated using reduced integration. The coefficient matrix associated with the mass is specified in Eq. 5.36 as

$$[M^0] = \int_{-1}^{+1}\int_{-1}^{+1}\rho[N]^T[N]|J|d\eta d\zeta \qquad (6.4)$$

with the mass density ρ and the shape functions $[N]$ (Eq. 5.26). Note that $[E^0]$ and $[M^0]$ are positive definite while $[\overline{\overline{E}}^0]$, $[E^2]$ and $[\overline{\overline{E}}^2]$ are semi-positive definite which is verified by inspection.

As for the static-stiffness and mass matrices, the coefficient matrices of the finite elements $[\overline{\overline{E}}^0], \ldots, [\overline{\overline{E}}^2]$ and $[M^0]$ are assembled to form those of the structure-medium interface. Again, to simplify the nomenclature, the same symbols are used for the assembled coefficient matrices in the following.

6.1.2 Consistent Infinitesimal Finite-element Cell Equation

For simplicity, the derivation of the consistent infinitesimal finite-element cell equation is performed in the frequency domain. This equation with the displacement dynamic-stiffness matrix of the unbounded medium $[S^\infty(\omega)]$ as the unknown is given in Eq. 5.113 as

$$([S^\infty(\omega)] + [E^1])[E^0]^{-1}([S^\infty(\omega)] + [E^1]^T) - (s-2)[S^\infty(\omega)]$$
$$- \omega[S^\infty(\omega)]_{,\omega} - [E^2] + \omega^2[M^0] = 0 \quad (6.5)$$

with the spatial dimension s ($=2$ or $=3$).

To be able to perform the limit $\alpha \to \infty$ analytically, $[E^0]^{-1}$ is determined by solving the following eigenvalue problem

$$[\overline{\overline{E}}^0][\Phi] = [E^0][\Phi]\lfloor\Lambda\rfloor \qquad (6.6)$$

with the diagonal matrix of eigenvalues

$$\lfloor\Lambda\rfloor = \begin{bmatrix} 0 & \\ & \lfloor\lambda\rfloor^{-1} \end{bmatrix} \qquad (6.7)$$

and the eigenvector matrix $[\Phi]$ normalized as (unit matrix $[I]$)

$$[\Phi]^T [E^0][\Phi] = [I] \tag{6.8}$$

yielding

$$[\Phi]^T [\overline{\overline{E}}^0][\Phi] = \lceil \Lambda \rfloor \tag{6.9}$$

Defining

$$\lceil e^0 \rfloor = [\Phi]^T [E^0][\Phi] \tag{6.10}$$

and substituting Eq. 6.2a results in the diagonal matrix

$$\lceil e^0 \rfloor = [I] + \alpha \lceil \Lambda \rfloor \tag{6.11}$$

From Eq. 6.10

$$[E^0]^{-1} = [\Phi] \lceil e^0 \rfloor^{-1} [\Phi]^T \tag{6.12}$$

follows. Substituting Eq. 6.7 in Eq. 6.11 leads to

$$\lceil e^0 \rfloor^{-1} = \begin{bmatrix} [I] & 0 \\ 0 & \end{bmatrix} + \begin{bmatrix} 0 & \\ & \left([I] + \alpha \lceil \lambda \rfloor^{-1}\right)^{-1} \end{bmatrix} \tag{6.13}$$

The inverse of the ith element of the diagonal matrix $[I] + \alpha \lceil \lambda \rfloor^{-1}$ equals

$$\frac{1}{1 + \dfrac{\alpha}{\lambda_i}} = \frac{\lambda_i}{\alpha} \frac{1}{1 + \dfrac{\lambda_i}{\alpha}} = \frac{\lambda_i}{\alpha} - \left(\frac{\lambda_i}{\alpha}\right)^2 + O(\alpha^{-3}) \tag{6.14}$$

where two terms of the Taylor expansion are kept for $1/\alpha \to 0$. Thus

$$\lceil e^0 \rfloor^{-1} = \begin{bmatrix} [I] & 0 \\ 0 & \end{bmatrix} + \frac{1}{\alpha}\begin{bmatrix} 0 & \\ & \lceil \lambda \rfloor \end{bmatrix} - \frac{1}{\alpha^2}\begin{bmatrix} 0 & \\ & \lceil \lambda \rfloor^2 \end{bmatrix} + O(\alpha^{-3}) \tag{6.15}$$

results. Substituting Eq. 6.12 in Eq. 6.5 which is premultiplied by $[\Phi]^T$ and postmultiplied by $[\Phi]$ yields

$$([s^\infty(\omega)] + [\varepsilon^1]) \lceil e^0 \rfloor^{-1} ([s^\infty(\omega)] + [\varepsilon^1]^T) - (s-2)[s^\infty(\omega)]$$
$$- \omega[s^\infty(\omega)]_{,\omega} - [\varepsilon^2] + \omega^2[m^0] = 0 \tag{6.16}$$

6.1 FUNDAMENTAL EQUATIONS

where

$$[s^\infty(\omega)] = [\Phi]^T [S^\infty(\omega)][\Phi] \qquad (6.17)$$

and

$$[\varepsilon^1] = [\bar{\varepsilon}^1] + \alpha[\bar{\bar{\varepsilon}}^1] = [\Phi]^T [\bar{E}^1][\Phi] + \alpha[\Phi]^T [\bar{\bar{E}}^1][\Phi] \qquad (6.18a)$$

$$[\varepsilon^2] = [\bar{\varepsilon}^2] + \alpha[\bar{\bar{\varepsilon}}^2] = [\Phi]^T [\bar{E}^2][\Phi] + \alpha[\Phi]^T [\bar{\bar{E}}^2][\Phi] \qquad (6.18b)$$

$$[m^0] = [\Phi]^T [M^0][\Phi] \qquad (6.19)$$

Typical terms in Eq. 6.16 are evaluated after substituting Eqs. 6.15 and 6.18 as

$$[\varepsilon^1][e^0]^{-1} = \alpha[\bar{\bar{\varepsilon}}^1]\begin{bmatrix}[I] & \\ & 0\end{bmatrix} + [\bar{\varepsilon}^1]\begin{bmatrix}[I] & \\ & 0\end{bmatrix} + [\bar{\varepsilon}^1]\begin{bmatrix}0 & \\ & \lceil\lambda\rfloor\end{bmatrix}$$

$$+ \frac{1}{\alpha}\left([\bar{\varepsilon}^1]\begin{bmatrix}0 & \\ & \lceil\lambda\rfloor\end{bmatrix} - [\bar{\bar{\varepsilon}}^1]\begin{bmatrix}0 & \\ & \lceil\lambda\rfloor^2\end{bmatrix}\right) + O(\alpha^{-2}) \quad (6.20a)$$

$$[\varepsilon^1][e^0]^{-1}[\varepsilon^1]^T - [\varepsilon^2] \approx \alpha^2[\bar{\bar{\varepsilon}}^1]\begin{bmatrix}[I] & \\ & 0\end{bmatrix}[\bar{\bar{\varepsilon}}^1]^T$$

$$+ \alpha\left([\bar{\bar{\varepsilon}}^1]\begin{bmatrix}[I] & \\ & 0\end{bmatrix}[\bar{\varepsilon}^1]^T + [\bar{\varepsilon}^1]\begin{bmatrix}[I] & \\ & 0\end{bmatrix}[\bar{\bar{\varepsilon}}^1]^T + [\bar{\bar{\varepsilon}}^1]\begin{bmatrix}0 & \\ & \lceil\lambda\rfloor\end{bmatrix}[\bar{\bar{\varepsilon}}^1]^T - [\bar{\bar{\varepsilon}}^2]\right)$$

$$+ [\bar{\varepsilon}^1]\begin{bmatrix}[I] & \\ & 0\end{bmatrix}[\bar{\varepsilon}^1]^T + [\bar{\varepsilon}^1]\begin{bmatrix}0 & \\ & \lceil\lambda\rfloor\end{bmatrix}[\bar{\bar{\varepsilon}}^1]^T + [\bar{\bar{\varepsilon}}^1]\begin{bmatrix}0 & \\ & \lceil\lambda\rfloor\end{bmatrix}[\bar{\varepsilon}^1]^T$$

$$- [\bar{\bar{\varepsilon}}^1]\begin{bmatrix}0 & \\ & \lceil\lambda\rfloor^2\end{bmatrix}[\bar{\bar{\varepsilon}}^1]^T - [\bar{\varepsilon}^2] + O(\alpha^{-1}) \quad (6.20b)$$

For the limit $\alpha \to \infty$, $[s^\infty(\omega)]$ in Eq. 6.16 must remain finite. Thus, the coefficient matrices of α^2 and of α in Eq. 6.20 must vanish, which results in

$$[\bar{\bar{\varepsilon}}^1]\begin{bmatrix}[I] & \\ & 0\end{bmatrix} = 0 \qquad (6.21a)$$

$$[\bar{\bar{\varepsilon}}^1]\begin{bmatrix}0 & \\ & \lceil\lambda\rfloor\end{bmatrix}[\bar{\bar{\varepsilon}}^1]^T - [\bar{\bar{\varepsilon}}^2] = 0 \qquad (6.21b)$$

Eq. 6.21 has also been verified numerically for a large number of cases. Enforcing Eq. 6.21 and neglecting terms in $1/\alpha$, $1/\alpha^2$, ... , Eq. 6.16 is formulated for incompressible elasticity with the limit $\nu = 0.5$ performed analytically as

$$[s^\infty(\omega)]\begin{bmatrix}[I] & \\ & 0\end{bmatrix}[s^\infty(\omega)] + [e^1][s^\infty(\omega)] + [s^\infty(\omega)][e^1]^T + 3[s^\infty(\omega)]$$

$$- \omega[s^\infty(\omega)]_{,\omega} - [e^2] + \omega^2[m^0] = 0 \quad (6.22)$$

where

$$[e^1] = [\bar{\varepsilon}^1]\begin{bmatrix}[I] & \\ & 0\end{bmatrix} + [\bar{\varepsilon}^1]\begin{bmatrix}0 & \\ & \lceil\lambda\rfloor\end{bmatrix} - \frac{s+1}{2}[I] \quad (6.23a)$$

$$[e^2] = -[\bar{\varepsilon}^1]\begin{bmatrix}[I] & \\ & 0\end{bmatrix}[\bar{\varepsilon}^1]^T - [\bar{\varepsilon}^1]\begin{bmatrix}0 & \\ & \lceil\lambda\rfloor\end{bmatrix}[\bar{\varepsilon}^1]^T \quad (6.23b)$$

$$- [\bar{\varepsilon}^1]\begin{bmatrix}0 & \\ & \lceil\lambda\rfloor\end{bmatrix}[\bar{\varepsilon}^1]^T + [\bar{\varepsilon}^1]\begin{bmatrix}0 & \\ & \lceil\lambda\rfloor^2\end{bmatrix}[\bar{\varepsilon}^1]^T + [\bar{\varepsilon}^2]$$

As a formulation of the unbounded medium's interaction forces based on convolution integrals of the accelerations is to be determined, the acceleration unit-impulse response matrix is calculated. Its Fourier transform is obtained from the dynamic-stiffness matrix as (Eq. 2.14)

$$[M^\infty(\omega)] = \frac{[S^\infty(\omega)]}{(i\omega)^2} \quad (6.24)$$

which using Eq. 6.17 yields

$$[m^\infty(\omega)] = \frac{[s^\infty(\omega)]}{(i\omega)^2} \quad (6.25)$$

Substituting Eq. 6.25 in Eq. 6.22 leads to

$$[m^\infty(\omega)]\begin{bmatrix}[I] & \\ & 0\end{bmatrix}[m^\infty(\omega)] + [e^1]\frac{[m^\infty(\omega)]}{(i\omega)^2} + \frac{[m^\infty(\omega)]}{(i\omega)^2}([e^1]^T + [I])$$

$$+ \frac{1}{\omega}[m^\infty(\omega)]_{,\omega} - \frac{1}{(i\omega)^4}[e^2] - \frac{1}{(i\omega)^2}[m^0] = 0 \quad (6.26)$$

After performing the Fourier transformation of Eq. 6.26, the *consistent infinitesimal finite-element cell equation of incompressible elasticity* results

$$\int_0^t [m^\infty(t-\tau)]\begin{bmatrix}[I] & \\ & 0\end{bmatrix}[m^\infty(\tau)]d\tau + [e^1]\int_0^t\int_0^\tau [m^\infty(\tau')]d\tau'd\tau$$

$$+ \int_0^t\int_0^\tau [m^\infty(\tau')]d\tau'd\tau[e^1]^T + t\int_0^t [m^\infty(\tau)]d\tau - \frac{t^3}{6}[e^2]H(t) - t[m^0]H(t) = 0 \quad (6.27)$$

6.1 FUNDAMENTAL EQUATIONS

Eq. 6.27 is the same as Eq. 5.118 of the compressible case with the exception of the coefficient matrix $\begin{bmatrix} [I] \\ 0 \end{bmatrix}$ of the first term.

In passing, it is worth mentioning that for a horizontally layered unbounded medium (Section 5.2.10) the corresponding equation for the compressible case Eq. 5.145 is modified for the incompressible case in the same way using the appropriate coefficient matrices (Eqs. 6.23 and 6.19).

The time discretization of Eq. 6.27 is the same as that described in Section 5.3. After determining $[m^\infty(t)]$ from Eq. 6.27, the acceleration unit-impulse response matrix follows as

$$[M^\infty(t)] = \left([\Phi]^{-1}\right)^T [m^\infty(t)][\Phi]^{-1} \tag{6.28}$$

The *response of an incompressible unbounded medium is instantaneous in the entire domain* owing to the infinite dilatational-wave velocity. This manifests itself by a mass which does not appear in the compressible case.

To discuss the difference between the behaviour of an unbounded medium in compressible elasticity and that in incompressible elasticity, the instantaneous response in the time domain is examined. This corresponds to the response at the high-frequency limit in the frequency domain. As is well known, the high-frequency limit of the dynamic-stiffness matrix of a compressible unbounded medium is proportional to $i\omega$ with the proportionality matrix representing the dashpot matrix (see discussion in connection with Eq. 2.7 and turn to Eq. A.5.29 for an example). This is easily verified from Eq. 5.113 formulated for compressible elasticity

$$[S^\infty(\omega)][E^0]^{-1}[S^\infty(\omega)] + [E^1][E^0]^{-1}[S^\infty(\omega)] + [S^\infty(\omega)][E^0]^{-1}[E^1]^T$$
$$- (s-2)[S^\infty(\omega)] - \omega[S^\infty(\omega)]_{,\omega} - [E^2] + [E^1][E^0]^{-1}[E^1]^T + \omega^2[M^0] = 0 \tag{6.29}$$

For $\omega \to \infty$, the first term, which is quadratic in $[S^\infty(\omega)]$, and the last term, which is proportional to ω^2, dominate. $[S^\infty(\omega)]$ will thus be proportional to $i\omega$. In contrast, for incompressible elasticity, the coefficient matrix of the first term in Eq. 6.22 equals $\begin{bmatrix} [I] \\ 0 \end{bmatrix}$. As the last term is positive definite, the elements of $[s^\infty(\omega)]$ corresponding to the zero submatrix in the first term cannot be determined based on just the first and last terms. These elements of $[s^\infty(\omega)]$ will be proportional to ω^2 as the other terms are linear in $[s^\infty(\omega)]$. The same also applies to $[S^\infty(\omega)]$ (Eq. 6.17) with the proportionality matrix representing the mass matrix. Thus, for an *incompressible unbounded medium, the high-frequency response is dominated by the concentrated masses located in the nodes on the structure-medium interface* whereby dashpots are also present.

6.2 ACCURACY

6.2.1 Spherical Cavity Embedded in Full-space

The spherical cavity embedded in a full-space with a uniform normal displacement prescribed on the structure-medium interface described in Appendix A.1 (Figure A-1) for compressible elasticity is calculated for the incompressible case.

To derive the analytical solution for the acceleration unit-impulse response coefficient $M^\infty(t)$, the corresponding expression in the frequency domain (Eq. A.1.47) is formulated as

$$M^\infty(\omega) = K^\infty \left(\frac{1}{(i\omega)^2} + \frac{1}{4c_s^2} \frac{r_0^2}{1 + i\frac{\omega r_0}{c_p}} \right) \tag{6.30}$$

after substituting $c_p^2/c_s^2 = 2(1-\nu)/(1-2\nu)$ (shear-wave velocity $c_s = \sqrt{G/\rho}$). For the limit of the incompressible case, $c_p \to \infty$ yields

$$M^\infty(\omega) = 4\pi r_0^3 \rho + K^\infty \frac{1}{(i\omega)^2} \tag{6.31}$$

with $K^\infty = 16\pi G r_0$ (Eq. A.1.17). The inverse Fourier transformation of Eq. 6.31 leads to the acceleration unit-impulse response coefficient

$$M^\infty(t) = 4\pi r_0^3 \rho \delta(t) + K^\infty t H(t) \tag{6.32}$$

The coefficient $4\pi r_0^3 \rho$ represents the mass. The interaction force (Eq. 2.11) equals after integration by parts

$$R(t) = 4\pi r_0^3 \rho \ddot{u}_0(t) + K^\infty u_0(t) \tag{6.33}$$

When applying the consistent infinitesimal finite-element cell method, only one octant of the structure-medium interface is discretized due to symmetry (Figure 5-7). Three 8-node isoparametric surface finite elements are used yielding a dynamic system with 33 degrees of freedom. After determining the unit-impulse response matrix $[M^\infty(t)]$ of order 33×33, the uniform radial nodal accelerations are enforced. This defines the spatial motion pattern $\{\phi\}$. The unit-impulse response coefficient $M^\infty(t)$ follows from Eq. A.0.1, which corresponds to summing all radial nodal forces for a unit radial acceleration. The analysis is performed with $\Delta t = 0.05 r_0/c_s$. To ease the representation of the result, the acceleration is specified as a step function $\ddot{u}_0(t) = \ddot{u}_0 H(t)$ with a constant acceleration $\ddot{u}_0$. The corresponding interaction force $R(t)$ follows from Eq. 2.11. $R(t)$ non-dimensionalized with $4\pi r_0^3 \rho \ddot{u}_0$ is plotted as a function of the dimensionless time $\bar{t} = tc_s/r_0$ in Figure 6-1. The agreement of the

6.2 ACCURACY

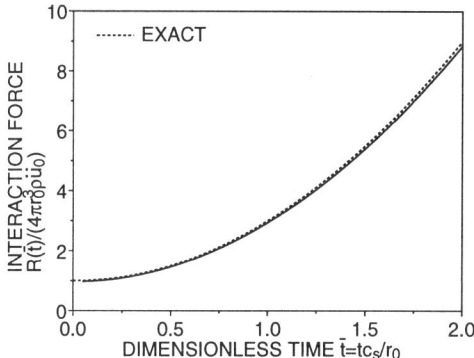

Figure 6-1 Interaction force of spherical cavity embedded in incompressible full-space caused by acceleration step function

result of the consistent infinitesimal finite-element cell method with that of the exact solution determined from Eq. 6.33

$$R(\bar{t}) = 4\pi r_0^3 \rho \ddot{u}_0 \left(1 + 2\bar{t}^2\right) H(\bar{t}) \tag{6.34}$$

is excellent.

6.2.2 In-plane Motion of Semi-infinite Layer of Constant Depth

The in-plane motion of a semi-infinite layer of constant depth is described in Appendix A.3.2 (Figure A-10). It is instructive to determine explicit equations for the dynamic-stiffness matrix selecting only one line element on the structure-medium interface. In particular, the high-frequency behaviour described by the dashpot and mass matrices is discussed.

The structure-medium interface is discretized with 1 2-node line element with linear shape functions resulting in 2 degrees of freedom after enforcing the fixed boundary condition at the base. The consistent infinitesimal finite-element cell equation for this case where the characteristic length d does not change is specified in Section 5.2.10. The coefficient matrices $[E^0]$, $[E^1]$, $[E^2]$ and $[M^0]$ are defined in Eqs. 5.137 and 5.138. Performing the decomposition of the elasticity matrix $[D]$ into the shear and volumetric parts (Eq. 6.1) yields Eqs. 6.2 and 6.3 with $[B^1]$ and $[B^2]$ defined in Eq. 5.131. The integrations for the coefficient matrices of the shear part $[E^0]$, $[E^1]$, $[E^2]$ and for $[M^0]$ are performed exactly. The integrations for the coefficient matrices of the volumetric part $[\overline{E}^0]$, $[\overline{E}^1]$, $[\overline{E}^2]$ are evaluated analytically with one Gauss point which corresponds to reduced integration. This yields

$$[E^0] = \frac{G}{3} \begin{bmatrix} 2 & 0 \\ 0 & 1 \end{bmatrix} \tag{6.35a}$$

$$[\overline{\overline{E}}^0] = \frac{G}{4}\begin{bmatrix} 1 & 0 \\ 0 & 0 \end{bmatrix} \tag{6.35b}$$

$$[\overline{E}^1] = -\frac{G}{2}\begin{bmatrix} 0 & 1 \\ 0 & 0 \end{bmatrix} \tag{6.35c}$$

$$[\overline{\overline{E}}^1] = -\frac{G}{2}\begin{bmatrix} 0 & 0 \\ 1 & 0 \end{bmatrix} \tag{6.35d}$$

$$[\overline{E}^2] = G\begin{bmatrix} 1 & 0 \\ 0 & 2 \end{bmatrix} \tag{6.35e}$$

$$[\overline{\overline{E}}^2] = G\begin{bmatrix} 0 & 0 \\ 0 & 1 \end{bmatrix} \tag{6.35f}$$

and

$$[M^0] = \frac{d^2\rho}{3}[I] \tag{6.36}$$

The eigenvalue problem (Eq. 6.6) yields

$$\lceil \Lambda \rfloor = \begin{bmatrix} 0 & \\ & \frac{3}{8} \end{bmatrix} \tag{6.37a}$$

$$[\Phi] = \sqrt{\frac{3}{G}}\begin{bmatrix} 0 & 1 \\ 1 & 0 \end{bmatrix} \tag{6.37b}$$

The identities in Eq. 6.21 are easily verified.

The equation for the horizontally layered unbounded medium (Eq. 5.143), but modified for the incompressible case as in Eq. 6.22, is constructed

$$[s^\infty(a_0)]\begin{bmatrix} 1 & 0 \\ 0 & 0 \end{bmatrix}[s^\infty(a_0)] - \frac{1}{2\sqrt{2}}\begin{bmatrix} 0 & 8 \\ 3 & 0 \end{bmatrix}[s^\infty(a_0)]$$

$$-[s^\infty(a_0)]\frac{1}{2\sqrt{2}}\begin{bmatrix} 0 & 8 \\ 3 & 0 \end{bmatrix} + \frac{a_0^2}{2}\begin{bmatrix} 2 & 0 \\ 0 & 1 \end{bmatrix} - \frac{1}{8}\begin{bmatrix} 64 & 0 \\ 0 & 9 \end{bmatrix} = 0 \tag{6.38}$$

with the dimensionless frequency $a_0 = \omega d/c_s$. Formulated in the elements $s_{11}^\infty(a_0)$, etc. of the symmetric matrix $[s^\infty(a_0)]$, the following three equations

$$(s_{11}^\infty(a_0))^2 - 4\sqrt{2}s_{12}^\infty(a_0) + a_0^2 - 8 = 0 \tag{6.39a}$$

$$s_{11}^\infty(a_0)s_{12}^\infty(a_0) - 2\sqrt{2}s_{22}^\infty(a_0) - \frac{3}{2\sqrt{2}}s_{11}^\infty(a_0) = 0 \tag{6.39b}$$

$$(s_{12}^\infty(a_0))^2 - \frac{3}{\sqrt{2}}s_{12}^\infty(a_0) + \frac{a_0^2}{2} - \frac{9}{8} = 0 \tag{6.39c}$$

6.2 ACCURACY

result. From Eq. 6.39c $s_{12}^\infty(a_0)$ follows, then from Eq. 6.39a $s_{11}^\infty(a_0)$ and finally from Eq. 6.39b $s_{22}^\infty(a_0)$

$$s_{11}^\infty(a_0) = \sqrt{14 + 2\sqrt{2}\sqrt{9-2a_0^2} - a_0^2} \quad (6.40a)$$

$$s_{12}^\infty(a_0) = \frac{1}{2}\left(\frac{3}{\sqrt{2}} + \sqrt{9-2a_0^2}\right) \quad (6.40b)$$

$$s_{22}^\infty(a_0) = \frac{1}{4\sqrt{2}}\sqrt{9-2a_0^2}\sqrt{14+2\sqrt{2}\sqrt{9-2a_0^2} - a_0^2} \quad (6.40c)$$

The first two terms of the high-frequency limit follow from Taylor expansions with $1/a_0 \ll 1$ as

$$[s^\infty(a_0)] = -\frac{a_0^2}{4}\begin{bmatrix} 0 & 0 \\ 0 & 1 \end{bmatrix} + \frac{ia_0}{2}\begin{bmatrix} 2 & \sqrt{2} \\ \sqrt{2} & 1 \end{bmatrix} + \frac{1}{16\sqrt{2}}\begin{bmatrix} 32\sqrt{2} & 24 \\ 24 & 29\sqrt{2} \end{bmatrix} \quad (6.41)$$

The dynamic-stiffness matrix $[S^\infty(\omega)]$ is obtained from Eq. 6.17 with $[\Phi]$ specified in Eq. 6.37b as

$$[S^\infty(\omega)] = -\frac{\rho d^2 \omega^2}{6}\begin{bmatrix} 1 & 0 \\ 0 & 0 \end{bmatrix} + \frac{i\omega\rho c_s d}{3}\begin{bmatrix} 1 & 1 \\ 1 & 1 \end{bmatrix} + \frac{G}{24}\begin{bmatrix} 29 & 12 \\ 12 & 16 \end{bmatrix} \quad (6.42)$$

The first term on the right-hand side represents the "mass" matrix. The mass coefficient $\rho d^2/6$ is only present for the horizontal degree of freedom. The second term represents the dashpot matrix with the finite velocity c_s. The off-diagonal terms in the dashpot matrix are non-zero which is not the case in compressible elasticity.

6.2.3 Circular Cavity Embedded in Full-plane

The in-plane motion of a circular cavity embedded in a full-plane described in Appendix A.5 (Figure A-14) is addressed. The analytical expression for the dynamic-stiffness coefficient in the frequency domain for incompressible elasticity is specified in Eq. A.5.32.

The same example is calculated for compressible elasticity in Section 5.4.7. The same spatial and temporal discretizations are used in the incompressible case.

For $\nu = 0.5$, the dynamic-stiffness coefficient in the frequency domain $S^\infty(\omega)$, non-dimensionalized with the shear modulus G, is decomposed in $k(a_0)$ and $c(a_0)$. Good agreement with the analytical solution results (Figure 6-2).

6.2.4 Prism Embedded in Half-space

The three-dimensional motion of a prism embedded in an inhomogeneous half-space for the vector wave equation described in Appendix A.7 (Figure A-17) is

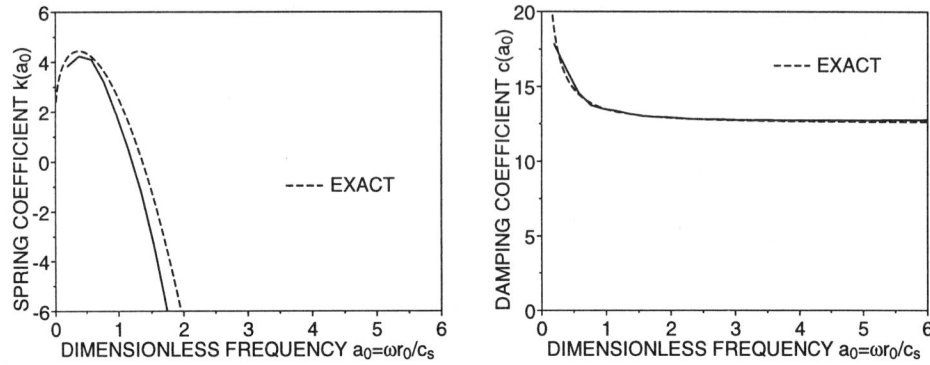

Figure 6-2 Dynamic-stiffness coefficient in frequency domain of circular cavity embedded in incompressible full-plane

calculated for incompressible elasticity. The embedment ratio $e/b = 2/3$ is selected. An inhomogeneous half-space compatible with similarity with the shear modulus G_1 of the unbounded medium adjacent to the walls and $G_2 = 4G_1$ adjacent to the base is calculated. The mass density ρ does not vary. The structure-medium interface is first discretized by finite elements. Owing to symmetry, only a quarter of the prism is analysed. The finite-element discretization of the three faces is shown in Figure 5-26. The corresponding unit-impulse response matrix $[M^\infty(t)]$ is then established with $\Delta t = 0.02b/c_{s1}$ ($c_{s1} = \sqrt{G_1/\rho}$).

To ease comparison, the constraints of motion corresponding to a rigid structure-medium interface are enforced. This yields the unit-impulse response coefficients of the horizontal, vertical, rocking and torsional motions $M_h^\infty(t)$, $M_v^\infty(t)$, $M_r^\infty(t)$ and $M_t^\infty(t)$ as described at the beginning of Section 5.4.12. The acceleration is specified for each degree of freedom as a step function with a constant acceleration $\ddot{u}_0$. The corresponding interaction forces $R_h(t)$ and $R_v(t)$ non-dimensionalized with $\rho b^3 \ddot{u}_0$ as well as $R_r(t)$ and $R_r(t)$ with $\rho b^5 \ddot{u}_0$ are plotted as a function of $\bar{t} = tc_s/b$ in Figure 6-3. An extended mesh of finite elements for $\nu = 0.49$ is also analysed with selective reduced integration. The element length in the radial direction equals $0.15b$ and $\Delta t = 0.02b/c_{s1}$. 20 rows of finite elements are chosen. The agreement is excellent.

6.2 ACCURACY

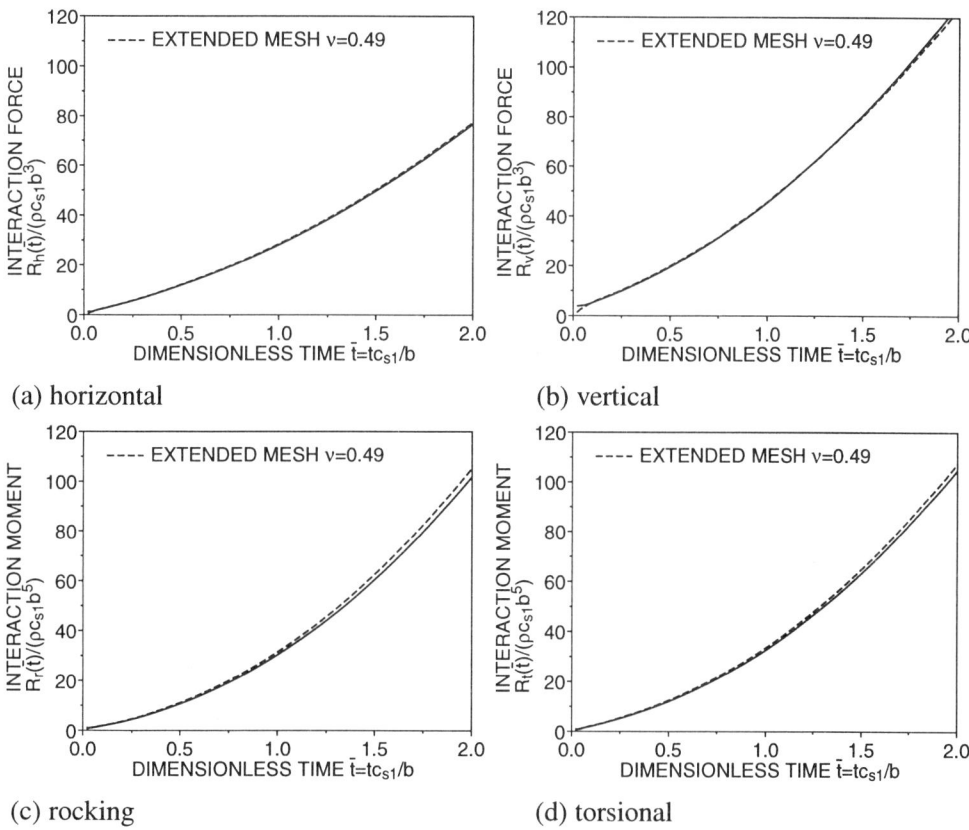

Figure 6-3 Interaction force of square prism embedded in inhomogeneous incompressible half-space caused by acceleration step function

7

CONSISTENT INFINITESIMAL FINITE-ELEMENT CELL METHOD IN FREQUENCY DOMAIN

7.1 OVERVIEW

In the previous chapters of the book the methods for performing dynamic analysis directly in the time domain are described. When the structure-unbounded medium system behaves linearly, the analysis in the frequency domain using a Fourier transformation is an alternative. For a transient load the dynamic stiffness throughout the frequency range of interest is needed. In particular, when the harmonic (steady-state) response to a load of a specific frequency is to be calculated, the frequency domain analysis is efficient. In this case the dynamic stiffness for a single frequency is sufficient.

This chapter addresses the calculation of the dynamic stiffness in the frequency domain. The consistent infinitesimal finite-element cell method is well suited to evaluating the dynamic stiffness for a large range of frequencies. It is, however, not competitive with other methods for determining the dynamic stiffness at a single frequency, as a system of nonlinear equations has to be solved (Reference [63]).

As the derivation of the consistent infinitesimal finite-element cell equation in the time domain is performed via the frequency domain in Section 5.2, the consistent infinitesimal finite-element cell equation in the frequency domain is established as an intermediate result (Eq. 5.113)

$$([S^\infty(\omega)] + [E^1])[E^0]^{-1}([S^\infty(\omega)] + [E^1]^T) - (s-2)[S^\infty(\omega)]$$
$$- \omega[S^\infty(\omega)]_{,\omega} - [E^2] + \omega^2[M^0] = 0 \quad (7.1)$$

This is a *system of first-order nonlinear ordinary differential equations for the displacement dynamic-stiffness matrix* $[S^\infty(\omega)]$ *in the independent variable* ω. If $[S^\infty(\omega)]$ is known at a specific frequency, starting from this boundary condition $[S^\infty(\omega)]$ can be calculated for increasing and decreasing frequencies by integrating Eq. 7.1 numerically. The integration of ordinary differential equations is well known and will not be repeated here. Powerful algorithms are e.g. listed in Reference [40].

It is worth mentioning that material damping of the unbounded medium is straightforwardly introduced in this frequency domain formulation. Using the correspondence principle it is sufficient to replace the real elastic moduli by the complex moduli in the elasticity matrix $[D]$.

The boundary condition must satisfy the radiation condition in the frequency domain (Section 1.2). Obviously, the static-stiffness matrix cannot be used. The radiation condition is formulated at an infinite distance from the structure-medium interface which corresponds to an infinite characteristic length r (see e.g. Eq. 1.3). The associated dimensionless frequency $a_0 = \omega r / c_s$ is thus infinite, which can also be achieved by an infinite ω. The boundary condition $[S^\infty(\omega)]$ satisfying the radiation condition is formulated at infinite ω. In an actual calculation, a large ω is selected. Thus, the high-frequency behaviour of $[S^\infty(\omega)]$ is studied examining the asymptotic expansion of Eq. 7.1 (Section 7.2).

In contrast to Eq. 7.1 of the consistent infinitesimal finite-element cell method with an analytical limit of the infinitesimal cell width, Eq. 5.101 expresses a *relationship between the dynamic-stiffness matrices at the interior boundary of the cell* $[S_i^\infty(\omega)]$ *and at the exterior boundary* $[S_e^\infty(\omega)]$ *for a finite cell width*. This equation can be used either to calculate $[S_i^\infty(\omega)]$ from $[S_e^\infty(\omega)]$ (*dynamic condensation method*) or to determine $[S_e^\infty(\omega)]$ from $[S_i^\infty(\omega)]$ (*substructure deletion method*). These procedures can be applied repeatedly covering either a decreasing or an increasing characteristic length (for a fixed frequency). For similar boundaries, the result can also be interpreted as either decreasing or increasing the frequency (at the structure-medium interface with a fixed characteristic length, see also Eq. 3.8). To start the procedure the boundary condition $[S^\infty(\omega)]$ at a specific frequency must be known (Section 7.3).

The accuracy of the consistent infinitesimal finite-element cell method is examined (Section 7.4). The spherical cavity embedded in a full-space with symmetric waves is used to study the propagation of errors (Section 7.4.1). It is also demonstrated analytically that the static-stiffness coefficient cannot be used as the boundary condition.

7.2 ASYMPTOTIC EXPANSION FOR HIGH FREQUENCY

The asymptotic behaviour at high frequency of the consistent infinitesimal finite-element cell equation (Eq. 7.1) is studied. The dynamic-stiffness matrix $[S^\infty(\omega)]$ at high frequency for compressible elasticity is expanded in a power series of $i\omega$ in descending order starting at one

$$[S^\infty(\omega)] \approx i\omega[C_\infty] + [K_\infty] + \sum_{j=1}^{m} \frac{1}{(i\omega)^j}[A_j] \qquad (7.2)$$

7.2 ASYMPTOTIC EXPANSION FOR HIGH FREQUENCY

The first two terms on the right-hand side represent the singular part $[S_s^\infty(\omega)]$ with the constant dashpot matrix $[C_\infty]$ and the constant spring matrix $[K_\infty]$ addressed in Eqs. 2.6 and 2.7. The third term denotes the asymptotic expansion of the regular part $[S_r^\infty(\omega)]$ with the unknown coefficient matrices $[A_j]$ ($j = 1,\ldots,m$). A concise formulation results when the transformation based on the following eigenvalue problem is introduced

$$[M^0][\Phi] = [E^0][\Phi]\lceil\Lambda^2\rfloor \tag{7.3}$$

$[M^0]$ and $[E^0]$ are positive definite matrices resulting in positive eigenvalues $\lceil\Lambda^2\rfloor$. The eigenvectors $[\Phi]$ are normalized as

$$[\Phi]^T[E^0][\Phi] = [I] \tag{7.4}$$

yielding

$$[\Phi]^T[M^0][\Phi] = \lceil\Lambda^2\rfloor \tag{7.5}$$

and

$$[E^0]^{-1} = [\Phi][\Phi]^T \tag{7.6}$$

Premultiplying Eq. 7.1 by $[\Phi]^T$ and postmultiplying by $[\Phi]$ results in

$$([s^\infty(\omega)] + [e^1])([s^\infty(\omega)] + [e^1]^T) - (s-2)[s^\infty(\omega)]$$
$$- \omega[s^\infty(\omega)]_{,\omega} - [e^2] + \omega^2\lceil\Lambda^2\rfloor = 0 \tag{7.7}$$

where

$$[s^\infty(\omega)] = [\Phi]^T[S^\infty(\omega)][\Phi] \tag{7.8}$$

and

$$[e^1] = [\Phi]^T[E^1][\Phi] \tag{7.9a}$$
$$[e^2] = [\Phi]^T[E^2][\Phi] \tag{7.9b}$$

Substituting Eq. 7.2 in Eq. 7.8 yields

$$[s^\infty(\omega)] \approx i\omega[c_\infty] + [k_\infty] + \sum_{j=1}^{m}\frac{1}{(i\omega)^j}[a_j] \tag{7.10}$$

with

$$[c_\infty] = [\Phi]^T[C_\infty][\Phi] \tag{7.11a}$$
$$[k_\infty] = [\Phi]^T[K_\infty][\Phi] \tag{7.11b}$$
$$[a_j] = [\Phi]^T[A_j][\Phi] \tag{7.11c}$$

Substituting Eq. 7.10 in Eq. 7.7 and rearranging in descending order of the power series of $i\omega$ leads to

$$(i\omega)^2 \left([c_\infty]^2 - \lceil \Lambda^2 \rfloor \right)$$
$$+ i\omega \left([c_\infty][k_\infty] + [k_\infty][c_\infty] + [c_\infty][e^1]^T + [e^1][c_\infty] - (s-1)[c_\infty] \right)$$
$$+ [c_\infty][a_1] + [a_1][c_\infty] + ([k_\infty] + [e^1])([k_\infty] + [e^1]^T) - (s-2)[k_\infty] - [e^2]$$
$$+ \frac{1}{i\omega} \left([c_\infty][a_2] + [a_2][c_\infty] + ([k_\infty] + [e^1])[a_1] + [a_1]([k_\infty] + [e^1]^T) - (s-3)[a_1] \right) \approx 0$$
(7.12)

In Eq. 7.12, $m = 2$ is selected. The generalization to any m leading to additional terms is straightforward.

The coefficient matrix of each term of the power series in $i\omega$ is set equal to zero in descending order. The first yields

$$[c_\infty]^2 = \lceil \Lambda^2 \rfloor \tag{7.13}$$

Selecting the positive roots of each element on the diagonal of $\lceil \Lambda^2 \rfloor$ leads to

$$[c_\infty] = \lceil \Lambda \rfloor \tag{7.14}$$

The dashpot matrix $[C_\infty]$ follows from Eq. 7.11a as

$$[C_\infty] = \left([\Phi]^{-1} \right)^T \lceil \Lambda \rfloor [\Phi]^{-1} \tag{7.15}$$

As each coefficient of $\lceil \Lambda \rfloor$ is positive, $[C_\infty]$ will be positive definite.

The second term in Eq. 7.12 results after substituting Eq. 7.14 in

$$\lceil \Lambda \rfloor [k_\infty] + [k_\infty] \lceil \Lambda \rfloor = -\lceil \Lambda \rfloor [e^1]^T - [e^1] \lceil \Lambda \rfloor + (s-1) \lceil \Lambda \rfloor \tag{7.16}$$

This linear equation for $[k_\infty]$ is a Lyapunov equation similar to Eq. 5.185 but with a diagonal coefficient matrix $\lceil \Lambda \rfloor$. Its solution for each element $k_{\infty kl}$ equals

$$k_{\infty kl} = \frac{1}{\Lambda_k + \Lambda_l} \left(-\Lambda_k e^1_{lk} - \Lambda_l e^1_{kl} + (s-1) \Lambda_k \delta_{kl} \right) \tag{7.17}$$

with the Kronecker delta δ_{kl} (=1 for $k = l$; =0 for $k \neq l$). The spring matrix $[K_\infty]$ is calculated from Eq. 7.11b as

$$[K_\infty] = \left([\Phi]^{-1} \right)^T [k_\infty][\Phi]^{-1} \tag{7.18}$$

The third term in Eq. 7.12 leads after substituting Eq. 7.14 with the known $[k_\infty]$ to

$$\lceil \Lambda \rfloor [a_1] + [a_1] \lceil \Lambda \rfloor = -([k_\infty] + [e^1])([k_\infty] + [e^1]^T) + (s-2)[k_\infty] + [e^2] \tag{7.19}$$

7.3 DYNAMIC CONDENSATION AND SUBSTRUCTURE DELETION

This equation for $[a_1]$ is in the same form as Eq. 7.16. Analogously, $[a_2]$ is determined from the fourth term in Eq. 7.12 with the known $[c_\infty] = \lceil \Lambda \rfloor$, $[k_\infty]$ and $[a_1]$

$$\lceil \Lambda \rfloor [a_2] + [a_2] \lceil \Lambda \rfloor = -\left([k_\infty] + [e^1]\right)[a_1] - [a_1]\left([k_\infty] + [e^1]^T\right) + (s-3)[a_1] \quad (7.20)$$

The coefficient matrices $[A_j]$ result from Eq. 7.11c as

$$[A_j] = \left([\Phi]^{-1}\right)^T [a_j][\Phi]^{-1} \quad (7.21)$$

After calculating $[C_\infty]$, $[K_\infty]$, $[A_1]$ and $[A_2]$, the asymptotic behaviour for $m = 2$ follows from Eq. 7.2.

In an actual application of the consistent infinitesimal finite-element cell method in the frequency domain the boundary condition at the specified high frequency ω_h $[S^\infty(\omega_h)]$ follows from Eq. 7.2 for a selected m. It is then used as the starting value to integrate the consistent infinitesimal finite-element cell equation (Eq. 7.1) for decreasing ω.

The consistent infinitesimal finite-element cell equation is processed for decreasing frequencies starting at a very large value ω_h which replaces $\omega \to \infty$. The starting value of $[S^\infty(\omega_h)]$ is determined based on an asymptotic expansion at high frequency (Eq. 7.2). The dashpot matrix $[C_\infty]$ is calculated after solving an eigenvalue problem (Eq. 7.15), and the spring matrix $[K_\infty]$ (Eq. 7.18) and the coefficient matrices of the asymptotic expansion $[A_j]$ (Eq. 7.21) follow from Lyapunov equations.

It is of interest to compare this frequency-domain procedure with the consistent infinitesimal finite-element cell method in the time domain. In this case the algorithm for increasing time starts at early time. The asymptotic behaviour of the acceleration unit-impulse response matrix $[M^\infty(t)]$ at early time is derived. Substituting Eq. 7.2 in Eq. 2.14 and performing the inverse Fourier transformation yields

$$[M^\infty(t)] = [C_\infty]H(t) + [K_\infty]tH(t) + \sum_{j=1}^m \frac{t^{j+1}}{(j+1)!}[A_j]H(t) \quad (7.22)$$

Substituting Eq. 7.22 in Eq. 5.115, processing analogously to Eqs. 7.3 to 7.11 and examining the orders of magnitude in t, i.e. $1, t, t^2, t^3$ lead to the same equations for $[C_\infty]$, $[K_\infty]$ and $[A_j]$ as specified in Eqs. 7.14 to 7.21. Thus, the asymptotic expansion of $[M^\infty(t)]$ at $t = 0$ results in the same coefficient matrices $[C_\infty]$, $[K_\infty]$ and $[A_j]$ as the asymptotic expansion of $[S^\infty(\omega)]$ at $\omega \to \infty$. This is a consequence of the initial value theorem.

7.3 DYNAMIC CONDENSATION AND SUBSTRUCTURE DELETION METHODS

A straightforward algorithm which uses only the standard static-stiffness and mass matrices of finite elements without performing any limit of the cell width in the radial direction is discussed. Assemblage of a cell of finite elements and the unbounded medium as described in Section 5.2.7 leads to the equation (Eq. 5.101)

$$[S_i^\infty(\omega)] = [S_{ii}(\omega)] - [S_{ie}(\omega)]([S_e^\infty(\omega)] + [S_{ee}(\omega)])^{-1}[S_{ei}(\omega)] \tag{7.23}$$

In this so-called *dynamic condensation* equation, the dynamic-stiffness matrix of the unbounded medium at the interface coinciding with the interior boundary of the cell $[S_i^\infty(\omega)]$ is expressed as a function of that at the interface coinciding with the exterior boundary $[S_e^\infty(\omega)]$ and of the dynamic-stiffness submatrices of the cell. For instance,

$$[S_{ie}(\omega)] = [K_{ie}] - \omega^2[M_{ie}] \tag{7.24}$$

applies with the corresponding static-stiffness submatrix $[K_{ie}]$ and mass submatrix $[M_{ie}]$.

In an efficient dynamic condensation algorithm a series of *similar* finite-element cells with a constant dimensionless cell width $w = (r_e - r_i)/r_i$, from the structure-medium interface up to a large distance, is introduced (Figure 7-1). For the *j*th cell,

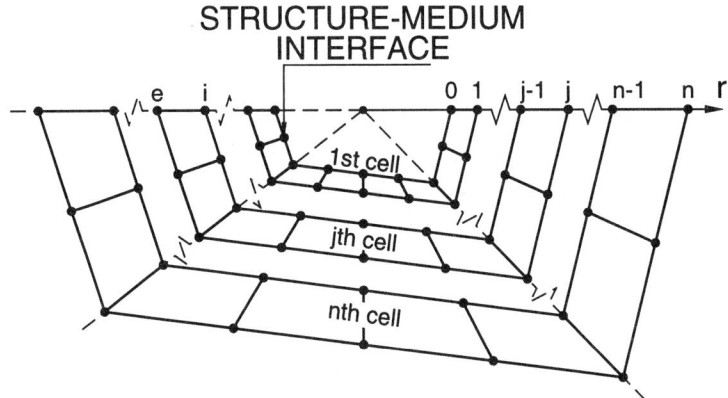

Figure 7-1 Finite-element cells of part of unbounded medium used in dynamic condensation method

Eq. 7.23 is formulated as

$$[S_i^{\infty j}(\omega)] = [S_{ii}^j(\omega)] - [S_{ie}^j(\omega)]([S_e^{\infty j}(\omega)] + [S_{ee}^j(\omega)])^{-1}[S_{ei}^j(\omega)] \tag{7.25}$$

with the characteristic length $r_i = r_{j-1}$ and $r_e = r_j$. The static-stiffness and mass matrices for the similar cells are determined by scaling. The dynamic-stiffness matrix of the *j*th cell is constructed as

$$[S^j(\omega)] = \left(\frac{r_{j-1}}{r_0}\right)^{s-2}[K^1] - \omega^2\left(\frac{r_{j-1}}{r_0}\right)^s[M^1] \tag{7.26}$$

7.3 DYNAMIC CONDENSATION AND SUBSTRUCTURE DELETION 157

with the static-stiffness matrix $[K^1]$ and mass matrix $[M^1]$ of the first cell and the characteristic length r_0 of the structure-medium interface. ω is fixed at first. The dynamic-stiffness matrix $[S^\infty(r_n, \omega)]$ of the unbounded medium at the interface with r_n is assumed to be known. Applying Eq. 7.25 for the nth cell with $[S_e^{\infty n}(\omega)] = [S^\infty(r_n, \omega)]$ yields $[S_i^{\infty n}(\omega)]$ which equals the dynamic-stiffness matrix $[S^\infty(r_{n-1}, \omega)]$ of the unbounded medium at the interface with r_{n-1}. Eq. 7.25 is applied repeatedly for $j = n-1, \ldots, 1$ resulting in $[S^\infty(r_{n-2}, \omega)], \ldots, [S^\infty(r_0, \omega)]$.

The relationship of the dynamic-stiffness matrices of the unbounded medium at similar interfaces derived in Section 3.1 is used to interpret $[S^\infty(r_j, \omega)]$ $(j = n, \ldots, 0)$ at the structure-medium interface with a fixed $r = r_0$ but varying ω. Eq. 3.6 is formulated at r_j for ω and at r_0 for $(r_j/r_0)\omega$ which yields the same $a_0 = \omega r_j/c_s$

$$[S^\infty(r_j, \omega)] = Gr_j^{s-2}[\bar{S}^\infty\left(\frac{\omega r_j}{c_s}\right)] \tag{7.27a}$$

$$[S^\infty(r_0, \frac{r_j}{r_0}\omega)] = Gr_0^{s-2}[\bar{S}^\infty\left(\frac{r_j}{r_0}\omega\frac{r_0}{c_s}\right)] \tag{7.27b}$$

Eliminating the common term $[\bar{S}^\infty(\omega r_j/c_s)]$ results in

$$[S^\infty(r_0, \frac{r_j}{r_0}\omega)] = \left(\frac{r_0}{r_j}\right)^{s-2}[S^\infty(r_j, \omega)] \tag{7.28}$$

Eq. 7.28 leads to the dynamic-stiffness matrix at the structure-medium interface at decreasing frequencies $(r_j/r_0)\omega$ for $j = n, \ldots, 0$. Thus, through an appropriate choice of the number of cells n and aspect ratio r_j/r_{j-1}, the dynamic-stiffness matrix can be calculated in the frequency range of interest.

As a practical procedure to determine the boundary condition which serves as the starting value for the dynamic condensation method, the dynamic-stiffness matrix on the structure-medium interface at the high frequency selected as $\omega_h = (r_n/r_0)\omega$ is expressed as

$$[S^\infty(r_0, \omega_h)] = i\omega_h[C_\infty] \tag{7.29}$$

From Eq. 7.28, the dynamic-stiffness matrix at r_n for ω is calculated as

$$[S^\infty(r_n, \omega_h)] = i\omega_h\left(\frac{r_n}{r_0}\right)^{s-2}[C_\infty] = i\omega\left(\frac{r_n}{r_0}\right)^{s-1}[C_\infty] \tag{7.30}$$

The dashpot matrix $[C_\infty]$ is calculated based on purely physical considerations formulating the law of conservation of momentum (see Section 1.2). In the dynamic condensation method the error of the dynamic-stiffness matrix at the structure-medium interface reduces for decreasing ω. This is verified in Section 7.4.1. Although the boundary condition in Eq. 7.30 is approximate, the error at the structure-medium interface for intermediate and low frequencies will be acceptable.

The same concepts of the dynamic condensation method are applied in the so-called *substructure deletion method*, with the exception that the boundary condition $[S^\infty(r_0,\omega)]$ is specified at the structure-medium interface. The spatial discretization using finite-element cells shown in Figure 7-1 is still valid. The substructure-deletion equation of the jth cell follows from Eq. 7.25 as

$$[S_e^{\infty j}(\omega)] = -[S_{ee}^j(\omega)] - [S_{ei}^j(\omega)]([S_i^{\infty j}(\omega)] - [S_{ii}^j(\omega)])^{-1}[S_{ie}^j(\omega)] \tag{7.31}$$

Applying Eq. 7.31 repeatedly yields $[S^\infty(r_1,\omega)],\ldots,[S^\infty(r_n,\omega)]$. Eq. 7.28 is again used to evaluate the dynamic-stiffness matrix at the structure-medium interface but for increasing ω.

No generally applicable simple method comparable with that of dynamic condensation exists to determine the boundary condition which is needed to start the substructure deletion method. In particular, the static-stiffness matrix cannot be used, as is illustrated in Section 7.4.1. In addition, the error of the dynamic-stiffness matrix at the structure-medium interface increases for increasing ω. Thus, the substructure deletion method is of limited value only.

7.4 ACCURACY

7.4.1 Spherical Cavity Embedded in Full-space

The spherical cavity embedded in a full-space with a uniform normal displacement prescribed on the structure-medium interface described in Appendix A.1 (Figure A-1) is addressed. The analytical solution of the dynamic-stiffness coefficient is specified in Eq. A.1.16.

Before the accuracy is addressed, it is demonstrated analytically that the static-stiffness coefficient cannot be used as the boundary condition in the consistent infinitesimal finite-element cell method and in the substructure deletion method and that the error decreases in the dynamic condensation method for decreasing ω and increases in the substructure deletion method for increasing ω.

Substituting the coefficients E^0, E^1, E^2 (Eq. 5.204) and M^0 (Eq. 5.205) into Eq. 7.1 yields

$$\frac{1}{8\pi}\frac{1-2\nu}{1-\nu}\frac{1}{Gr_0}(S^\infty(\omega))^2 - \frac{1-5\nu}{1-\nu}S^\infty(\omega) - \omega S^\infty(\omega)_{,\omega} + 4\pi r_0^3\rho\omega^2 - 16\pi\frac{1+\nu}{1-\nu}Gr_0 = 0 \tag{7.32}$$

or with $a_0 = \omega r_0/c_p$

$$\left(\frac{S^\infty(a_0)}{E^0}\right)^2 - \frac{1-5\nu}{1-\nu}\frac{S^\infty(a_0)}{E^0} - a_0\left(\frac{S^\infty(a_0)}{E^0}\right)_{,a_0} + a_0^2 - \frac{2(1+\nu)(1-2\nu)}{(1-\nu)^2} = 0 \tag{7.33}$$

7.4 ACCURACY

Its solution equals

$$S^\infty(a_0) = E^0 \left(\frac{2(1-2\nu)}{1-\nu} - a_0^2 \frac{c_1 e^{+ia_0} + c_2 e^{-ia_0}}{c_1(1+ia_0)e^{+ia_0} + c_2(1-ia_0)e^{-ia_0}} \right) \quad (7.34)$$

Eq. 7.34 can be derived starting from Eq. A.1.14 by including both the outgoing and incoming waves and by using Eq. A.1.9. The integration constant of the first-order differential equation (Eq. 7.32) equals c_2/c_1 for $c_1 \neq 0$ and c_1/c_2 for $c_2 \neq 0$. $c_1 = 0$ and $c_2 = 0$ correspond to outgoing and incoming waves, respectively.

The limit of $S^\infty(a_0)$ for $a_0 \to 0$ is investigated. Eq. 7.34 results in

$$\lim_{a_0 \to 0} S^\infty(a_0) = 16\pi G r_0 - 24\pi \frac{1-\nu}{1-2\nu} G r_0 \quad \text{for} \quad c_1 = -c_2 \quad (7.35a)$$

$$= 16\pi G r_0 \quad \text{for} \quad c_1 \neq -c_2 \quad (7.35b)$$

Eq. 7.35a does not correspond to the static-stiffness coefficient $K^\infty = 16\pi G r_0$ (Eq. A.1.17). As all other c_1, c_2 in Eq. 7.34 lead to the static-stiffness coefficient (Eq. 7.35b), K^∞ cannot be used to determine the integration constant. Thus, K^∞ is useless as a starting value.

The propagation of the error present in the boundary condition used as the starting value is examined. First, the dynamic condensation and substructure deletion methods both working with a finite cell width are addressed. A cell with one finite element based on a linear shape function in the radial direction is selected. For the first cell adjacent to the structure-medium interface of the spherical cavity, the static-stiffness matrix $[K^1]$ and the mass matrix $[M^1]$ to be used in Eq. 7.26 equal

$$[K^1] = 8\pi \frac{1-\nu}{1-2\nu} \frac{G r_0}{w} \begin{bmatrix} 1 + \frac{1-3\nu}{1-\nu} w + w^2 & -1 - w \\ -1 - w & 1 + \frac{1+\nu}{1-\nu} w + \frac{1+\nu}{1-\nu} w^2 \end{bmatrix} \quad (7.36a)$$

$$[M^1] = \frac{\pi}{15} \rho r_0^3 w \begin{bmatrix} 20 + 10w + 2w^2 & 10 + 10w + 3w^2 \\ 10 + 10w + 3w^2 & 20 + 30w + 12w^2 \end{bmatrix} \quad (7.36b)$$

with the dimensionless cell width $w = r_1/r_0 - 1$. Poisson's ratio $\nu = 0.25$ is chosen. For the dynamic condensation method $\omega = 0.1 c_p/r_0$ is selected, and the boundary condition which is formulated at $r_n = 200 r_0$ equals (Eq. 7.30)

$$S^\infty(r_n, \omega) = i\omega \left(\frac{r_n}{r_0}\right)^2 C_\infty \quad (7.37)$$

with (Eq. A.1.23a)

$$C_\infty = 4\pi r_0^2 \rho c_p \quad (7.38)$$

Applying Eq. 7.27b with $j = n$ yields on the left-hand side $S^\infty(r_0, \omega_h)$ with $\omega_h = 20c_p/r_0$ at the structure-medium interface, yielding $a_0 = 20$. The exact dynamic-stiffness coefficient at r_n for $a_0 = 20$ (Eqs. A.1.18 and A.1.19) equals $K^\infty(0.252 + i14.963)$ which compares to the applied boundary condition (Eqs. 7.37 and 7.38) of $K^\infty(0 + i15)$. With $w = 0.0406$, which yields $1/8$ wavelengths in the nth cell, $n = 133$ results. The dynamic condensation method is applied down to r_0 yielding $K^\infty(0.9926 + i0.0007356)$ which corresponds to $a_0 = 0.1$. The exact solution equals $K^\infty(0.9926 + i0.0007426)$. The dimensionless spring coefficient $k(a_0)$ and damping coefficient $c(a_0)$ (Eq. A.1.18) are plotted in Figure 7-2. Note that a logarithmic scale

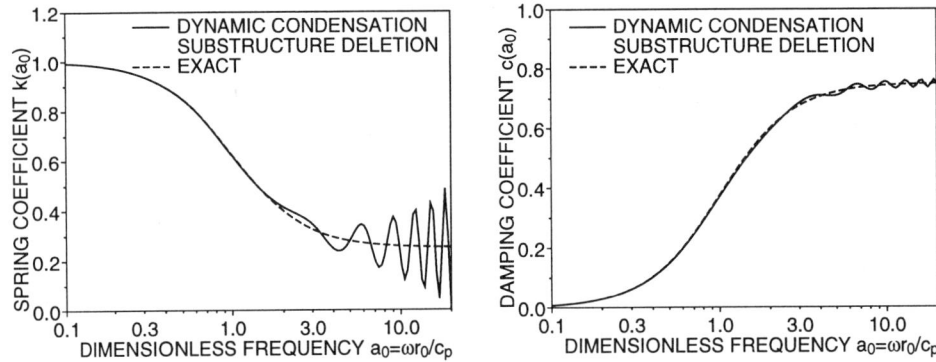

Figure 7-2 Dynamic-stiffness coefficient of spherical cavity calculated with dynamic condensation and substructure deletion methods

is used on the a_0-axis. The violent oscillations present in the high-frequency range diminish drastically in the low-frequency range, yielding results of high accuracy for $a_0 < 3$. The substructure deletion method leads to the same curve as the dynamic condensation method, indicating that the error increases for increasing a_0.

Second, the same study of the propagation of the error for the consistent infinitesimal finite-element cell method, for which the limit of the infinitesimal cell width is incorporated in the formulation analytically, with the same boundary condition (Eq. 7.37) is performed. From a practical point of view, identical results as for dynamic condensation are obtained for this case (results not shown). Thus, the error of the consistent infinitesimal finite-element cell method decreases for decreasing a_0.

The accuracy of the consistent infinitesimal finite-element cell method is investigated. First, the asymptotic expansion for high frequency, which provides the boundary condition, is examined. For this one-dimensional case an asymptotic expansion can be derived from the analytical solution and used in a comparison with that of the consistent infinitesimal finite-element cell equation. The analytical solution of $S^\infty(\omega)$ is specified in Eqs. A.1.21 to A.1.23 with C_∞ and K_∞ listed. The asymptotic

7.4 ACCURACY

expansion of the regular part $S_r^\infty(\omega)$ (Eq. A.1.23c) for high frequency

$$S_r^\infty(\omega) \approx \frac{1}{i\omega}A_1 + \frac{1}{(i\omega)^2}A_2 \tag{7.39}$$

yields

$$A_1 = 4\pi\rho c_p^3 \tag{7.40a}$$

$$A_2 = -4\pi\rho c_p^4 \frac{1}{r_0} \tag{7.40b}$$

The coefficients of the consistent infinitesimal finite-element cell equation E^0, E^1, E^2 and M^0 are listed for the spherical cavity in Eqs. 5.204 and 5.205. From Eqs. 7.3 and 7.4

$$\Lambda^2 = \frac{M^0}{E^0} \tag{7.41}$$

$$\Phi = (E^0)^{-0.5} \tag{7.42}$$

result. The dashpot coefficient C_∞ follows from Eq. 7.15 as

$$C_\infty = \frac{\Lambda}{\Phi^2} = 4\pi r_0^2 \rho c_p \tag{7.43}$$

and the spring coefficient K_∞ from Eqs. 7.16 and 7.18 as

$$K_\infty = E^0 - E^1 = 8\pi \frac{1-3\nu}{1-2\nu} G r_0 \tag{7.44}$$

The coefficients A_1 and A_2 result from Eqs. 7.19, 7.20 and 7.21 as

$$A_1 = 4\pi\rho c_p^3 \tag{7.45a}$$

$$A_2 = -4\pi\rho c_p^4 \frac{1}{r_0} \tag{7.45b}$$

As expected, the coefficients of the asymptotic expansion $C_\infty, K_\infty, A_1, A_2$ determined from the consistent infinitesimal finite-element cell equation are identical to those of the analytical solution (Eqs. A.1.23a, A.1.23b, 7.40).

The consistent infinitesimal finite-element cell equation Eq. 7.32 is solved starting from the boundary condition evaluated at different frequencies ω_h

$$S^\infty(\omega_h) \approx i\omega_h C_\infty + K_\infty + \sum_{j=1}^{m} \frac{1}{(i\omega_h)^j} A_j \tag{7.46}$$

The analysis is performed for Poisson's ratio $\nu = 0.25$ using the Bulirsch-Stoer method described in Reference [40], starting at $\omega_h = a_{0h}c_p/r_0$ for $a_{0h} = 3$ and $= 6$. The dynamic-stiffness coefficient decomposed as in Eq. A.0.3 is plotted in Figure 7-3. The (small) error of the boundary condition decreases for decreasing a_0. Even for $a_{0h} = 3$ with the boundary condition $K^\infty(0.333 + i2)$ (the analytical solution equals $K^\infty(0.325 + i2.025)$), the results are highly accurate.

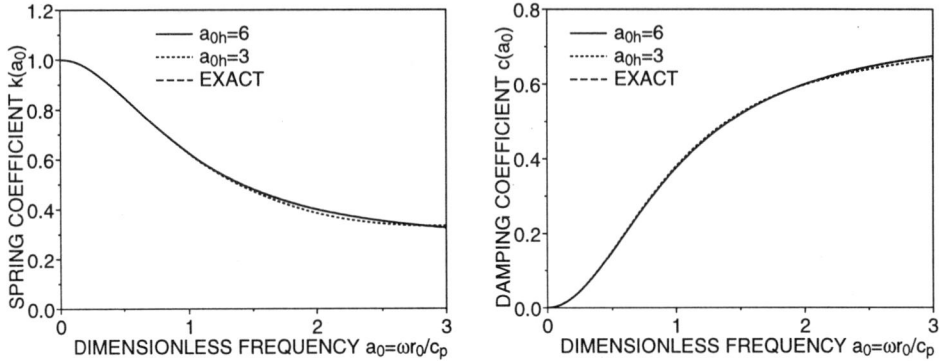

Figure 7-3 Dynamic-stiffness coefficient of spherical cavity calculated with consistent infinitesimal finite-element cell method for different starting values

7.4.2 Circular Cavity Embedded in Full-plane

The in-plane motion of a circular cavity embedded in a full-plane described in Appendix A.5 (Figure A-14) is addressed. The analytical expression for the dynamic-stiffness coefficient in the frequency domain is specified in Eq. A.5.22 with Eq. A.5.19. Poisson's ratio is selected as $\nu = 1/3$.

Owing to symmetry only one quarter of the structure-medium interface is discretized with 4 3-node line elements of equal length. The boundary condition of the dynamic-stiffness matrix $[S^\infty(\omega_h)]$ is determined from the asymptotic expansion with $m = 2$ (Eq. 7.2) at $\omega_h = a_{0h}c_s/r_0$ for $a_{0h} = 6$. $[S^\infty(\omega)]$ of order 16×16 is calculated for decreasing ω with the consistent infinitesimal finite-element cell method based on the Bulirsch-Stoer method of Reference [40]. The dynamic-stiffness coefficient $S^\infty(\omega)$ follows from Eq. A.0.2 with the vector $\{\phi\}$ determined from the translational motion of the rigid structure-medium interface (Eq. A.5.5). $S^\infty(\omega)$ non-dimensionalized with the shear modulus G is decomposed in $k(a_0)$ and $c(a_0)$. Excellent agreement with the analytical solution of Eq. A.5.22 results (Figure 7-4).

7.4.3 Prism Embedded in Half-space

The three-dimensional motion of a prism embedded in a half-space for the vector wave equation addressed in Appendix A.7 (Figure A-17) is calculated. The embedment ratio $e/b = 2/3$ is selected. The isotropic homogeneous half-space with a common shear modulus G and Poisson's ratio $\nu = 1/3$ for the 2 zones is analysed. The finite-element discretization of one quarter of the structure-medium interface is shown in Figure 5-26. The dynamic-stiffness matrix $[S^\infty(\omega)]$ is calculated for decreasing ω using the consistent infinitesimal finite-element cell method based on the Bulirsch-Stoer method of Reference [40], starting from the boundary condition $[S^\infty(\omega_h)]$ determined from the asymptotic expansion with $m = 1$ (Eq. 7.2) at $\omega_h =$

7.4 ACCURACY

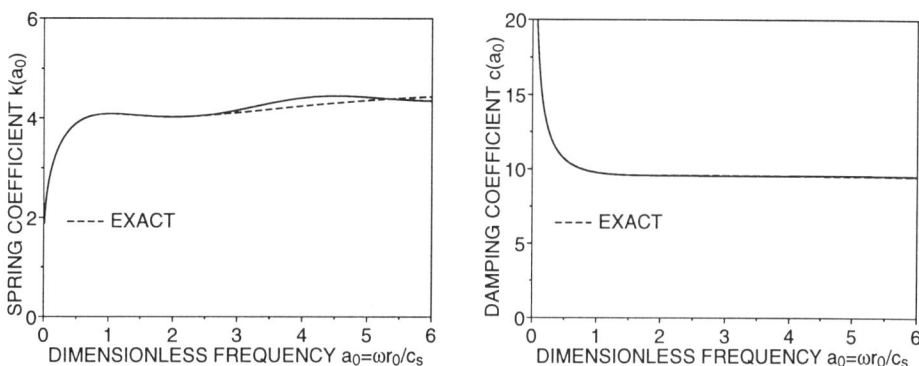

Figure 7-4 Dynamic-stiffness coefficient of circular cavity calculated with consistent infinitesimal finite-element cell method

$a_{0h}c_s/r_0$ for $a_{0h} = 30$. A rigid interface is introduced. $\{\phi_h\}$, $\{\phi_v\}$, $\{\phi_r\}$ and $\{\phi_t\}$ correspond to the motion patterns of the nodes on the structure-medium interface associated with the horizontal, vertical, rocking and torsional motions, respectively. The dynamic-stiffness coefficients $S_h^\infty(\omega)$, $S_v^\infty(\omega)$, $S_r^\infty(\omega)$ and $S_t^\infty(\omega)$ are calculated from Eq. A.0.2. After non-dimensionalization with the static-stiffness coefficients, a decomposition in the dimensionless spring coefficients and damping coefficients is performed as specified in Eq. A.0.3. The dynamic-stiffness coefficients plotted in Figure 7-5 are compared with the results of the boundary-element method for the horizontal, vertical and rocking degrees of freedom in the frequency domain [15] and with those of the hybrid method for the torsional degree of freedom [35].

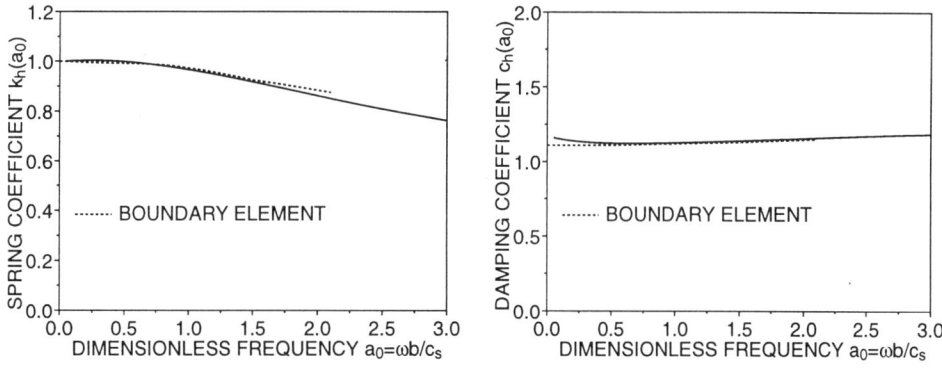

(a) horizontal

Figure 7-5 Dynamic-stiffness coefficients of rigid prism calculated with consistent infinitesimal finite-element cell method

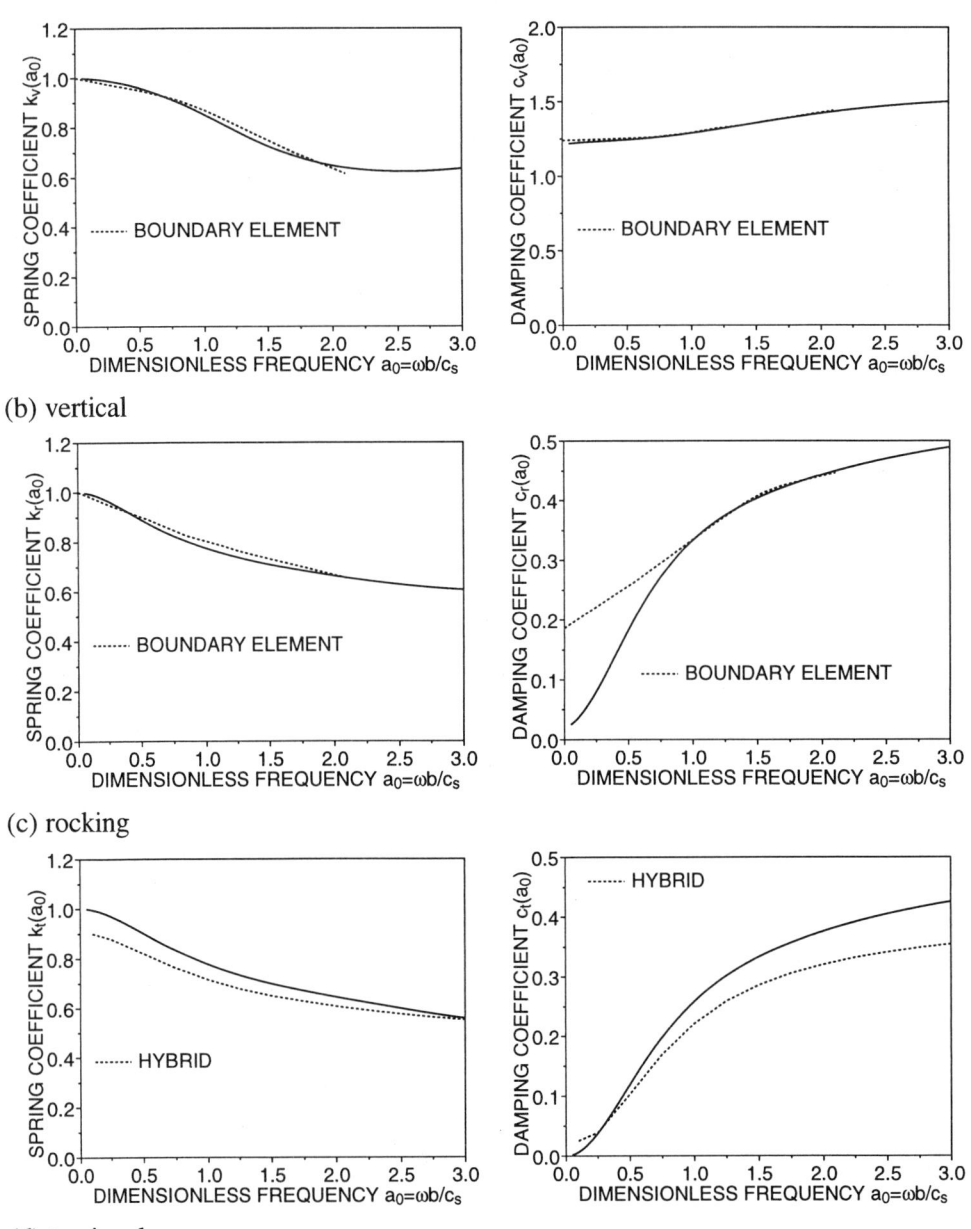

(b) vertical

(c) rocking

(d) torsional

Figure 7-5 Dynamic-stiffness coefficients of rigid prism calculated with consistent infinitesimal finite-element cell method

8

CONSISTENT INFINITESIMAL FINITE-ELEMENT CELL METHOD FOR STATICS

The consistent infinitesimal finite-element cell method for statics is contained as a special case in that for dynamics. It is, in principle, sufficient to set $\omega = 0$ or $t \to \infty$. However, a derivation of the key equations for the static case without the variable ω or t is simpler. It is also straightforward to derive explicit expressions for the displacements and strains at internal points in the unbounded medium. The advantages of the consistent infinitesimal finite-element cell method examined in Section 1.5.1 of course also apply to the static case. The fundamental equations are summarized in Section 8.1. Displacements and strains at internal points are determined in Section 8.2. The accuracy is examined in Section 8.3. Sections 8.1 and 8.3 are based on Reference [69].

8.1 SUMMARY OF FUNDAMENTAL EQUATIONS

The consistent infinitesimal finite-element cell method is based on two relationships of the static-stiffness matrices at similar structure-medium interfaces of an unbounded medium. One is based on similarity and the other on equilibrium and compatibility of the assemblage of finite elements.

The relationship based on similarity (Figure 3-1) is verified as follows. The static-stiffness matrix $[K^\infty(r)]$ is proportional to the shear modulus G and is a function of the characteristic length r and Poisson's ratio ν which is dimensionless. Examining the dimensions yields

$$[K^\infty(r)] = Gr^{s-2}[\overline{K}^\infty] \tag{8.1}$$

with the spatial dimension s ($= 2$ or $= 3$). The dimensionless matrix $[\overline{K}^\infty]$ is a function of ν only. Eq. 8.1 also follows from Eq. 3.6 for $\omega = 0$. The static-stiffness matrices of the unbounded medium at two similar structure-medium interfaces characterized by

r_i (interior) and r_e (exterior) are calculated from Eq. 8.1. Eliminating $[\overline{K}^\infty]$ leads to

$$[K_e^\infty] = \left(\frac{r_e}{r_i}\right)^{s-2} [K_i^\infty] \tag{8.2}$$

Eq. 8.2 corresponds to Eq. 3.8.

The relationship based on the assemblage of the finite-element cell and the unbounded medium (Figure 5-4) is derived for dynamics in Section 5.2.7. For statics, the static stiffness replaces the dynamic stiffness which results in (Eq. 5.99)

$$\begin{bmatrix} [K_{ii}] & [K_{ie}] \\ [K_{ei}] & [K_{ee}] \end{bmatrix} \begin{Bmatrix} \{u_i\} \\ \{u_e\} \end{Bmatrix} = \begin{bmatrix} [K_i^\infty] & 0 \\ 0 & -[K_e^\infty] \end{bmatrix} \begin{Bmatrix} \{u_i\} \\ \{u_e\} \end{Bmatrix} \tag{8.3}$$

and (Eq. 5.101)

$$[K_i^\infty] = [K_{ii}] - [K_{ie}]([K_e^\infty] + [K_{ee}])^{-1}[K_{ei}] \tag{8.4}$$

The limit of infinitesimal cell width is performed analytically as described in Section 5.2.8 with the static-stiffness matrix of the cell instead of the dynamic-stiffness matrix. Eq. 8.4 is then formulated as (Eq. 5.110)

$$([K_e^\infty] + [E^1])[E^0]^{-1}([K_i^\infty] + [E^1]^T) - \frac{[K_e^\infty] - [K_i^\infty]}{w} - [E^2] = O(w) \tag{8.5}$$

The coefficient matrices $[E^0]$, $[E^1]$ and $[E^2]$ for three-dimensional elasticity are defined in Eq. 5.22. The dimensionless cell width equals $w = r_e/r_i - 1$. To perform the limit of the second term of Eq. 8.5 for $w \to 0$, Eq. 8.2 is substituted leading to

$$\lim_{w \to 0} \frac{[K_e^\infty] - [K_i^\infty]}{w} = \lim_{w \to 0} \frac{(1+w)^{s-2} - 1}{w}[K_i^\infty] = (s-2)[K^\infty] \tag{8.6}$$

with $[K^\infty] = [K_i^\infty]$. The limit of Eq. 8.5 with $[K_e^\infty] = [K^\infty] + O(w)$ results in

$$([K^\infty] + [E^1])[E^0]^{-1}([K^\infty] + [E^1]^T) - (s-2)[K^\infty] - [E^2] = 0 \tag{8.7}$$

which is the *consistent infinitesimal finite-element cell equation for statics*. It corresponds to Eq. 5.113 in the frequency domain. Eq. 8.7 can also be written as

$$[K^\infty][E^0]^{-1}[K^\infty] + \left([E^1][E^0]^{-1} - \frac{s-2}{2}[I]\right)[K^\infty]$$
$$+ [K^\infty]\left([E^0]^{-1}[E^1]^T - \frac{s-2}{2}[I]\right) - [E^2] + [E^1][E^0]^{-1}[E^1]^T = 0 \tag{8.8}$$

This is the algebraic Riccati equation with the unknown static-stiffness matrix $[K^\infty]$. Such an equation is also encountered in the first time step of the consistent

8.2 DISPLACEMENT AND STRAIN AT INTERNAL POINT

infinitesimal finite-element cell method in the time domain (Section 5.3.1). The solution procedure for the Riccati equation is outlined in Eqs. 5.177 to 5.181.

In the three-dimensional case $[K^\infty]$ is positive definite. In the two-dimensional case, as the static-stiffness coefficients vanish for the two translational directions, $[K^\infty]$ is semi-positive definite. To facilitate the solution in the two-dimensional case, $[E^2]$ in Eq. 8.8 is replaced by $[E^2] + \varepsilon[I]$ where ε is a very small constant with the dimension of G. The matrix $[K^\infty]$ will then be positive definite. Physically, this operation corresponds to connecting a flexible support with a very small spring constant to each node on the structure-medium interface. This is verified as follows. Distributed springs with a spring constant per (dimensionless) length in the radial direction 0.5ε are assumed to act, resulting in the additional terms $0.5w\varepsilon[I]$ in $[K_{ii}]$ and $[K_{ee}]$. This corresponds in Eq. 5.20 to additional terms $0.5\varepsilon[I]$ in $[K_{ii}^2]$ and $[K_{ee}^2]$. It follows from Eq. 5.37c that the term $\varepsilon[I]$ is added to $[E^2]$.

For incompressible elasticity in statics the limit of Poisson's ratio $\nu = 0.5$ is performed as in Section 6.1. This yields

$$[k^\infty] \begin{bmatrix} [I] & \\ & 0 \end{bmatrix} [k^\infty] + [e^1][k^\infty] + [k^\infty][e^1]^T + 3[k^\infty] - [e^2] = 0 \tag{8.9}$$

which corresponds to setting $\omega = 0$ in Eq. 6.22. The coefficient matrices are defined in Section 6.1. The equation for the static-stiffness matrix is derived from Eq. 6.17

$$[K^\infty] = ([\Phi]^{-1})^T [k^\infty][\Phi]^{-1} \tag{8.10}$$

8.2 DISPLACEMENT AND STRAIN AT INTERNAL POINT

For statics the displacements and strains at internal points of the unbounded medium, located at any finite distance from the structure-medium interface, are calculated concisely after solving the Riccati equation. Conceptually, a surface finite-element mesh at the characteristic length r is introduced which is similar to that on the structure-medium interface. This permits the calculation of the displacements in the nodes.

For the known static-stiffness matrix $[K^\infty]$ the first set of equations in Eq. 8.3 is written as

$$[K_{ii}]\{u_i\} + [K_{ie}]\{u_e\} = [K_i^\infty]\{u_i\} \tag{8.11}$$

For an infinitesimal cell width

$$\{u_e\} = \{u_i\} + \{u\}_{,r}|_{r=r_i} dr \tag{8.12}$$

applies with $dr = wr_i$ (Eq. 5.1). Substituting Eq. 8.12 in Eq. 8.11 yields

$$([K_{ii}] + [K_{ie}])\{u_i\} + [K_{ie}]dr\{u\}_{,r}|_{r=r_i} = [K_i^\infty]\{u_i\} \tag{8.13}$$

Substituting Eqs. 5.37b and 5.20 with Eq. 5.37a in Eq. 8.13 and performing the limit $w \to 0$ leads to

$$r\{u\}_{,r} = -[E^0]^{-1}([K^\infty] + [E^1]^T)\{u\} \qquad (8.14)$$

with the subscript i omitted. This is a differential equation in $\{u\}$ with the coefficients $[E^0]$, $[E^1]$ and $[K^\infty]$ being functions of r. The latter are non-dimensionalized as

$$[\bar{E}^0] = \frac{1}{Gr^{s-2}}[E^0] \qquad (8.15a)$$

$$[\bar{E}^1] = \frac{1}{Gr^{s-2}}[E^1] \qquad (8.15b)$$

$$[\bar{E}^2] = \frac{1}{Gr^{s-2}}[E^2] \qquad (8.15c)$$

$$[\bar{K}^\infty] = \frac{1}{Gr^{s-2}}[K^\infty] \qquad (8.15d)$$

Substituting Eq. 8.15 in Eq. 8.14 results in

$$r\{u\}_{,r} = -[\bar{E}^0]^{-1}([\bar{K}^\infty] + [\bar{E}^1]^T)\{u\} \qquad (8.16)$$

where the coefficient matrices on the right-hand side are independent of r.

Substituting Eq. 8.15 in Eq. 8.8 yields

$$[\bar{K}^\infty][\bar{E}^0]^{-1}[\bar{K}^\infty] + \left([\bar{E}^1][\bar{E}^0]^{-1} - \frac{s-2}{2}[I]\right)[\bar{K}^\infty]$$

$$+ [\bar{K}^\infty]\left([\bar{E}^0]^{-1}[\bar{E}^1]^T - \frac{s-2}{2}[I]\right) - [\bar{E}^2] + [\bar{E}^1][\bar{E}^0]^{-1}[\bar{E}^1]^T = 0 \qquad (8.17)$$

Note that Eq. 8.17 is independent of r and applies to any interface with the characteristic length r within the unbounded medium (Eq. 8.15d).

For an elegant derivation the solution of the Riccati equation Eq. 8.17 is expanded slightly. Instead of the Schur decomposition described in Reference [26] the eigenvalue problem is solved

$$\begin{bmatrix} -[\bar{E}^0]^{-1}[\bar{E}^1]^T + \frac{s-2}{2}[I] & -[\bar{E}^0]^{-1} \\ -[\bar{E}^2] + [\bar{E}^1][\bar{E}^0]^{-1}[\bar{E}^1]^T & [\bar{E}^1][\bar{E}^0]^{-1} - \frac{s-2}{2}[I] \end{bmatrix} \begin{bmatrix} [\Phi_{11}] & [\Phi_{12}] \\ [\Phi_{21}] & [\Phi_{22}] \end{bmatrix}$$

$$= \begin{bmatrix} [\Phi_{11}] & [\Phi_{12}] \\ [\Phi_{21}] & [\Phi_{22}] \end{bmatrix} \begin{bmatrix} \lfloor\lambda\rfloor & \\ & -\lfloor\lambda\rfloor \end{bmatrix} \qquad (8.18)$$

with the eigenvalues $\lfloor\lambda\rfloor$ and the eigenvectors $[\Phi]$. The real part of λ_i is smaller than zero ($\text{Re}(\lambda_i) < 0$). Analogously to Eq. 5.181 the dimensionless static-stiffness matrix equals

$$[\bar{K}^\infty] = [\Phi_{21}][\Phi_{11}]^{-1} \qquad (8.19)$$

8.2 DISPLACEMENT AND STRAIN AT INTERNAL POINT

The first set of equation in Eq. 8.18 is written as

$$-[\bar{E}^0]^{-1}\left([\bar{K}^\infty]+[\bar{E}^1]^T\right) = [\Phi_{11}]\lceil\lambda\rfloor[\Phi_{11}]^{-1} - \frac{s-2}{2}[I] \quad (8.20)$$

Substituting Eq. 8.20 in Eq. 8.16 results in

$$r\{u\}_{,r} = \left([\Phi_{11}]\lceil\lambda\rfloor[\Phi_{11}]^{-1} - \frac{s-2}{2}[I]\right)\{u\} \quad (8.21)$$

Transforming the displacements $\{u\}$ to $\{\bar{u}\}$ as

$$\{\bar{u}\} = [\Phi_{11}]^{-1}\{u\} \quad (8.22)$$

yields from Eq. 8.21

$$r\{\bar{u}\}_{,r} = \left(\lceil\lambda\rfloor - \frac{s-2}{2}[I]\right)\{\bar{u}\} \quad (8.23)$$

This system of equations is decomposed as

$$r\bar{u}_{i,r} = \left(\lambda_i - \frac{s-2}{2}\right)\bar{u}_i \quad (8.24)$$

Its solution equals

$$\bar{u}_i = c_i r^{\lambda_i - \frac{s-2}{2}} \quad (8.25)$$

with the integration constant c_i, leading to

$$\{\bar{u}\} = \left[r^{\lambda_i - \frac{s-2}{2}}\right]\{c\} \quad (8.26)$$

The internal displacements $\{u\}$ follow from Eqs. 8.22 and 8.26 as

$$\{u\} = [\Phi_{11}]\left[r^{\lambda_i - \frac{s-2}{2}}\right]\{c\} \quad (8.27)$$

The internal interaction forces are equal to

$$\{R\} = [K^\infty]\{u\} \quad (8.28)$$

Substituting Eqs. 8.15d, 8.19 and 8.27 in Eq. 8.28 results in

$$\{R\} = G[\Phi_{21}]\left[r^{\lambda_i + \frac{s-2}{2}}\right]\{c\} \quad (8.29)$$

$\{c\}$ follows from the boundary condition enforced at the structure-medium interface with the characteristic length r_0.

$$\{u(r=r_0)\} = \{u_0\} \quad (8.30)$$

Formulating Eq. 8.27 at the structure-medium interface yields

$$\{c\} = \left[r_0^{-\left(\lambda_i - \frac{s-2}{2}\right)} \right] [\Phi_{11}]^{-1} \{u_0\} \tag{8.31}$$

Substituting Eq. 8.31 in Eq. 8.27 and in Eq. 8.29 leads to the internal displacements

$$\{u\} = [\Phi_{11}] \left[\left(\frac{r}{r_0}\right)^{\lambda_i - \frac{s-2}{2}} \right] [\Phi_{11}]^{-1} \{u_0\} \tag{8.32}$$

and the internal interaction forces

$$\{R\} = Gr^{s-2}[\Phi_{21}] \left[\left(\frac{r}{r_0}\right)^{\lambda_i - \frac{s-2}{2}} \right] [\Phi_{11}]^{-1} \{u_0\} \tag{8.33}$$

As the differential equations in the radial direction are solved analytically (Eq. 8.27), the displacements at any radial coordinate r can be determined directly. No additional approximations are introduced.

Stresses at internal points are also of interest. The stresses on the structure-medium interface can be important and are calculated as a special case. They follow straightforwardly from the strains.

In the finite-element method the strains $\{\hat{\varepsilon}\}$ within an element are calculated from the nodal displacements $\{\hat{u}\}$ as

$$\{\hat{\varepsilon}\} = [B]\{\hat{u}\} = [\,[B_i]\ [B_e]\,] \left\{ \begin{array}{c} \{u_i\} \\ \{u_e\} \end{array} \right\} \tag{8.34}$$

with the strain-nodal displacement matrix $[B]$ (Eq. 5.16). The circumflex $\hat{}$ denotes the three-dimensional finite element of the cell, and the subscripts i and e refer to the interior and exterior boundaries. $[B_i]$ and $[B_e]$ are calculated from Eq. 5.17 with $\xi_i = -1$ and $\xi_e = +1$, respectively. Substituting in Eq. 8.34 and using Eq. 8.12 yields

$$\{\hat{\varepsilon}\} = [B^1] r_i \{u\}_{,r}|_{r=r_i} + \frac{1}{2\left(1 + \frac{w}{2}(1+\xi)\right)} [B^2] \left(2\{u_i\} + (1+\xi)wr_i \{u\}_{,r}|_{r=r_i}\right) \tag{8.35}$$

with $[B^1]$ and $[B^2]$ defined in Eq. 5.15. Performing the limit of the dimensionless cell width $w \to 0$ leads to the strains $\{\varepsilon\}$ at the radial coordinate $r = r_i$

$$\{\varepsilon\} = \lim_{w \to 0} \{\hat{\varepsilon}\} = [B^1] r \{u\}_{,r} + [B^2]\{u\} \tag{8.36}$$

$\{u\}$ and $\{u\}_{,r}$ are calculated at the radial coordinate r as described in Eq. 8.32. Eq. 8.36 can then be applied for each (surface) element.

8.3 ACCURACY

8.3.1 Spherical Cavity Embedded in Full-space

The spherical cavity embedded in a full-space with a uniform normal displacement enforced on the structure-medium interface described in Appendix A.1 (Figure A-1) is addressed.

First, the one-dimensional problem is examined. The coefficients $[E^0]$, $[E^1]$ and $[E^2]$ are specified in Eq. 5.204. After substitution in Eq. 8.8 the quadratic equation in the static-stiffness coefficient K^∞ equals

$$\frac{1}{8\pi}\frac{1-2\nu}{1-\nu}\frac{1}{Gr_0}(K^\infty)^2 - \frac{1-5\nu}{1-\nu}K^\infty - 16\pi\frac{1+\nu}{1-\nu}Gr_0 = 0 \tag{8.37}$$

The two solutions are

$$K_1^\infty = 16\pi Gr_0 \tag{8.38a}$$

$$K_2^\infty = -8\pi\frac{1+\nu}{1-2\nu}Gr_0 \tag{8.38b}$$

The positive K_1^∞ is selected which represents the exact solution (Eq. A.1.17).

To calculate the internal displacement, the eigenvalue problem of Eq. 8.18 is solved

$$\begin{bmatrix} \dfrac{1-5\nu}{2(1-\nu)} & -\dfrac{1}{8\pi}\dfrac{1-2\nu}{1-\nu} \\ -16\pi\dfrac{1+\nu}{1-\nu} & -\dfrac{1-5\nu}{2(1-\nu)} \end{bmatrix}\begin{Bmatrix} \Phi_{1i} \\ \Phi_{2i} \end{Bmatrix} = \lambda_i \begin{Bmatrix} \Phi_{1i} \\ \Phi_{2i} \end{Bmatrix} \tag{8.39}$$

This yields the eigenvalues

$$\lambda_1 = -\frac{3}{2} \tag{8.40a}$$

$$\lambda_2 = -\lambda_1 \tag{8.40b}$$

and the eigenvectors

$$\begin{bmatrix} \Phi_{11} & \Phi_{12} \\ \Phi_{21} & \Phi_{22} \end{bmatrix} = \begin{bmatrix} 1 & 1 \\ 16\pi & -8\pi\dfrac{1+\nu}{1-2\nu} \end{bmatrix} \tag{8.41}$$

As expected from Eqs. 8.15d and 8.19

$$K^\infty = Gr\frac{\Phi_{21}}{\Phi_{11}} = 16\pi Gr \tag{8.42}$$

holds, which agrees with Eq. 8.38a. The internal displacement of the full-space on the outside of the structure-medium interface follows from Eq. 8.32 as

$$u = \left(\frac{r_0}{r}\right)^2 u_0 \tag{8.43}$$

This agrees with the exact solution in Eq. A.1.15 for $\omega = 0$.

To calculate the internal stresses, Eq. 8.36 is used to determine the strains with $\{B^1\}$ and $\{B^2\}$ at the coordinate r specified as (Eq. 5.202)

$$\{B^1\} = \frac{1}{r}\begin{Bmatrix} 1 \\ 0 \\ 0 \end{Bmatrix} \tag{8.44a}$$

$$\{B^2\} = \frac{1}{r}\begin{Bmatrix} 0 \\ 1 \\ 1 \end{Bmatrix} \tag{8.44b}$$

u and $ru_{,r}$ follow from Eq. 8.43. After substitution, the strains are equal to

$$\begin{Bmatrix} \varepsilon_r \\ \varepsilon_\theta \\ \varepsilon_\varphi \end{Bmatrix} = \frac{r_0^2}{r^3}\begin{Bmatrix} -2 \\ 1 \\ 1 \end{Bmatrix} u_0 \tag{8.45}$$

Again, the strains are exact as can be verified using Eq. A.1.2. The stresses follow from Eq. A.1.3.

Second, the static-stiffness coefficient K^∞ is calculated as a three-dimensional problem with the same material properties and spatial discretization of the structure-medium interface (Figure 5-7) as described in the second part of Section 5.4.1. The calculated K^∞ using the consistent infinitesimal finite-element cell method equals $15.74\pi G r_0$, which is very close to the analytical solution $K^\infty = 16\pi G r_0$ (Eq. A.1.17).

8.3.2 Semi-infinite Wedge

The out-of-plane motion of the semi-infinite wedge is described in Appendix A.4.1 (Figure A-11). The static case is investigated.

The analytical solution for statics is easily rederived. Eqs. A.4.1 to A.4.17 apply for $\omega = 0$. The solution of Eq. A.4.12b for $\omega = 0$ equals

$$R_i(r) = c_i r^{-\lambda_i} \tag{8.46}$$

where the boundary condition at infinity is enforced. The internal displacement is written as

$$u = \sum_{i=0}^{\infty} c_i r^{-\lambda_i} \sin \lambda_i \theta \tag{8.47}$$

The integration coefficients follow from the boundary condition (Eq. A.4.7) as

$$c_i = \frac{(-1)^i 2}{\alpha^2 \lambda_i^2 r_0^{-\lambda_i}} u_0 \tag{8.48}$$

8.3 ACCURACY

This yields

$$u = u_0 \frac{2}{\alpha^2} \sum_{i=0}^{\infty} \frac{(-1)^i}{\lambda_i^2} \left(\frac{r}{r_0}\right)^{-\lambda_i} \sin \lambda_i \theta \qquad (8.49)$$

The shear stresses follow from Eqs. A.4.1 and A.4.2 as

$$\tau_{zr} = G u_0 \frac{2}{\alpha^2 r_0} \sum_{i=0}^{\infty} \frac{(-1)^{i+1}}{\lambda_i} \left(\frac{r}{r_0}\right)^{-\lambda_i - 1} \sin \lambda_i \theta \qquad (8.50a)$$

$$\tau_{z\theta} = G u_0 \frac{2}{\alpha^2 r_0} \sum_{i=0}^{\infty} \frac{(-1)^i}{\lambda_i} \left(\frac{r}{r_0}\right)^{-\lambda_i - 1} \cos \lambda_i \theta \qquad (8.50b)$$

The semi-infinite wedge with the opening angle $\alpha = 30°$ is calculated using the consistent infinitesimal finite-element cell method. The structure-medium interface is discretized with 10 3-node line elements of equal length. The static-stiffness matrix $[K^\infty]$ of order 20×20 is determined. A linear variation of the displacement on the structure-medium interface is enforced (Eq. A.4.7), which determines the vector $\{\phi\}$. The static-stiffness coefficient follows from Eq. A.0.2 as $K^\infty = 0.5427G$ (Eq. A.4.31). The non-dimensionalized displacement u and shear stresses τ_{zx} and τ_{zy} are plotted as functions of the circumferential coordinate θ and radial coordinate r/r_0 on the left-hand side of Figure 8-1. On the right-hand side the error determined as the difference of the results of the consistent infinitesimal finite-element cell method and of the analytical solution of Eq. 8.49 and of Eq. 8.50 transformed into the x-y coordinate system is represented. While the displacement is highly accurate, the stress concentration at the corner ($\theta = 30°, r/r_0 = 1$) leads to deviations in the stresses which are, however, limited to the first two elements on the structure-medium interface.

8.3.3 Hemispherical Foundation Embedded in Half-space

A rigid hemispherical foundation with radius r_0 embedded in a half-space with shear modulus G is examined. Because of symmetry, only a quarter is analysed with the same finite-element discretization of the structure-medium interface as for an octant of the spherical cavity (Figure 5-7). After enforcing the rigid-body constraint, the torsional static-stiffness coefficient equals $K_t^\infty = 3.96\pi G r_0^3$ which agrees well with the analytical solution $4\pi G r_0^3$ of Reference [28].

8.3.4 Prism Embedded in Half-space

As a truly three-dimensional problem, a square prism of length $2b$ embedded with depth e in a half-space described in Appendix A.7 (Figure A-17) is addressed. Three embedment ratios $e/b = 2/3$, $=1$ and $=2$ are examined. Besides the homogeneous half-space, an inhomogeneous half-space compatible with similarity is also analysed. The

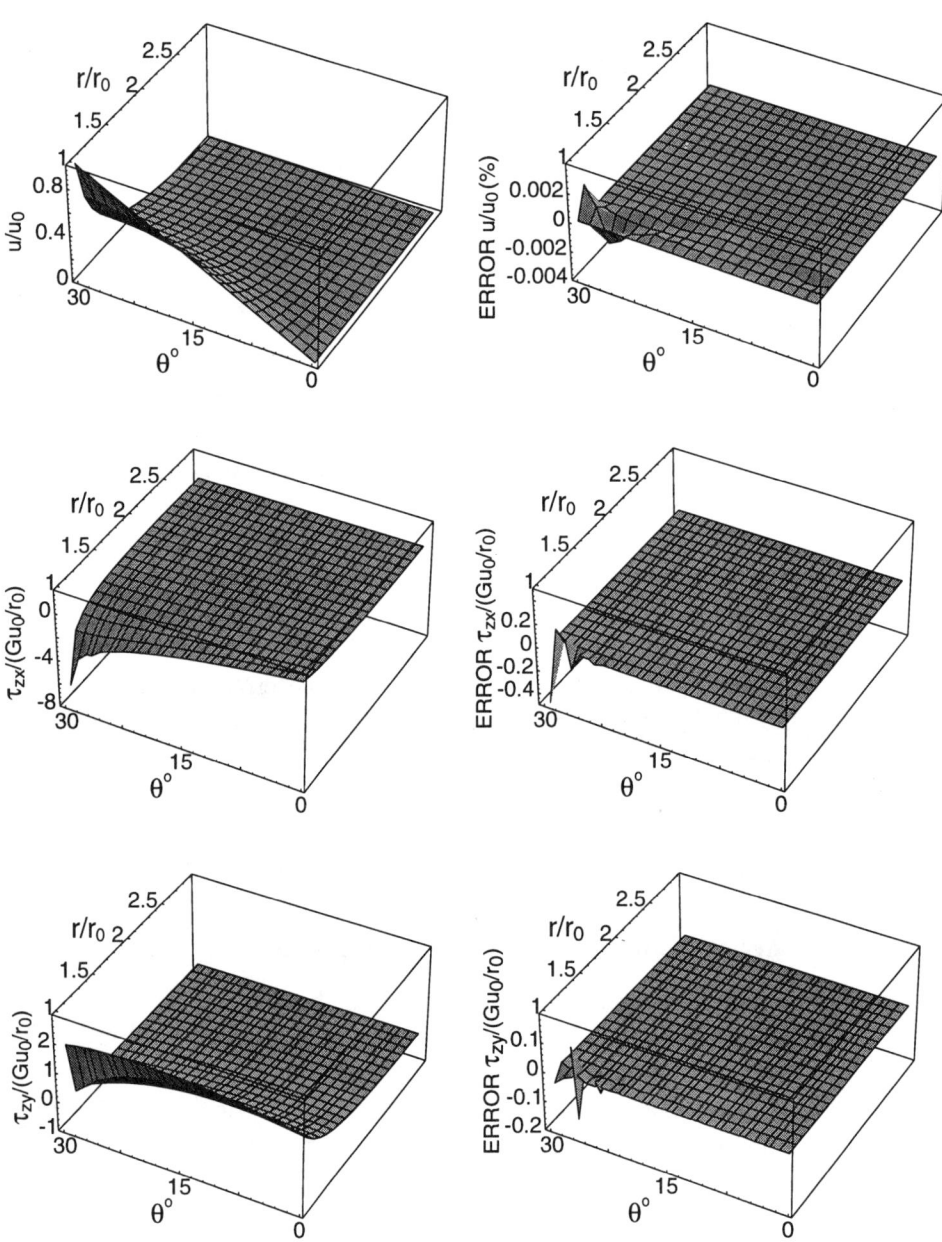

Figure 8-1 Displacement and shear stresses at internal points of semi-infinite wedge

8.3 ACCURACY

Table 8.1 Convergence of static-stiffness coefficients of rigid square prism embedded in homogenous half-space with embedment ratio $e/b = 2/3$ and Poisson's ratio $\nu = 1/3$

	FINITE-ELEMENT DISCRETIZATION PER FACE				REFERENCE [15]
	1×1	2×2	3×3	4×4	
$K_h^\infty/(Gb)$	10.07	9.77	9.71	9.69	9.50
$K_v^\infty/(Gb)$	10.04	9.42	9.31	9.27	9.20
$K_r^\infty/(Gb^3)$	17.60	15.39	14.98	14.82	14.16
$K_t^\infty/(Gb^3)$	27.66	25.51	25.00	24.78	24.23

structure-medium interface is first discretized with finite elements. The corresponding static-stiffness matrix is then established. A rigid interface is introduced. $\{\phi_h\}$, $\{\phi_v\}$, $\{\phi_r\}$ and $\{\phi_t\}$ correspond to the displacement patterns of the nodes on the structure-medium interface associated with the horizontal, vertical, rocking and torsional degrees of freedom, respectively. The static-stiffness coefficients, K_h^∞, K_v^∞, K_r^∞ and K_t^∞ are then calculated from Eq. A.0.2. Owing to symmetry, only a quarter of the foundation is analysed. A typical finite-element discretization of the three faces with 3×3 finite elements per face (for $e/b = 2/3$ and $=1$) is shown in Figure 5-26.

The static-stiffness coefficients are calculated for the homogeneous half-space for $e/b = 2/3$ and $\nu = 1/3$. The behaviour refining the finite-element mesh is studied (Table 8.1). The results seem to converge to those of Eq. 7.7 of Reference [15] which are determined using curve fitting of boundary-element results. Note that convergence for the consistent infinitesimal finite-element cell method is monotonic from the "stiff side".

Excellent accuracy is also achieved for the larger embedment ratios (Table 8.2). Note that for $e/b = 2$ the base and the two side faces are discretized with 3×3 and 3×6 finite elements, respectively.

The incompressible case is also calculated (Table 8.3).

The case of an inhomogeneity compatible with similarity, with the shear modulus G_1 of the unbounded soil adjacent to the walls and G_2 adjacent to the base (Figure A-17), is finally calculated using the discretization in Figure 5-26. Poisson's ratio $\nu = 1/3$ does not vary. The analysis is performed for $e/b = 2/3$ and $G_2/G_1 = 4$. The following static-stiffness coefficients result: $K_h^\infty = 19.18G_1b$, $K_v^\infty = 30.39G_1b$, $K_r^\infty = 37.30G_1b^3$, $K_t^\infty = 43.72G_1b^3$. As no results for this inhomogeneous case exist in the literature, an extended mesh consisting of 60 rows of finite elements with an aspect ratio of 1.08 extending to a distance of $101b$ where the boundary is fixed is analysed. The extended mesh leads to the following values: $K_h^\infty = 19.66G_1b$, $K_v^\infty = 31.34G_1b$, $K_r^\infty = 38.25G_1b^3$, $K_t^\infty = 44.39G_1b^3$.

Table 8.2 Static-stiffness coefficients of rigid square prism embedded in homogeneous half-space with embedment ratios $e/b = 1$, $e/b = 2$ and Poisson's ratio $v = 1/3$

	$e/b = 1$		$e/b = 2$	
	3×3, 3×3 3×3	REFERENCE [15]	3×3, 3×6 3×6	REFERENCE [15]
$K_h^\infty/(Gb)$	11.14	11.12	14.85	14.79
$K_v^\infty/(Gb)$	10.14	10.12	12.43	12.40
$K_r^\infty/(Gb^3)$	20.81	20.43	49.28	49.14
$K_t^\infty/(Gb^3)$	31.80	31.44	52.56	50.87

Table 8.3 Static-stiffness coefficients of rigid square prism embedded in homogeneous half-space with embedment ratios $e/b = 1$, $e/b = 2$ and Poisson's ratio $v = 1/2$

	$e/b = 1$		$e/b = 2$	
	3×3, 3×3 3×3	REFERENCE [15]	3×3, 3×6 3×6	REFERENCE [15]
$K_h^\infty/(Gb)$	12.52	12.36	16.74	16.44
$K_v^\infty/(Gb)$	12.34	13.49	14.72	16.53
$K_r^\infty/(Gb^3)$	24.54	27.24	56.69	65.52
$K_t^\infty/(Gb^3)$	34.36	31.44	55.96	50.87

8.3.5 Tunnel in Inhomogeneous Transversely Isotropic Unbounded Rock

As an example from actual practice, a railway tunnel intersecting a fault is addressed (Figure 8-2a). The geometry of the cross-section is specified in the figure. The fault (Zone C) exhibits isotropic behaviour with shear modulus $G^C = 0.1$ GPa and Poisson's ratio $v^C = 0.35$. The two zones of unbounded rock on both sides of the fault are modelled as transversely isotropic with a horizontal plane of isotropy (Appendix A.8). In Zone A: moduli of elasticity in the plane of isotropy $E_{hh}^A = 25$ GPa and perpendicular to it $E_{hv}^A = 15$ GPa, shear modulus in a plane perpendicular to the plane of isotropy $G_{hv}^A = 5$ GPa, Poisson's ratios in the plane of isotropy $v_{hh}^A = 0.2$ and perpendicular to it $v_{hv}^A = 0.15$; Region B: $E_{hh}^B = 5$ GPa, $E_{hv}^B = 3$ GPa, $G_{hv}^B = 1$ GPa, $v_{hh}^B = 0.2$, $v_{hv}^B = 0.15$. A plane strain condition is assumed. The railway loads result in concentrated forces P as shown in Figure 8-2a.

The centre of similarity is selected on the middle line of the fault. As similarity is not satisfied exactly, small deviations of the locations of the interfaces between the fault and the rock occur (dashed lines in Figure 8-2a). The surface is discretized with

8.3 ACCURACY

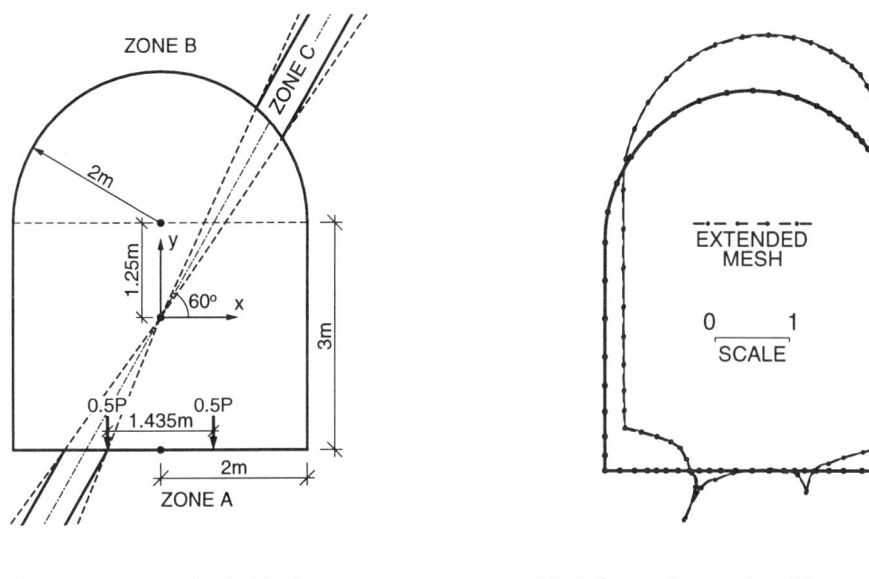

(a) geometry with similarity (b) deformed tunnel wall

Figure 8-2 Tunnel in inhomogeneous transversely isotropic unbounded rock

31 isoparametric 3-node line elements (Figure 8-2b). To determine the static-stiffness matrix $[K^\infty]$ from Eq. 8.8, $\varepsilon[I]$ is added to $[E^2]$ with $\varepsilon = 10^{-9} G_{hv}^A$. The corresponding displacements of the tunnel wall $\{u\}$ follow from

$$[K^\infty]\{u\} = \{R\} \tag{8.51}$$

with the load vector $\{R\}$ containing the two elements $P/2$. The displacements relative to the centre of the track, non-dimensionalized by P/G_{hv}^A, are plotted in Figure 8-2b. The results using an extended mesh consisting of 100 rows of finite element with the aspect ratio 1.06 in the radial direction which models the rock up to a distance of 679m from the tunnel are also presented. From a practical point of view they coincide with those of the consistent infinitesimal finite-element cell method shown as a solid line.

9

CONSISTENT INFINITESIMAL FINITE-ELEMENT CELL METHOD FOR DIFFUSION

The previous chapters (Chapters 2 to 8 of this Part on similarity-based formulations) address the *wave propagation* equation. In the time domain a *hyperbolic* partial differential equation is solved leading to the unit-impulse response function. In the frequency domain with statics as a special case an *elliptic* partial differential equation is processed resulting in the dynamic stiffness. This chapter examines the time- and frequency-domain formulations of the consistent infinitesimal finite-element cell method applied to the *diffusion* equation which is a *parabolic* partial differential equation in the time domain. The diffusion equation occurs in many fields of engineering. Heat conduction is used for illustration in this chapter. The concept described in Section 5.1 and the characteristics outlined in Section 1.5.1 still apply.

In contrast to the wave equation with a second derivative of the function with respect to time (see Eq. 5.41), the diffusion equation exhibits only a first derivative of the function with respect to time which leads to an instantaneous response over the entire medium. The spatial derivatives are in the same form as in the scalar wave equation. Thus, the spatial discretization for the diffusion equation will be the same as that for the scalar wave equation while the time discretization will differ. In principle, this Chapter 9 stands alone as far as is feasible and can thus be read without extensive study of Chapter 5. However, to avoid unnecessary duplication the spatial discretization is not repeated and precise references to sections and equations are provided.

The fundamental equations starting from the problem definition up to the consistent infinitesimal finite-element cell equation in the time domain for the diffusion equation are developed in Section 9.1. The time discretization is addressed in Section 9.2. The dynamic stiffness is calculated from the consistent infinitesimal finite-element cell equation in the frequency domain in Section 9.3. The accuracy is evaluated in Section 9.4. Sections 9.1, 9.2 and 9.4 are based on Reference [52].

9.1 FUNDAMENTAL EQUATIONS

9.1.1 Definition of Problem

The homogeneous diffusion equation in three dimensions is formulated as

$$q_{x,x} + q_{y,y} + q_{z,z} + \mu \dot{u} = 0 \tag{9.1}$$

$u = u(x,y,z,t)$ denotes the function to be determined, q_x, q_y, q_z are the components of the flux vector $\{q\}$, and μ is a material constant. $\{q\}$ is related to the gradient of u as

$$\{q\} = -[\kappa] \begin{Bmatrix} u_{,x} \\ u_{,y} \\ u_{,z} \end{Bmatrix} \tag{9.2}$$

with the anisotropic material matrix $[\kappa]$, which is positive definite. Besides vanishing initial conditions, boundary conditions are prescribed as

$$u = \bar{u} \qquad \text{on } S_u \tag{9.3a}$$
$$-q_n = \bar{q} \qquad \text{on } S_q \tag{9.3b}$$

$\bar{u}$ is the prescribed function on the surface S_u and $\bar{q}$ the prescribed normal flux on S_q.

For heat conduction u corresponds to the temperature, $\{q\}$ to the heat flux, $[\kappa]$ to the thermal conductivity and μ to the product of the mass density ρ and the specific heat c of the material.

The spatial discretization of the three-dimensional finite-element method leads to the static-stiffness matrix $[K]$ and mass matrix $[M]$ of an element of volume V

$$[K] = \int_V [B]^T [\kappa][B] dV \tag{9.4}$$

$$[M] = \int_V \mu [\hat{N}]^T [\hat{N}] dV \tag{9.5}$$

$[B]$ denotes the gradient-nodal function matrix and $[N]$ the shape function matrix.

For simplicity, the same symbols $[K]$ and $[M]$ are used for the assembled global system. The semi-discrete diffusion equation of the global system is formulated as

$$[M]\{\dot{u}(t)\} + [K]\{u(t)\} = \{P(t)\} \tag{9.6}$$

$\{u(t)\}$ denotes the vector of the nodal values of the function, and $\{P(t)\}$ is the nodal flux vector. For heat conduction, $\{u(t)\}$ represents the nodal temperature vector, $[K]$ is the thermal conductivity matrix, $[M]$ the capacity matrix and $\{P(t)\}$ denotes the nodal heat flux vector.

9.1 FUNDAMENTAL EQUATIONS

In the frequency domain, Eq. 9.6 is transformed to

$$[S(\omega)]\{u(\omega)\} = \{P(\omega)\} \tag{9.7}$$

with the dynamic-stiffness matrix

$$[S(\omega)] = [K] + i\omega[M] \tag{9.8}$$

9.1.2 Unit-impulse Response Matrices

The dynamic properties of an unbounded medium can be expressed in the nodes on the structure-medium interface by the response matrices to various unit impulses. In Chapter 2 the corresponding relationships for the wave equation are derived, linking the displacements, the velocities or the accelerations to the interaction forces. The key equations which are used in solving the diffusion equation are summarized.

The nodal fluxes on the structure-medium interface of the unbounded medium $\{R(t)\}$ are expressed as a function of the nodal values of the function $\{u(t)\}$. In the frequency domain their amplitudes are related as

$$\{R(\omega)\} = [S^\infty(\omega)]\{u(\omega)\} \tag{9.9}$$

with the dynamic-stiffness matrix in the frequency domain $[S^\infty(\omega)]$ (superscript ∞ for the unbounded medium). Eq. 9.9 corresponds to Eq. 2.1. In contrast to the wave equation no decomposition into the singular and regular parts as in Eq. 2.6 is performed, which simplifies the expressions. In the time domain this relationship is written as a convolution integral with the function $\{u(t)\}$ (Eq. 2.2)

$$\{R(t)\} = \int_0^t [S^\infty(t-\tau)]\{u(\tau)\}d\tau \tag{9.10}$$

$[S^\infty(t)]$ is the response matrix to a unit impulse of the function $\{u(t)\}$. $[S^\infty(t)]$ and $[S^\infty(\omega)]$ form a Fourier transform pair. Alternatively, Eq. 9.9 is reformulated as (Eq. 2.28)

$$\{R(\omega)\} = [V^\infty(\omega)]\{\dot{u}(\omega)\} \tag{9.11}$$

using the amplitudes of the time derivative of the nodal values of the function $\{\dot{u}(\omega)\} = i\omega\{u(\omega)\}$ with (Eq. 2.29)

$$[V^\infty(\omega)] = \frac{[S^\infty(\omega)]}{i\omega} \tag{9.12}$$

Transformed to the time domain, Eq. 9.11 is written as a convolution integral with the time derivative of the nodal values of the function $\{\dot{u}(t)\}$ (Eq. 2.26)

$$\{R(t)\} = \int_0^t [V^\infty(t-\tau)]\{\dot{u}(\tau)\}d\tau \tag{9.13}$$

$[V^\infty(t)]$ is the response matrix to a unit impulse of the time derivative of the function $\{\dot{u}(t)\}$ and is equal to the response matrix to a unit-step of $\{u(t)\}$. For short, $[V^\infty(t)]$ is called the velocity unit-impulse response matrix as in Section 2.3. $[V^\infty(t)]$ follows from Eq. 9.12 as (Eq. 2.38)

$$[V^\infty(t)] = \int_0^t [S^\infty(\tau)]d\tau \tag{9.14}$$

The descriptions of the unbounded medium using the unit-impulse response matrix $[S^\infty(t)]$ and the velocity unit-impulse response matrix $[V^\infty(t)]$ are equivalent. The consistent infinitesimal finite-element cell method calculates $[V^\infty(t)]$. The nodal flux $\{R(t)\}$ follows either directly from Eq. 9.13 or from Eq. 9.10, whereby

$$[S^\infty(t)] = [\dot{V}^\infty(t)] \tag{9.15}$$

applies (Eq. 9.14).

9.1.3 Dynamic Stiffness at Similar Structure-medium Interfaces

In the derivation of the consistent infinitesimal finite-element cell equation similar structure-medium interfaces of the unbounded medium are addressed. The only independent variable which fully defines a specific structure-medium interface is the characteristic length r (Figure 3-1). The procedure of Section 3.1 for the wave equation is modified for the diffusion equation.

To derive the relationship of the dynamic-stiffness matrices at similar interfaces for the diffusion equation, heat conduction is used as an example. The result is valid for all cases described by the diffusion equation. The nodal heat flux-nodal temperature relationship of an unbounded medium at similar structure-medium interfaces characterized by the variable r is discussed

$$\{R(\omega)\} = [S^\infty(r,\omega)]\{u(\omega)\} \tag{9.16}$$

To determine the independent dimensionless variables of which $[S^\infty(r,\omega)]$ is a function, a dimensional analysis is performed. The characteristic length r, the thermal conductivity κ, the product μ of the mass density ρ and the specific heat c of the material, and the frequency ω are sufficient to determine $[S^\infty(r,\omega)]$. A bracket denotes the dimension; $[L]$, $[M]$, $[T]$ and $[\mathcal{T}]$ are the dimensions of length, mass, time and temperature. The dimensions of the variables are

$$[[S^\infty]] = [L]^{s-1}[M][T]^{-2}[\mathcal{T}]^{-1} \tag{9.17a}$$
$$[r] = [L] \tag{9.17b}$$
$$[\kappa] = [L][M][T]^{-2}[\mathcal{T}]^{-1} \tag{9.17c}$$
$$[\mu] = [L]^{-1}[M][T]^{-1}[\mathcal{T}]^{-1} \tag{9.17d}$$
$$[\omega] = [T]^{-1} \tag{9.17e}$$

9.1 FUNDAMENTAL EQUATIONS

with the spatial dimension s ($= 2$ or $= 3$). The product of all these variables raised to unknown powers n_i ($i = 1, 2, \ldots, 5$) must be dimensionless.

$$[[S^\infty]]^{n_1} [r]^{n_2} [\kappa]^{n_3} [\mu]^{n_4} [\omega]^{n_5} =$$
$$[L]^{(s-1)n_1+n_2+n_3-n_4} [M]^{n_1+n_3+n_4} [T]^{-2n_1-2n_3-n_4-n_5} [T]^{-n_1-n_3-n_4} \quad (9.18)$$

This yields

$$(s-1)n_1 + n_2 + n_3 - n_4 = 0 \quad (9.19a)$$
$$n_1 + n_3 + n_4 = 0 \quad (9.19b)$$
$$-2n_1 - 2n_3 - n_4 - n_5 = 0 \quad (9.19c)$$
$$-n_1 - n_3 - n_4 = 0 \quad (9.19d)$$

The rank of the coefficient matrix of the system of equations (Eq. 9.19) with 5 unknowns equals 3, which permits the two unknowns n_1, n_5 to be chosen arbitrarily. For $n_1 = 1$, $n_5 = 0$, the other unknowns are $n_2 = 2 - s$, $n_3 = -1$, $n_4 = 0$, yielding the first dimensionless variable $[S^\infty] r^{2-s} \kappa^{-1}$. For $n_1 = 0$, $n_5 = 1$, the other unknowns are $n_2 = 2$, $n_3 = -1$, $n_4 = 1$, resulting in the second dimensionless variable $r^2 \kappa^{-1} \mu \omega$, which is the dimensionless frequency

$$a_0 = \frac{\omega \mu r^2}{\kappa} \quad (9.20)$$

The first dimensionless variable $[S^\infty] r^{2-s} \kappa^{-1}$ will be a function $[\overline{S}^\infty]$ of the second dimensionless variable a_0 yielding

$$[S^\infty(r, \omega)] = \kappa r^{s-2} [\overline{S}^\infty(a_0)] \quad (9.21)$$

The a_0 of the diffusion equation in Eq. 9.20 should be compared to that of the wave equation in Eq. 3.5. In Eq. 9.21, $[\overline{S}^\infty(a_0)]$ is a function of one independent variable, i.e. a_0. For constant ω, $[\overline{S}^\infty(a_0)]$ is a function of r, or just as valid, for constant r, $[\overline{S}^\infty(a_0)]$ is a function of ω. The same change in a_0 can be achieved by varying the values of either r or ω with the other fixed. The derivative thus follows for a constant ω varying r as

$$[\overline{S}^\infty(a_0)]_{,a_0} = \frac{\kappa}{2 \omega \mu r} [\overline{S}^\infty(a_0)]_{,r} \quad (9.22a)$$

and the same result is calculated for a constant r but varying ω as

$$[\overline{S}^\infty(a_0)]_{,a_0} = \frac{\kappa}{2 \mu r} [\overline{S}^\infty(a_0)]_{,\omega} \quad (9.22b)$$

Setting the two right-hand sides of Eq. 9.22 equal yields

$$r [\overline{S}^\infty(a_0)]_{,r} = 2\omega [\overline{S}^\infty(a_0)]_{,\omega} \quad (9.23)$$

The partial derivative of $[\overline{S}^\infty(a_0)]$ with respect to r can thus be replaced by that with respect to ω for similar interfaces. The partial derivatives with respect to r and ω are calculated from Eq. 9.21 as

$$[\overline{S}^\infty(a_0)]_{,r} = \frac{1}{\kappa r^{s-2}}\left(-\frac{s-2}{r}[S^\infty(r,\omega)] + [S^\infty(r,\omega)]_{,r}\right) \quad (9.24a)$$

$$[\overline{S}^\infty(a_0)]_{,\omega} = \frac{1}{\kappa r^{s-2}}[S^\infty(r,\omega)]_{,\omega} \quad (9.24b)$$

Substituting Eq. 9.24 in Eq. 9.23 leads to

$$r[S^\infty(r,\omega)]_{,r} = (s-2)[S^\infty(r,\omega)] + 2\omega[S^\infty(r,\omega)]_{,\omega} \quad (9.25)$$

Eq. 9.25 expresses the relationship of the dynamic-stiffness matrices of the unbounded medium at similar interfaces based on similarity. *Compared with the corresponding relationship for the wave equation (Eq. 3.12), the only difference is the factor 2 in the last term in Eq. 9.25, which results from r^2 instead of r appearing in a_0 (Eq. 9.20).* This relationship permits the partial derivative of $[S^\infty(r,\omega)]$ with respect to r to be replaced by that with respect to ω.

9.1.4 Coefficient Matrices of Finite-element Cell

In the derivation of the consistent infinitesimal finite-element cell equation starting from the mesh of surface finite elements on the structure-medium interface, a three-dimensional finite-element cell satisfying similarity with a single element in the radial direction is constructed (Section 5.2.1 for the vector wave equation). The static-stiffness matrix and the mass matrix of the finite-element cell are decomposed with respect to the dimensionless cell width. This defines the coefficient matrices $[E^0]$, $[E^1]$, $[E^2]$ and $[M^0]$. Comparing the definitions of the static-stiffness matrix $[K]$ (Eq. 9.4) and the mass matrix $[M]$ (Eq. 9.5) for the diffusion equation with those of the scalar wave equations (Eqs. 5.47 and 5.32), it becomes apparent that $[\kappa]$ replaces G and μ replaces ρ. This yields for the three-dimensional diffusion equation (Eqs. 5.49 and 5.50)

$$[E^0] = \int_{-1}^{+1}\int_{-1}^{+1}[B^1]^T[\kappa][B^1]|J|d\eta d\zeta \quad (9.26a)$$

$$[E^1] = \int_{-1}^{+1}\int_{-1}^{+1}[B^2]^T[\kappa][B^1]|J|d\eta d\zeta \quad (9.26b)$$

$$[E^2] = \int_{-1}^{+1}\int_{-1}^{+1}[B^2]^T[\kappa][B^2]|J|d\eta d\zeta \quad (9.26c)$$

and

$$[M^0] = \int_{-1}^{+1}\int_{-1}^{+1}\mu[N]^T[N]|J|d\eta d\zeta \quad (9.27)$$

9.1 FUNDAMENTAL EQUATIONS

[N] denotes the shape functions of the surface finite element, as described in Eq. 5.51 (see as an example Eq. 5.2). $|J|$ and $[B^1]$, $[B^2]$ are defined in Eqs. 5.9 and 5.46. For the two-dimensional diffusion equation (Eqs. 5.66 and 5.67)

$$[E^0] = \int_{-1}^{+1} [B^1]^T [\kappa][B^1]|J|d\eta \tag{9.28a}$$

$$[E^1] = \int_{-1}^{+1} [B^2]^T [\kappa][B^1]|J|d\eta \tag{9.28b}$$

$$[E^2] = \int_{-1}^{+1} [B^2]^T [\kappa][B^2]|J|d\eta \tag{9.28c}$$

and

$$[M^0] = \int_{-1}^{+1} \mu[N]^T [N]|J|d\eta \tag{9.29}$$

result.

9.1.5 Assemblage of Finite-element Cell and Unbounded Medium

The assemblage enforcing continuity of the function and of the nodal flux links the dynamic-stiffness matrix of the unbounded medium at the structure-medium interface coinciding with the interior boundary of the finite-element cell to that at the interface corresponding to the exterior boundary, as described for the wave equation in Section 5.2.7 (Figure 5-4).

Partitioning the nodal flux-nodal function relationship of the finite-element cell (Eq. 9.7) into the nodes lying on the interior (subscript i) and exterior (subscript e) boundaries yields

$$\begin{bmatrix} [S_{ii}(\omega)] & [S_{ie}(\omega)] \\ [S_{ei}(\omega)] & [S_{ee}(\omega)] \end{bmatrix} \begin{Bmatrix} \{u_i(\omega)\} \\ \{u_e(\omega)\} \end{Bmatrix} = \begin{Bmatrix} \{P_i(\omega)\} \\ \{P_e(\omega)\} \end{Bmatrix} \tag{9.30}$$

The dynamic-stiffness matrix $[S(\omega)]$ is defined in Eq. 9.8

The nodal flux-nodal function relationship of the unbounded medium at the interfaces corresponding to the interior and exterior boundaries is formulated as (Eq. 9.9)

$$\{R_i(\omega)\} = [S_i^\infty(\omega)]\{u_i(\omega)\} \tag{9.31a}$$

$$\{R_e(\omega)\} = [S_e^\infty(\omega)]\{u_e(\omega)\} \tag{9.31b}$$

Formulating the continuity of the nodal flux at the interior and exterior boundaries relates the nodal flux amplitudes of the unbounded medium to the nodal flux amplitudes of the cell

$$\{P_i(\omega)\} = \{R_i(\omega)\} \tag{9.32a}$$

$$\{P_e(\omega)\} = -\{R_e(\omega)\} \tag{9.32b}$$

Eliminating $\{P(\omega)\}$, $\{R(\omega)\}$ and $\{u_e(\omega)\}$ from Eqs. 9.30 to 9.32 yields for an arbitrary $\{u_i(\omega)\}$

$$[S_i^\infty(\omega)] = [S_{ii}(\omega)] - [S_{ie}(\omega)]([S_e^\infty(\omega)] + [S_{ee}(\omega)])^{-1}[S_{ei}(\omega)] \tag{9.33}$$

This represents the relationship of the dynamic-stiffness matrices of the unbounded medium at the interfaces coinciding with the interior and exterior boundaries of the finite-element cell based on finite-element assemblage.

9.1.6 Consistent Infinitesimal Finite-element Cell Equation in Frequency Domain

The consistent infinitesimal finite-element cell equation is derived using the relationship of the dynamic-stiffness matrices based on *similarity*, which is in the continuous form (Eq. 9.25), and the relationship based on *assemblage* (Eq. 9.33) for which the *limit of infinitesimal-cell width* must be performed analytically.

For the wave equation the analytical limit is enforced in Section 5.2.8. Eq. 5.101 is identical to Eq. 9.33 of the diffusion equation with the exception of the dynamic-stiffness matrix of the finite-element cell. Replacing $-\omega^2$ in $[S(\omega)]$ for the wave equation (Eq. 5.95) by $+i\omega$ yields the $[S(\omega)]$ for the diffusion equation (Eq. 9.8). The limit of the infinitesimal cell width $w \to 0$ can be performed identically as described in Eqs. 5.102 to 5.109 with $+i\omega$ replacing $-\omega^2$. This leads to (Eq. 5.110)

$$([S_e^\infty(\omega)] + [E^1])[E^0]^{-1}([S_i^\infty(\omega)] + [E^1]^T) - \frac{[S_e^\infty(\omega)] - [S_i^\infty(\omega)]}{w}$$
$$- [E^2] - i\omega[M^0] = O(w) \tag{9.34}$$

The limit of $w \to 0$ can now be performed. With $w = (r_e - r_i)/r_i$

$$\lim_{w \to 0} \frac{[S_e^\infty(\omega)] - [S_i^\infty(\omega)]}{w} = \lim_{r_e \to r_i} r_i \frac{[S_e^\infty(\omega)] - [S_i^\infty(\omega)]}{r_e - r_i} = r[S^\infty(\omega)]_{,r} \tag{9.35}$$

results with $[S^\infty(\omega)] = [S_i^\infty(\omega)]$ and $r = r_i$. The last term in Eq. 9.35 is expressed using Eq. 9.25 based on similarity. With $[S_e^\infty(\omega)] = [S_i^\infty(\omega)] + O(w)$, the limit of Eq. 9.34 equals

$$([S^\infty(\omega)] + [E^1])[E^0]^{-1}([S^\infty(\omega)] + [E^1]^T) - (s-2)[S^\infty(\omega)]$$
$$- 2\omega[S^\infty(\omega)]_{,\omega} - [E^2] - i\omega[M^0] = 0 \tag{9.36}$$

In an actual application a specific structure-medium interface is addressed which fixes r. The dynamic-stiffness matrix thus becomes a function of ω only. The partial derivative $[S^\infty(r,\omega)]_{,\omega}$ in Eq. 9.25 is replaced by $[S^\infty(\omega)]_{,\omega}$.

This represents the *consistent infinitesimal finite-element cell equation for the diffusion equation in the frequency domain*. It is a system of non-linear ordinary differential equations of first order. Its solution in the frequency domain is discussed in Section 9.3.

9.1 FUNDAMENTAL EQUATIONS

9.1.7 Consistent Infinitesimal Finite-element Cell Equation in Time Domain

The transformation of the consistent infinitesimal finite-element cell equation in the frequency domain to the time domain is analogous to that for the wave equation in Section 5.2.9.

As a formulation of the unbounded medium's nodal flux on the structure-medium interface based on a convolution integral of the time derivative of the function is to be determined (Eq. 9.13), the velocity unit-impulse response matrix $[V^\infty(t)]$ is calculated. The (symmetric) positive definite coefficient matrix $[E^0]$ is decomposed by Cholesky's method as

$$[E^0] = [U]^T[U] \tag{9.37}$$

where $[U]$ is an upper-triangular matrix. Substituting Eqs. 9.12 and 9.37 in Eq. 9.36, which is premultiplied by $([U]^{-1})^T$ and postmultiplied by $[U]^{-1}$, yields

$$[v^\infty(\omega)]^2 + [e^1]\frac{[v^\infty(\omega)]}{i\omega} + \frac{[v^\infty(\omega)]}{i\omega}[e^1]^T + 2i[v^\infty(\omega)]_{,\omega} - \frac{1}{(i\omega)^2}[e^2] - \frac{1}{i\omega}[m^0] = 0 \tag{9.38}$$

where

$$[v^\infty(\omega)] = ([U]^{-1})^T [V^\infty(\omega)][U]^{-1} \tag{9.39}$$

and the coefficient matrices are equal to

$$[e^1] = ([U]^{-1})^T [E^1][U]^{-1} - \frac{s}{2}[I] \tag{9.40a}$$

$$[e^2] = ([U]^{-1})^T ([E^2] - [E^1][E^0]^{-1}[E^1]^T)[U]^{-1} \tag{9.40b}$$

$$[m^0] = ([U]^{-1})^T [M^0][U]^{-1} \tag{9.40c}$$

Note that $[e^2]$ is symmetric, and $[m^0]$ is positive definite. Applying the inverse Fourier transformation to Eq. 9.38 results in the *consistent infinitesimal finite-element cell equation in the time domain*

$$\int_0^t [v^\infty(t-\tau)][v^\infty(\tau)]d\tau + [e^1]\int_0^t [v^\infty(\tau)]d\tau + \int_0^t [v^\infty(\tau)]d\tau[e^1]^T$$
$$+ 2t[v^\infty(t)] - t[e^2]H(t) - [m^0]H(t) = 0 \tag{9.41}$$

After determining $[v^\infty(t)]$ from Eq. 9.41, the velocity unit-impulse response matrix follows from Eq. 9.39 as

$$[V^\infty(t)] = [U]^T[v^\infty(t)][U] \tag{9.42}$$

9.2 TIME DISCRETIZATION

After discretization with respect to time, Eq. 9.41 yields an equation for the velocity unit-impulse response matrix at each time station n. It is assumed that the latter is piece-wise constant over each time step, i.e. $[v^\infty]_n$ applies at time $t = (n - 1/2)\Delta t$ ($n \geq 1$). As a consequence, the integral terms in Eq. 9.41 are discretized as

$$[I]_n = \int_0^{n\Delta t} [v^\infty(\tau)]d\tau = [I]_{n-1} + \Delta t [v^\infty]_n \tag{9.43}$$

$$\int_0^t [v^\infty(t-\tau)][v^\infty(\tau)]d\tau = \Delta t \sum_{j=1}^n [v^\infty]_{n-j+1}[v^\infty]_j \tag{9.44}$$

9.2.1 First Time Step

As in Eq. 9.41 the convolution-integral term leads to a *quadratic* equation in the unknown matrix for the first time step and a special treatment is necessary. Substituting Eqs. 9.43 and 9.44 with $n = 1$ in Eq. 9.41 yields

$$[v^\infty]_1^2 + ([e^1] + [I])[v^\infty]_1 + [v^\infty]_1([e^1]^T + [I]) - [e^2] - \frac{[m^0]}{\Delta t} = 0 \tag{9.45}$$

Eq. 9.45 is the algebraic Riccati equation. It has a unique symmetric semi-positive definite solution for $[v^\infty]_1$, which can be computed by the algorithm outlined in Section 5.3.1. The velocity unit-impulse response matrix at the first time station follows from Eq. 9.42 as

$$[V^\infty]_1 = [U]^T [v^\infty]_1 [U] \tag{9.46}$$

It is obvious that $[V^\infty]_1$ is symmetric semi-positive definite.

9.2.2 *n*th Time Step

For $n \geq 2$ the convolution-integral term in Eq. 9.41 results in *linear* terms in the unknown $[v^\infty]_n$. Discretizing Eq. 9.41 leads to the equation for the transformed velocity unit-impulse response matrix $[v^\infty]_n$

$$\left([v^\infty]_1 + [e^1]\right)[v^\infty]_n + [v^\infty]_n \left([v^\infty]_1 + [e^1]^T\right) + 2n[v^\infty]_n =$$
$$-\sum_{j=2}^{n-1} [v^\infty]_{n-j+1}[v^\infty]_j - \frac{1}{\Delta t}\left([e^1][I]_{n-1} + [I]_{n-1}[e^1]^T\right) + n[e^2] + \frac{[m^0]}{\Delta t} \tag{9.47}$$

Eq. 9.47 is the Lyapunov equation in the form

$$[A][X] + [X][A]^T + 2n[X] = [C] \tag{9.48}$$

9.3 DYNAMIC STIFFNESS

with $[X] = [v^\infty]_n$ and the other matrices defined straightforwardly. The solution procedure is described in Section 5.3.2.

To prove that $[v^\infty]_n$ is symmetric, the transpose of equation Eq. 9.47 is subtracted from Eq. 9.47 with the symmetric matrices $[e^2]$, $[m^0]$ and $[v^\infty]_j$, $(j = 1, \ldots, n-1)$ leading to

$$([v^\infty]_1 + [e^1])([v^\infty]_n - [v^\infty]_n^T) + ([v^\infty]_n - [v^\infty]_n^T)([v^\infty]_1 + [e^1]^T)$$
$$+ 2n([v^\infty]_n - [v^\infty]_n^T) = 0 \quad (9.49)$$

This linear equation is satisfied for $[v^\infty]_n = [v^\infty]_n^T$.

After determining $[v^\infty]_n$, the velocity unit-impulse response matrix is calculated (Eq. 9.42)

$$[V^\infty]_n = [U]^T [v^\infty]_n [U] \quad (9.50)$$

$[V^\infty]_n$ is also symmetric.

9.3 DYNAMIC STIFFNESS

This section addresses the calculation of the dynamic-stiffness matrix in the frequency domain. The consistent infinitesimal finite-element cell method is well suited to determining the dynamic-stiffness matrix for a large range of frequencies.

It is shown that *the consistent infinitesimal finite-element cell equation for the diffusion equation can be transformed using a simple change of variable to that for the wave equation* examined in Chapter 7. Thus, all procedures developed for the wave equation can also be directly applied for the diffusion equation.

For the diffusion equation the consistent infinitesimal finite-element cell equation in the frequency domain (Eq. 9.36) is repeated here for convenience

$$([S^\infty(\omega)] + [E^1])[E^0]^{-1}([S^\infty(\omega)] + [E^1]^T) - (s-2)[S^\infty(\omega)]$$
$$- 2\omega[S^\infty(\omega)]_{,\omega} - [E^2] - i\omega[M^0] = 0 \quad (9.51)$$

As already discussed in Section 9.1.6, $-\omega^2 = (i\omega)^2$ in the wave equation corresponds to $i\omega$ in the diffusion equation. This leads to the selection of the new variable Ω with $-\Omega^2 = i\omega$ or

$$\omega = i\Omega^2 \quad (9.52)$$

Substituting Eq. 9.52 in Eq. 9.51 yields

$$([S^\infty(\Omega)] + [E^1])[E^0]^{-1}([S^\infty(\Omega)] + [E^1]^T) - (s-2)[S^\infty(\Omega)]$$
$$- \Omega[S^\infty(\Omega)]_{,\Omega} - [E^2] + \Omega^2[M^0] = 0 \quad (9.53)$$

This equation is identical to the consistent infinitesimal finite-element cell equation in the frequency domain for the wave equation (Eq. 7.1) with the independent variable ω replaced by Ω. As discussed in Section 7.1, in order to solve the ordinary differential equation of the first order a boundary condition is necessary. This is calculated on the basis of an *asymptotic expansion for high frequency*. The procedure developed in Section 7.2 is still valid for Eq. 9.53 with ω replaced by Ω. The asymptotic expression (Eq. 7.2) is formulated as

$$[S^\infty(\Omega)] \approx i\Omega[C_\infty] + [K_\infty] + \sum_{j=1}^{m} \frac{1}{(i\Omega)^j}[A_j] \tag{9.54}$$

The coefficient matrices $[C_\infty]$, $[K_\infty]$ and $[A_j]$ follow from Eqs. 7.14 to 7.21 with the eigenvalue problem in Eq. 7.3.

Substituting

$$i\Omega = \sqrt{i\omega} \tag{9.55}$$

from Eq. 9.52 in Eq. 9.54 results in the asymptotic expansion for the diffusion equation

$$[S^\infty(\omega)] \approx \sqrt{i\omega}[C_\infty] + [K_\infty] + \sum_{j=1}^{m} \frac{1}{(\sqrt{i\omega})^j}[A_j] \tag{9.56}$$

Eq. 9.56 is used to calculate the boundary condition $[S^\infty(\omega_h)]$ at the specified high frequency ω_h for a selected m, which serves as the starting value for integrating the consistent infinitesimal finite-element cell equation in the frequency domain for decreasing ω using Eq. 9.51.

Obviously, the static case called the steady-state response corresponds to the frequency-domain solution with $\omega = 0$. For $\omega = 0$, the consistent infinitesimal finite-element cell equations for the diffusion and wave equations are identical (Eqs. 9.51 and 7.1), with the corresponding coefficient matrices $[E^0]$, $[E^1]$, $[E^2]$. Thus, Chapter 8 also applies directly for the diffusion equation. In particular, the steady-state stiffness matrix $[K^\infty]$ follows from Eq. 8.8. The function at internal points is calculated in Section 8.2. The application of the consistent infinitesimal finite-element cell method to a bounded medium discussed in Chapter 10 is also valid for the diffusion equation.

9.4 ACCURACY

9.4.1 Spherical Cavity Embedded in Full-space

Heat conduction in a spherical cavity of radius r_0 embedded in an isotropic full-space with the thermal conductivity coefficient κ and the product of mass density ρ and specific heat c denoted as μ is used for illustration. To obtain a one-dimensional

9.4 ACCURACY

problem where the isothermal surfaces are concentric spheres, a uniform normal heat flux $\bar{q}(t)$ is prescribed on the wall, which forms the structure-medium interface. The temperature u thus depends only on the radial coordinate r and time t. The heat flux $R(t)$-time derivative of temperature $\dot{u}_0(t)$ relationship at $r = r_0$ with $R(t) = 4\pi r_0^2 \bar{q}(t)$ equals

$$R(t) = \int_0^t V^\infty(t - \tau)\dot{u}_0(\tau)d\tau \tag{9.57}$$

Using spherical coordinates, the analytical solution for the velocity unit-impulse response function $V^\infty(t)$ is derived. The heat-conduction equation equals

$$\kappa\left(u_{,rr} + \frac{2}{r}u_{,r}\right) - \mu\dot{u} = 0 \tag{9.58}$$

which is written in the frequency domain as

$$\kappa\left(u(\omega)_{,rr} + \frac{2}{r}u(\omega)_{,r}\right) - i\omega\mu u(\omega) = 0 \tag{9.59}$$

or

$$(ru(\omega))_{,rr} - i\omega\frac{\mu}{\kappa}ru(\omega) = 0 \tag{9.60}$$

Its solution equals

$$u(\omega) = c_1 \frac{1}{r}e^{+\sqrt{i\omega\frac{\mu}{\kappa}}r} + c_2 \frac{1}{r}e^{-\sqrt{i\omega\frac{\mu}{\kappa}}r} \tag{9.61}$$

with the integration constants c_1 and c_2. Substituting $\sqrt{i} = (1+i)/\sqrt{2}$ yields

$$e^{+\sqrt{i\omega\frac{\mu}{\kappa}}r} = e^{+\frac{1}{\sqrt{2}}\sqrt{\omega\frac{\mu}{\kappa}}r + \frac{i}{\sqrt{2}}\sqrt{\omega\frac{\mu}{\kappa}}r} \tag{9.62}$$

As the temperature remains finite for $r \to \infty$, $c_1 = 0$ results. For the prescribed temperature amplitude $u_0(\omega)$ at $r = r_0$

$$u(\omega) = u_0(\omega)\frac{r_0}{r}e^{-\sqrt{i\omega\frac{\mu}{\kappa}}(r-r_0)} \tag{9.63}$$

follows. The amplitude of the heat flux at $r = r_0$ equals

$$R(\omega) = -4\pi r_0^2 \kappa u(\omega)_{,r}\big|_{r=r_0} \tag{9.64}$$

Substituting Eq. 9.63 in Eq. 9.64 yields

$$R(\omega) = 4\pi\kappa r_0\left(1 + \sqrt{i\omega\frac{\mu}{\kappa}}r_0\right)u_0(\omega) \tag{9.65}$$

Introducing the dimensionless frequency

$$a_0 = \frac{\omega \mu r_0^2}{\kappa} \tag{9.66}$$

results in

$$R(\omega) = 4\pi\kappa r_0 \left(1 + \sqrt{ia_0}\right) u_0(\omega) \tag{9.67}$$

The dynamic-stiffness coefficient in the frequency domain defined in Eq. 9.9 equals

$$S^\infty(a_0) = 4\pi\kappa r_0 \left(1 + \sqrt{ia_0}\right) \tag{9.68}$$

Eq. 9.68 is in the form of Eq. 9.21 with $s = 3$. The static-stiffness coefficient (steady-state) follows with $a_0 = 0$ as

$$K^\infty = 4\pi\kappa r_0 \tag{9.69}$$

Using Eq. 9.12 leads to

$$V^\infty(\omega) = K^\infty \left(\frac{1}{i\omega} + \sqrt{\frac{\mu}{\kappa}} r_0 \frac{1}{\sqrt{i\omega}}\right) \tag{9.70}$$

Its inverse Fourier transform yields

$$V^\infty(\omega) = K^\infty \left(1 + \sqrt{\frac{\mu}{\pi\kappa t}} r_0\right) H(t) \tag{9.71}$$

or with the dimensionless time

$$\bar{t} = t \frac{\kappa}{\mu r_0^2} \tag{9.72}$$

the velocity unit-impulse response function equals

$$V^\infty(\bar{t}) = K^\infty \left(1 + \frac{1}{\sqrt{\pi \bar{t}}}\right) H(\bar{t}) \tag{9.73}$$

The steady-state response $K^\infty = 4\pi\kappa r_0$ represents the heat flux at $t \to \infty$ for a velocity unit-impulse temperature, i.e. a temperature step function.

When applying the consistent infinitesimal finite-element cell method, only one octant of the structure-medium interface is discretized due to symmetry (Figure 5-7). Three 8-node isoparametric surface finite elements are used yielding a system with 16 degrees of freedom. The velocity unit-impulse response matrix $[V^\infty(t)]$ of order 16×16 is determined. The uniform spatial distribution of temperature, selected as unity, is enforced, which defines the spatial temperature pattern $\{\phi\}$. Analogous to

9.4 ACCURACY

Eq. A.0.1, the corresponding velocity unit-impulse response coefficient $V^\infty(t)$ follows as

$$V^\infty(t) = \{\phi\}^T [V^\infty(t)]\{\phi\} \tag{9.74}$$

The analysis is performed with $\Delta t = 0.04 \mu r_0^2 / \kappa$. The coefficient $V^\infty(t)$ non-dimensionalized with K^∞ is plotted as a function of $\bar{t}$ in Figure 9-1. The agreement with the exact result (Eq. 9.73) is excellent.

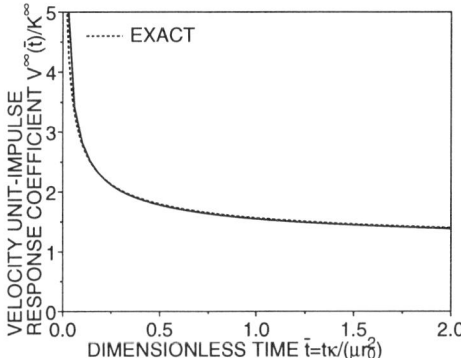

Figure 9-1 Velocity unit-impulse response coefficient for spherical cavity

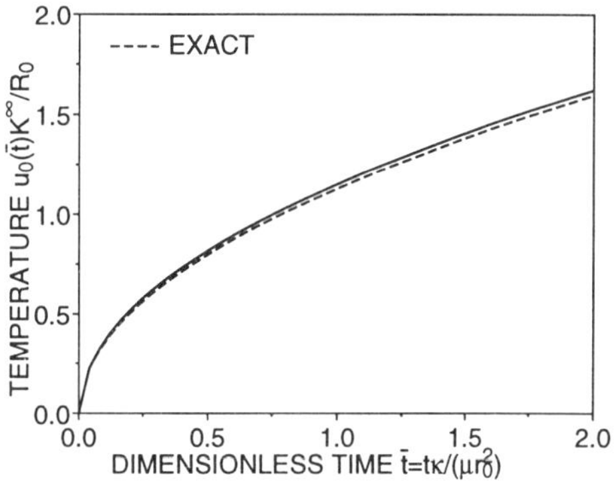

Figure 9-2 Temperature response to prescribed heat flux applied to wall of spherical cavity

As an application using the velocity unit-impulse response coefficient calculated above with the consistent infinitesimal finite-element cell method, the temperature response for an applied heat flux is determined. Discretizing Eq. 9.57 at time station n yields

$$R_n = \sum_{j=1}^{n} V_{n-j+1}^\infty \int_{t_{j-1}}^{t_j} \dot{u}_0(\tau) d\tau = \sum_{j=1}^{n} V_{n-j+1}^\infty (u_{0j} - u_{0j-1}) \tag{9.75}$$

u_{0n} follows from Eq. 9.75 with the zero initial condition $u_{00} = 0$.

For a spherical cavity with vanishing initial temperature the following heat flux is applied to its wall, the structure-medium interface

$$R(t) = R_0 \left(H(t) + 2\sqrt{\frac{\kappa t}{\pi \mu r_0^2}} \right) \tag{9.76}$$

The exact solution of the corresponding temperature

$$u_0(t) = \frac{2R_0}{K^\infty} \sqrt{\frac{\kappa t}{\pi \mu r_0^2}} \tag{9.77}$$

is easily verified by substituting Eqs. 9.77 and 9.73 in Eq. 9.57. The temperature determined from Eq. 9.75 agrees well with the exact solution (Figure 9-2).

9.4.2 Semi-infinite Wedge

As a two-dimensional problem with an analytical solution available in the frequency domain, heat conduction in an isotropic semi-infinite wedge with an opening angle α is addressed (Figure 9-3). The material constants of heat conduction are the thermal

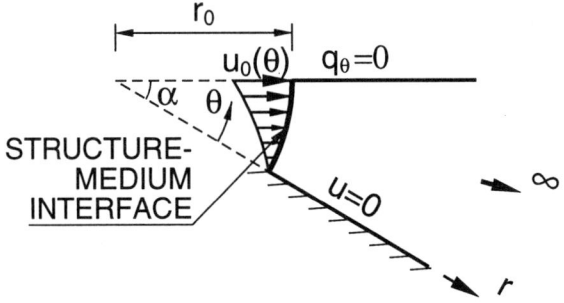

Figure 9-3 Heat conduction of semi-infinite wedge with prescribed temperature

conductivity coefficient κ and the product of mass density ρ and specific heat c, denoted as μ. On one of the boundaries extending to infinity in the radial direction a zero heat flux q_θ is enforced, and on the other a zero temperature u is prescribed

$$q_\theta(\theta = \alpha) = -\kappa \frac{1}{r} u_{,\theta} = 0 \qquad (9.78a)$$

$$u(\theta = 0) = 0 \qquad (9.78b)$$

The structure-medium interface coincides with the arc determined by the radius r_0, where a linear temperature distribution proportional to the angle θ is specified

$$u(r = r_0) = u_0(\theta) = \frac{\theta}{\alpha} u_0 \qquad (9.79)$$

The homogeneous heat conduction equation in polar coordinates r, θ follows from Eqs. 9.1 and 9.2 as

$$u_{,rr} + \frac{1}{r} u_{,r} + \frac{1}{r^2} u_{,\theta\theta} - \frac{\mu}{\kappa} \dot{u} = 0 \qquad (9.80)$$

A solution is derived in the frequency domain. The corresponding heat conduction equation in the temperature amplitude $u(\omega)$ equals

$$u(\omega)_{,rr} + \frac{1}{r} u(\omega)_{,r} + \frac{1}{r^2} u(\omega)_{;\theta\theta} - i\omega \frac{\mu}{\kappa} u(\omega) = 0 \qquad (9.81)$$

The method of separation of variables is applied to Eq. 9.81. $u(\omega)$ is written as the product of a function $R(r)$, which is independent of θ, and a function $\Theta(\theta)$, which is

9.4 ACCURACY

independent of r

$$u(\omega) = R(r)\Theta(\theta) \tag{9.82}$$

Substituting Eq. 9.82 in Eq. 9.81 yields after division by $R(r)\Theta(\theta)/r^2$

$$\frac{r^2 R(r)_{,rr} + rR(r)_{,r}}{R(r)} - i\omega\frac{\mu}{\kappa}r^2 = -\frac{\Theta(\theta)_{,\theta\theta}}{\Theta(\theta)} \tag{9.83}$$

The left-hand side is a function of r only and the right-hand side of θ only. To be able to satisfy Eq. 9.83, the two sides must be equal to the same constant, denoted as λ^2. Setting

$$\frac{\Theta(\theta)_{,\theta\theta}}{\Theta(\theta)} = -\lambda^2 \tag{9.84}$$

leads to the following two ordinary differential equations

$$\Theta(\theta)_{,\theta\theta} + \lambda^2 \Theta(\theta) = 0 \tag{9.85a}$$

$$r^2 R(r)_{,rr} + rR(r)_{,r} + \left(-i\omega\frac{\mu}{\kappa}r^2 - \lambda^2\right)R(r) = 0 \tag{9.85b}$$

The general solution of Eq. 9.85a equals

$$\Theta(\theta) = c_1 \cos\lambda\theta + c_2 \sin\lambda\theta \tag{9.86}$$

with the integration constants c_1 and c_2. Enforcing the boundary conditions of Eq. 9.78

$$\Theta(\alpha)_{,\theta} = 0 \tag{9.87a}$$
$$\Theta(0) = 0 \tag{9.87b}$$

results in

$$\cos\lambda\alpha = 0 \tag{9.88}$$

The eigenvalues λ follow as

$$\lambda_i = \frac{(2i+1)\pi}{2\alpha} \quad (i = 0, 1, \dots) \tag{9.89}$$

and the eigenfunctions as

$$\Theta_i(\theta) = \sin\lambda_i\theta \tag{9.90}$$

The general solution of the Bessel equation Eq. 9.85b for λ_i is

$$R_i(r) = c_{1i} H^{(1)}_{\lambda_i}\left(\sqrt{-i\omega\frac{\mu}{\kappa}}r\right) + c_{2i} H^{(2)}_{\lambda_i}\left(\sqrt{-i\omega\frac{\mu}{\kappa}}r\right) \tag{9.91}$$

with the first and second kind Hankel functions of order λ_i and the integration constants c_{1i}, c_{2i}. To introduce the boundary condition at $r \to \infty$, the asymptotic behaviour is examined with $\sqrt{-i} = (1-i)/\sqrt{2}$

$$H_{\lambda_i}^{(1)}\left(\sqrt{-i\omega\frac{\mu}{\kappa}}r\right) \approx \sqrt{\frac{2}{\pi r}}\sqrt{\frac{-\kappa}{i\omega\mu}}e^{+i\left(\frac{1-i}{\sqrt{2}}\sqrt{\omega\frac{\mu}{\kappa}}r - \frac{\lambda_i\pi}{2} - \frac{\pi}{4}\right)} \tag{9.92}$$

For $r \to \infty$, $H_{\lambda_i}^{(1)}\left(\sqrt{-i\omega\mu/\kappa r^2}\right) \to \infty$ results. To obtain a finite temperature, $c_{1i} = 0$ follows. This leads to

$$R_i(r) = c_i H_{\lambda_i}^{(2)}\left(\sqrt{-i\omega\frac{\mu}{\kappa}}r\right) \tag{9.93}$$

where the subscript 2 in the integration constant is omitted.

Using Eqs. 9.93 and 9.90 together with Eq. 9.82 permits the displacement amplitude to the expressed as

$$u(\omega) = \sum_{i=0}^{\infty} c_i H_{\lambda_i}^{(2)}\left(\sqrt{-i\omega\frac{\mu}{\kappa}}r\right) \sin\lambda_i\theta \tag{9.94}$$

The integration constants c_i are determined by enforcing the boundary condition Eq. 9.79

$$\sum_{i=0}^{\infty} c_i H_{\lambda_i}^{(2)}\left(\sqrt{-ia_0}\right) \sin\lambda_i\theta = \frac{\theta}{\alpha} u_0(\omega) \tag{9.95}$$

with the dimensionless frequency

$$a_0 = \frac{\omega\mu r_0^2}{\kappa} \tag{9.96}$$

Multiplying Eq. 9.95 by $\sin\lambda_i\theta$ and integrating from 0 to α yield

$$c_i = \frac{(-1)^i 2 u_0(\omega)}{\alpha^2 \lambda_i^2 H_{\lambda_i}^{(2)}\left(\sqrt{-ia_0}\right)} \tag{9.97}$$

Thus, Eq. 9.94 is reformulated as

$$u(\omega) = \frac{2u_0(\omega)}{\alpha^2} \sum_{i=0}^{\infty} \frac{(-1)^i}{\lambda_i^2 H_{\lambda_i}^{(2)}\left(\sqrt{-ia_0}\right)} H_{\lambda_i}^{(2)}\left(\sqrt{-i\omega\frac{\mu}{\kappa}}r\right) \sin\lambda_i\theta \tag{9.98}$$

The heat flux amplitude $q_r(\omega) = -\kappa u(\omega)_{,r}$ follows from Eq. 9.98 as

$$q_r(\omega) = \frac{2\kappa u_0(\omega)}{\alpha^2 r} \sum_{i=0}^{\infty} \frac{(-1)^i}{\lambda_i^2 H_{\lambda_i}^{(2)}\left(\sqrt{-ia_0}\right)}$$
$$\left(\sqrt{-i\omega\frac{\mu}{\kappa}}r H_{\lambda_i-1}^{(2)}\left(\sqrt{-i\omega\frac{\mu}{\kappa}}r\right) - \lambda_i H_{\lambda_i}^{(2)}\left(\sqrt{-i\omega\frac{\mu}{\kappa}}r\right)\right) \sin\lambda_i\theta \tag{9.99}$$

9.4 ACCURACY

where

$$H^{(2)}_{\lambda_i}\left(\sqrt{-i\omega\frac{\mu}{\kappa}}r\right)_{,r} = \frac{1}{r}\left(\sqrt{-i\omega\frac{\mu}{\kappa}}rH^{(2)}_{\lambda_i-1}\left(\sqrt{-i\omega\frac{\mu}{\kappa}}r\right) - \lambda_i H^{(2)}_{\lambda_i}\left(\sqrt{-i\omega\frac{\mu}{\kappa}}r\right)\right) \quad (9.100)$$

is substituted. Based on virtual work considerations, the normal heat flux amplitude $R(\omega)$ is calculated as

$$R(\omega) = -\int_0^\alpha q_r(\omega, r = r_0)\frac{\theta}{\alpha}d\theta \quad (9.101)$$

Substituting Eq. 9.99 in Eq. 9.101 yields

$$R(\omega) = \frac{2\kappa}{\alpha^3}\sum_{i=0}^\infty \frac{1}{\lambda_i^3}\left(1 - \frac{\sqrt{-ia_0}H^{(2)}_{\lambda_i-1}(\sqrt{-ia_0})}{\lambda_i H^{(2)}_{\lambda_i}(\sqrt{-ia_0})}\right)u_0(\omega) \quad (9.102)$$

The ratio of $R(\omega)$ and $u_0(\omega)$ is equal to the dynamic-stiffness coefficient

$$S^\infty(a_0) = \frac{2\kappa}{\alpha^3}\sum_{i=0}^\infty \frac{1}{\lambda_i^3}\left(1 - \frac{\sqrt{-ia_0}H^{(2)}_{\lambda_i-1}(\sqrt{-ia_0})}{\lambda_i H^{(2)}_{\lambda_i}(\sqrt{-ia_0})}\right) \quad (9.103)$$

Setting $a_0 = 0$ in Eq. 9.103 results in the steady-state coefficient

$$K^\infty = \frac{2\kappa}{\alpha^3}\sum_{i=0}^\infty \frac{1}{\lambda_i^3} \quad (9.104)$$

For $\alpha = 30°$, $K^\infty = 0.5427\kappa$ results. As explained in Section 9.3, replacing $-\omega^2$ for the wave equation Eq. A.4.8 by $i\omega$, i.e. ω by $\sqrt{-i\omega}$, yields the diffusion equation Eq. 9.81. The same rule applies to the dimensionless frequency a_0. Applying this substitution to the dynamic-stiffness coefficient of the semi-infinite wedge Eq. A.4.30 and changing G to κ results in Eq. 9.103.

In the consistent infinitesimal finite-element cell method the arc on the structure-medium interface is discretized with 3 3-node line elements of equal length. The boundary condition of the dynamic-stiffness matrix $[S^\infty(\omega_h)]$ is calculated from the asymptotic expansion in Eq. 9.56 with $m = 2$ at $\omega_h = a_{0h}\kappa/(\mu r_0^2)$ for $a_{0h} = 20$. $[S^\infty(\omega)]$ of order 6×6 is calculated for decreasing ω based on the Bulirsch-Stoer method of Reference [40]. The dynamic-stiffness coefficient $S^\infty(\omega)$ follows from Eq. A.0.2 with the vector $\{\phi\}$ determined from the boundary condition defined on the arc (Eq. 9.79). $S^\infty(\omega)$ calculated for $\alpha = 30°$ non-dimensionalized with the steady-state coefficient K^∞ is decomposed in the real and imaginary parts. Excellent agreement with the analytical solution of Eq. 9.103 results (Figure 9-4).

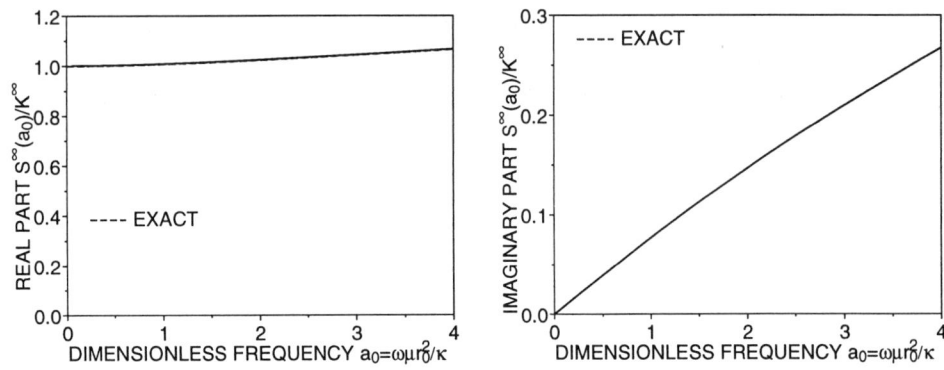

Figure 9-4 Dynamic-stiffness coefficient of semi-infinite wedge

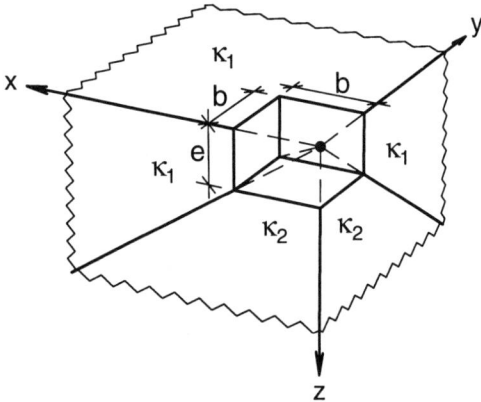

Figure 9-5 One quarter of square prism embedded in inhomogeneous half-space

9.4.3 Prism Embedded in Inhomogeneous Half-space

As a truly three-dimensional problem, a square prism of length $2b$ embedded with depth e in a half-space is addressed (Figure 9-5). The embedment ratio $e/b = 2/3$ is selected. Besides the homogeneous half-space, an inhomogeneous half-space compatible with similarity is also examined. The structure-medium interface is first discretized with finite elements. Owing to symmetry, only a quarter of the prism is analysed. The finite-element discretization of the three faces is shown in Figure 5-26. The corresponding velocity unit-impulse response matrix $[V^\infty(t)]$ of order 100×100 is then established with $\Delta t = 0.04 \mu b^2 / \kappa$.

To ease comparison, a spatial variation of the nodal values of the temperature is represented by the vector $\{\phi\}$. The corresponding velocity unit-impulse response coefficient $V^\infty(t)$ follows from Eq. 9.74. For the homogeneous half-space with κ

9.4 ACCURACY

and ρc two spatial variations of the temperature are investigated. For a uniform temperature on the structure-medium interface all elements in $\{\phi\}$ are equal to 1. $V^\infty(t)$ non-dimensionalized by the corresponding static-stiffness coefficient (steady-state) $K^\infty = 7.36\kappa b$ is plotted as a function of $\bar{t} = t\kappa/(\mu b^2)$ in Figure 9-6a. For comparison, the result using an extended mesh with 50 elements of width $0.1b$ in the radial direction and $\Delta t = 0.02\mu b^2/\kappa$ is also shown. $V^\infty(t)$ of the consistent infinitesimal finite-element cell method agrees well with the smoothed result of the extended mesh. For a temperature distribution proportional to the coordinate z and equal to 1 at the base ($z = e$), the same high accuracy is achieved (Figure 9-6b). In this case $K^\infty = 7.47\kappa b$ results.

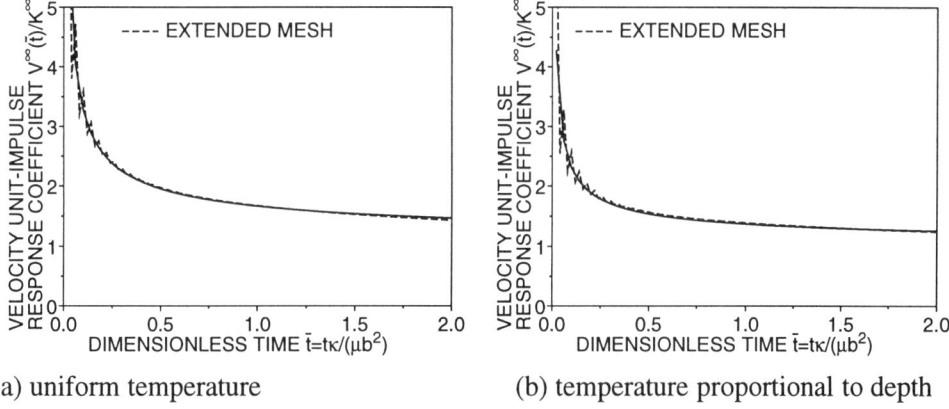

(a) uniform temperature   (b) temperature proportional to depth

Figure 9-6 Velocity unit-impulse response coefficient for square prism embedded in homogeneous half-space

For the inhomogeneous half-space with κ_1 adjacent to the walls and $\kappa_2 = 4\kappa_1$ adjacent to the base (Figure 9-5) a uniform temperature is calculated. The constant μ does not vary. The analysis is performed with $\Delta t = 0.04\mu b^2/\kappa_1$. The velocity unit-impulse response coefficient is plotted in Figure 9-7 with $K^\infty = 17.37\kappa_1 b$. For the extended mesh, 50 elements of width $0.1b$ in the radial direction and $\Delta t = 0.01\mu b^2/\kappa_1$ are introduced.

As well as for heat conduction as discussed above, the examples can also be interpreted for many other physical problems. For example, in flow in porous media u represents pressure, q seepage velocity, κ permeability and μ the specific storage. In this case, the velocity unit-impulse response matrix $[V^\infty(t)]$ permits the fluid flowing through the structure-medium interface for any pressure distribution in space and time to be analysed.

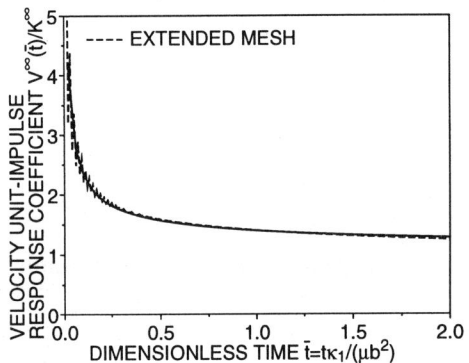

Figure 9-7 Velocity unit-impulse response coefficient for square prism embedded in inhomogeneous half-space for uniform temperature

10

CONSISTENT INFINITESIMAL FINITE-ELEMENT CELL METHOD APPLIED TO BOUNDED MEDIUM

The consistent infinitesimal finite-element cell method can also be used to analyse a *bounded* medium with a discretization on the boundary only. The corresponding consistent infinitesimal finite-element cell equation differs only slightly from that of the unbounded medium. The advantages of the method also apply for a bounded medium. In particular, plane interfaces between different materials or plane boundaries passing through the similarity centre are taken into consideration without any spatial discretization on the interfaces or the boundaries. This unique feature permits an accurate analysis of a crack in fracture mechanics without discretizing the crack faces. The consistent infinitesimal finite-element cell method can be regarded as a "boundary-element" method based on finite elements. This *boundary finite-element method* as a general procedure can be applied to bounded and unbounded media.

For a dynamic analysis in contrast to the unit-impulse response matrix calculated for the unbounded medium, the *static-stiffness* and *mass matrices* with respect to the degrees of freedom on the boundary are determined for the bounded medium. This mass matrix is determined without introducing any additional approximation, in contrast to the boundary-element method. The static-stiffness and mass matrices are then used straightforwardly in a dynamic analysis which leads to an efficient procedure.

After determining the static-stiffness matrix, it is possible to calculate the internal displacements and stresses with practically no additional effort. In contrast, in the boundary-element method the displacements and stresses in internal points are evaluated by integrating their fundamental solutions over the boundary.

The fundamental equations for the bounded medium are summarized in Section 10.1. The static-stiffness matrix is determined in Section 10.2 and the mass matrix in Section 10.3. The accuracy is discussed in Section 10.4.

10.1 SUMMARY OF FUNDAMENTAL EQUATIONS

The consistent infinitesimal finite-element cell method can be formulated for the dynamic-stiffness matrix in the frequency domain $[S^b(\omega)]$ (subscript b for bounded) of a bounded medium with degrees of freedom on the boundary only. The discretization is limited to the boundary. The force-displacement relationship equals

$$\{R(\omega)\} = [S^b(\omega)]\{u(\omega)\} \tag{10.1}$$

$\{R(\omega)\}$ and $\{u(\omega)\}$ are the amplitudes of the nodal forces and nodal displacements on the boundary.

The derivation of the consistent infinitesimal finite-element cell equation for the bounded medium is very similar to that for the unbounded medium discussed in Section 5.2. Similarity and assemblage enforcing compatibility and equilibrium are again invoked. The concept for the bounded medium is illustrated in Figure 10-1, which should be compared with that of the unbounded medium shown in Figure 5-1. A similarity centre O within the bounded medium is selected (Figure 10-1a). The characteristic length of the boundary equals r_e (and not r_i as for the unbounded medium). A similar fictitious boundary with the characteristic length r_i within the

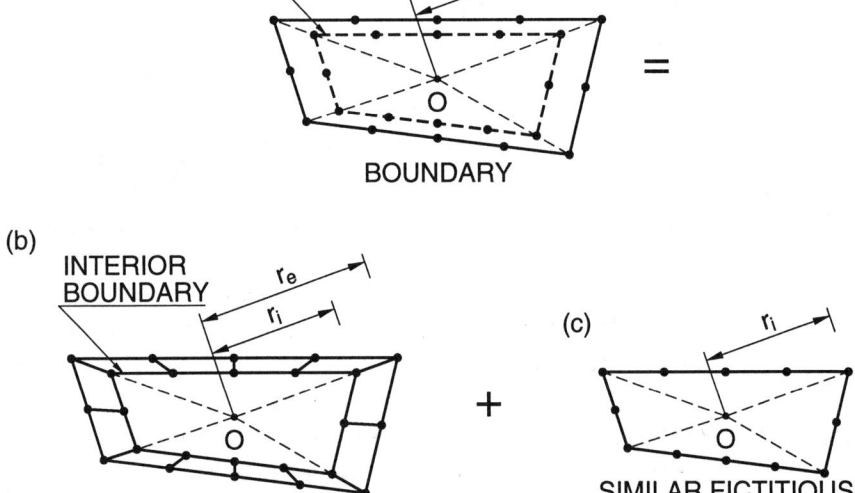

Figure 10-1 Concept of consistent infinitesimal finite-element cell method with infinitesimal cell width leading to finite-element discretization of boundary only

10.1 SUMMARY OF FUNDAMENTAL EQUATIONS

bounded medium ($r_i < r_e$) is constructed, where

$$r_i = \frac{r_e}{1+w} \tag{10.2}$$

holds with the dimensionless cell width w. The region between the boundary and the fictitious boundary is a cell of infinitesimal width which is discretized with finite elements (Figure 10-1b). Its interior and exterior boundaries coincide with the fictitious boundary and the boundary, respectively. The arrangement of the nodes on the two boundaries must satisfy similarity. The force-displacement relationship of the finite-element cell is written as (Eq. 5.96)

$$\begin{bmatrix} [S_{ii}(\omega)] & [S_{ie}(\omega)] \\ [S_{ei}(\omega)] & [S_{ee}(\omega)] \end{bmatrix} \begin{Bmatrix} \{u_i(\omega)\} \\ \{u_e(\omega)\} \end{Bmatrix} = \begin{Bmatrix} \{P_i(\omega)\} \\ \{P_e(\omega)\} \end{Bmatrix} \tag{10.3}$$

with $\{u_i(\omega)\}$ and $\{P_i(\omega)\}$ denoting the amplitudes of the nodal displacements and nodal forces on the interior boundary and $[S(\omega)]$ representing its dynamic-stiffness matrix (Eq. 5.95). Deleting the infinitesimal finite-element cell from the bounded medium defined by the boundary results in the bounded medium defined by the fictitious boundary (Figure 10-1c). The force-displacement relationship of the bounded medium at the boundaries corresponding to the interior and exterior boundaries of the finite-element cell is written as (Eq. 5.97)

$$\{R_i(\omega)\} = [S_i^b(\omega)]\{u_i(\omega)\} \tag{10.4a}$$
$$\{R_e(\omega)\} = [S_e^b(\omega)]\{u_e(\omega)\} \tag{10.4b}$$

The same deleting procedure also applies to the dynamic-stiffness matrices by applying finite-element assemblage. By using the same displacement amplitudes at the boundaries (Eqs. 10.3 and 10.4) compatibility is enforced. Formulating equilibrium at the two boundaries relates the amplitudes of the nodal forces of the bounded medium to those of the cell

$$\{P_i(\omega)\} = -\{R_i(\omega)\} \tag{10.5a}$$
$$\{P_e(\omega)\} = \{R_e(\omega)\} \tag{10.5b}$$

Note that the signs in Eq. 10.5 are changed compared with these for the unbounded medium specified in Eq. 5.98 and Eqs. 10.3 and 10.4 correspond to Eqs. 5.96 and 5.97 without a sign change. Eliminating $\{P(\omega)\}$ and $\{R(\omega)\}$ from Eqs. 10.3, 10.4 and 10.5 yields

$$\begin{bmatrix} [S_{ii}(\omega)] & [S_{ie}(\omega)] \\ [S_{ei}(\omega)] & [S_{ee}(\omega)] \end{bmatrix} \begin{Bmatrix} \{u_i(\omega)\} \\ \{u_e(\omega)\} \end{Bmatrix} = \begin{bmatrix} -[S_i^b(\omega)] & 0 \\ 0 & [S_e^b(\omega)] \end{bmatrix} \begin{Bmatrix} \{u_i(\omega)\} \\ \{u_e(\omega)\} \end{Bmatrix} \tag{10.6}$$

Eliminating $\{u_e(\omega)\}$ and choosing an arbitrary $\{u_i(\omega)\}$ results in the relationship of the dynamic-stiffness matrices at two boundaries of a bounded medium based on finite-element assemblage enforcing compatibility and equilibrium

$$-[S_i^b(\omega)] = [S_{ii}(\omega)] - [S_{ie}(\omega)](-[S_e^b(\omega)] + [S_{ee}(\omega)])^{-1}[S_{ei}(\omega)] \qquad (10.7)$$

Note that Eq. 10.7 is the same as Eq. 5.101 for the unbounded medium, but with the sign of the dynamic-stiffness matrices of the bounded medium $[S_i^b(\omega)]$, $[S_e^b(\omega)]$ reversed from those of the unbounded medium $[S_i^\infty(\omega)]$, $[S_e^\infty(\omega)]$.

In addition to Eq. 10.7 expressing the finite-element assemblage, the relationship based on *similarity* must be included. The equation for a bounded medium is the same as for an unbounded medium (Eq. 3.12)

$$r[S^b(r,\omega)]_{,r} = (s-2)[S^b(r,\omega)] + \omega[S^b(r,\omega)]_{,\omega} \qquad (10.8)$$

with the spatial dimension $s\ (= 2 \text{ or } = 3)$.

The limit of the infinitesimal dimensionless cell width w is performed analytically as for the unbounded medium in Section 5.2.8. Conceptually replacing $-[S_i^b(\omega)]$ by $[S_i^\infty(\omega)]$ and $-[S_e^b(\omega)]$ by $[S_e^\infty(\omega)]$ transforms Eq. 10.7 to Eq. 5.101. The derivation then yields the consistent infinitesimal finite-element cell equation (Eq. 5.113). Converting $[S^\infty(\omega)]$ back to the static-stiffness matrix $-[S^b(\omega)]$ of the bounded medium results in the *consistent infinitesimal finite-element cell equation in the frequency domain for a bounded medium*

$$(-[S^b(\omega)] + [E^1])[E^0]^{-1}(-[S^b(\omega)] + [E^1]^T) + (s-2)[S^b(\omega)]$$
$$+ \omega[S^b(\omega)]_{,\omega} - [E^2] + \omega^2[M^0] = 0 \qquad (10.9)$$

The coefficient matrices $[E^0]$, $[E^1]$, $[E^2]$ and $[M^0]$ for three-dimensional elasticity are specified in Eqs. 5.22 and 5.36.

Eq. 10.9 forms the starting point for calculating the static-stiffness and mass matrices.

10.2 STATICS

10.2.1 Static-stiffness Matrix

Setting $\omega = 0$ in Eq. 10.9 yields the consistent infinitesimal finite-element cell equation for statics with the unknown static-stiffness matrix $[K^b]$ of a bounded medium with degrees of freedom on the boundary only

$$[K^b][E^0]^{-1}[K^b] - \left([E^1][E^0]^{-1} - \frac{s-2}{2}[I]\right)[K^b]$$
$$- [K^b]\left([E^0]^{-1}[E^1]^T - \frac{s-2}{2}[I]\right) - [E^2] + [E^1][E^0]^{-1}[E^1]^T = 0 \qquad (10.10)$$

10.2 STATICS

Eq. 10.10 also follows from the consistent infinitesimal finite-element cell equation for the unbounded medium (Eq. 8.8) by replacing $[K^\infty]$ by $-[K^b]$.

When only the static-stiffness matrix $[K^b]$ is of interest, the algebraic Riccati equation (Eq. 10.10) is solved as described in Section 5.3.1 for the first time step of the consistent infinitesimal finite-element cell method. The Schur decomposition yields

$$\begin{bmatrix} [E^0]^{-1}[E^1]^T - \frac{s-2}{2}[I] & -[E^0]^{-1} \\ -[E^2]+[E^1][E^0]^{-1}[E^1]^T & -[E^1][E^0]^{-1}+\frac{s-2}{2}[I] \end{bmatrix} \begin{bmatrix} [V_{11}] & [V_{12}] \\ [V_{21}] & [V_{22}] \end{bmatrix}$$

$$= \begin{bmatrix} [V_{11}] & [V_{12}] \\ [V_{21}] & [V_{22}] \end{bmatrix} \begin{bmatrix} [S_{11}] & [S_{12}] \\ 0 & [S_{22}] \end{bmatrix} \quad (10.11)$$

$[S]$ is arranged in such a way that the real parts of the eigenvalues $[S_{11}]$ are negative and those of $[S_{22}]$ are positive. The static-stiffness matrix $[K^b]$ follows as

$$[K^b] = [V_{21}][V_{11}]^{-1} \quad (10.12)$$

As $[K^b]$ for a bounded medium is semi-positive definite, the term $\varepsilon[I]$ with a very small constant ε with the dimension of Gr_0^{s-2} is added to $[E^2]$. $[K^b]$ will then be positive definite as explained for the two-dimensional unbounded medium at the end of Section 8.1. The algorithm selects the positive definite solution of the matrix equation.

It is of interest to note that changing the sign of the negative definite solution of the Riccati equation in Eq. 10.10 (which makes the solution positive definite) leads to the static-stiffness matrix of the corresponding unbounded medium $[K^\infty]$. This follows from the fact that changing the sign of the unknown $[K^b]$ in Eq. 10.10 yields the equation for $[K^\infty]$ (Eq. 8.8). The Riccati equation needs to be solved only once to determine $[K^\infty]$ and $[K^b]$.

Assembling $[K^\infty]$ and $[K^b]$ leads to the static-stiffness matrix of the "free-field" system consisting of the unbounded medium and the bounded medium discretized on the surface between these two media.

The advantages of the consistent infinitesimal finite-element cell method discussed in Section 1.5.1 apply also to the bounded medium. In particular, plane interfaces between different materials or plane boundaries passing through the similarity centre are taken into consideration without any spatial discretization on the interfaces or the boundaries. Only the boundary of the bounded medium is discretized, illustrating that the consistent infinitesimal finite-element cell method is actually a *"boundary-element" method based on finite elements.*

10.2.2 Displacement and Strain at Internal Point

The derivation of the equation to calculate the displacements at an internal point for a bounded medium follows closely that for the unbounded medium described in Section 8.2. Only the key equations with changes are listed.

Comparing Eq. 10.6 for the bounded medium with $\omega = 0$ to Eq. 8.3 for the unbounded medium, it follows that the two equations are identical when $[K_i^b]$ and $[K_e^b]$ are replaced by $-[K_i^\infty]$ and $-[K_e^\infty]$, respectively. This represents the only source of the changes. Processing the first set of equations of Eq. 10.6 for $\omega = 0$ leads to (Eq. 8.16)

$$r\{u\}_{,r} = -[\bar{E}^0]^{-1}(-[\bar{K}^b] + [\bar{E}^1]^T)\{u\} \tag{10.13}$$

with (Eq. 8.15d)

$$[\bar{K}^b] = \frac{1}{Gr^{s-2}}[K^b] \tag{10.14}$$

The corresponding eigenvalue problem of the Riccati equation (Eq. 10.10) is in the same form as Eq. 8.18 with the signs of the coefficient submatrices on the diagonal changed

$$\begin{bmatrix} [\bar{E}^0]^{-1}[\bar{E}^1]^T - \frac{s-2}{2}[I] & -[\bar{E}^0]^{-1} \\ -[\bar{E}^2] + [\bar{E}^1][\bar{E}^0]^{-1}[\bar{E}^1]^T & -[\bar{E}^1][\bar{E}^0]^{-1} + \frac{s-2}{2}[I] \end{bmatrix} \begin{bmatrix} [\Phi_{11}] & [\Phi_{12}] \\ [\Phi_{21}] & [\Phi_{22}] \end{bmatrix}$$
$$= \begin{bmatrix} [\Phi_{11}] & [\Phi_{12}] \\ [\Phi_{21}] & [\Phi_{22}] \end{bmatrix} \begin{bmatrix} \lceil \lambda \rfloor & \\ & -\lceil \lambda \rfloor \end{bmatrix} \tag{10.15}$$

Again, $\text{Re}(\lambda_i) < 0$ applies. Introducing (Eq. 8.19)

$$[\bar{K}^b] = [\Phi_{21}][\Phi_{11}]^{-1} \tag{10.16}$$

the first set of equations of Eq. 10.15 is written as (Eq. 8.20)

$$-[\bar{E}^0]^{-1}\left(-[\bar{K}^b] + [\bar{E}^1]^T\right) = -[\Phi_{11}]\lceil \lambda \rfloor [\Phi_{11}]^{-1} - \frac{s-2}{2}[I] \tag{10.17}$$

Substituting Eq. 10.17 in Eq. 10.13 yields

$$r\{u\}_{,r} = \left(-[\Phi_{11}]\lceil \lambda \rfloor [\Phi_{11}]^{-1} - \frac{s-2}{2}[I]\right)\{u\} \tag{10.18}$$

Eq. 10.18 can also be obtained from Eq. 8.21 by replacing $\lceil \lambda \rfloor$ by $-\lceil \lambda \rfloor$. The solution of Eq. 10.17 is specified as (Eq. 8.27)

$$\{u\} = [\Phi_{11}]\left\lceil r^{-\lambda_i - \frac{s-2}{2}} \right\rfloor \{c\} \tag{10.19}$$

10.3 MASS MATRIX

Substituting the integration constants $\{c\}$ determined from enforcing the boundary condition (Eq. 8.30)

$$\{u(r=r_0)\} = \{u_0\} \tag{10.20}$$

in Eq. 10.19 leads to (Eq. 8.32)

$$\{u\} = [\Phi_{11}] \left[\left(\frac{r}{r_0}\right)^{-\lambda_i - \frac{s-2}{2}} \right] [\Phi_{11}]^{-1} \{u_0\} \tag{10.21}$$

The analysis of the strains at an internal point is based on Eq. 8.36. $\{u\}$ and $r\{u\}_{,r}$ follow from Eqs. 10.21 and 10.18 for the bounded medium.

The *explicit* displacements in the interior of the bounded medium (Eq. 10.21) can be used to calculate equivalent nodal loads on the boundary for a loading applied in the interior by integration.

10.3 MASS MATRIX

To determine the mass matrix $[M^b]$ of a bounded medium with degrees of freedom on the boundary only, the low-frequency behaviour of the consistent infinitesimal finite-element cell equation is addressed (Eq. 10.9). As in structural dynamics the dynamic-stiffness matrix $[S^b(\omega)]$ of the bounded medium is postulated as

$$[S^b(\omega)] = [K^b] - \omega^2 [M^b] \tag{10.22}$$

Terms of order in $i\omega$ equal to or higher than 4 are neglected in this expansion valid for low frequencies. The static-stiffness matrix $[K^b]$ is determined using Eq. 10.10.

Substituting Eq. 10.22 in Eq. 10.9 leads to a constant term independent of $i\omega$, a term in $(i\omega)^2$ and higher-order terms in $i\omega$ which are neglected. The constant term, which is equal to the left-had side of Eq. 10.10, vanishes. The coefficient matrix of the remaining term in $(i\omega)^2$ is written as

$$\left((-[K^b] + [E^1])[E^0]^{-1} - \frac{s}{2}[I] \right) [M^b]$$
$$+ [M^b] \left([E^0]^{-1}(-[K^b] + [E^1]^T) - \frac{s}{2}[I] \right) + [M^0] = 0 \tag{10.23}$$

This is the equation to calculate the mass matrix $[M^b]$. Eq. 10.23 is the Lyapunov equation

$$[A][X] + [X][A]^T = [C] \tag{10.24}$$

with the unknown $[X] = [M^b]$ and the other matrices defined straightforwardly. Proceeding as in connection with Eqs. 5.184 and 5.191, $[M^b]$ is proven to be symmetric as $[M^0]$ is symmetric. The solution of a Lyapunov equation is addressed in Section 5.3.2.

As an alternative procedure to solve Eq. 10.23, use is made of intermediate results of the Schur decomposition (Eq. 10.11) performed to calculate the static-stiffness matrix $[K^b]$. The first submatrix of the product in Eq. 10.11, postmultiplied by $[V_{11}]^{-1}$, leads to

$$[E^0]^{-1}(-[K^b]+[E^1]^T) - \frac{s-2}{2}[I] = [V_{11}][S_{11}][V_{11}]^{-1} \qquad (10.25)$$

with $[K^b]$ defined in Eq. 10.12. Substituting Eq. 10.25 in Eq. 10.23 results in

$$-([V_{11}]^{-1})^T[S_{11}]^T[V_{11}]^T[M^b] - [M^b][V_{11}][S_{11}][V_{11}]^{-1} + 2[M^b] = [M^0] \qquad (10.26)$$

Premultiplying Eq. 10.26 by $[V_{11}]^T$ and postmultiplying by $[V_{11}]$ yields

$$([I] - [S_{11}]^T)[m^b] + [m^b]([I] - [S_{11}]) = [V_{11}]^T[M^0][V_{11}] \qquad (10.27)$$

with the transformation

$$[m^b] = [V_{11}]^T[M^b][V_{11}] \qquad (10.28)$$

As $[S_{11}]$ is a quasi-upper triangular matrix consisting of 2×2 blocks or 1×1 blocks on the diagonal, Eq. 10.27 is solved by simple back substitution (Reference [5]). As the real parts of the eigenvalues of $[S_{11}]$ are negative, a solution of Eq. 10.27 always exists. The mass matrix is obtained from Eq. 10.28 as

$$[M^b] = ([V_{11}]^{-1})^T[m^b][V_{11}]^{-1} \qquad (10.29)$$

As $[M^0]$ is positive definite, $[M^b]$ will have the same property. The additional computational effort starting from $[K^b]$ to calculate $[M^b]$ is small.

Note that in the consistent infinitesimal finite-element cell method the *mass matrix follows from the low-frequency behaviour of the equation without any additional assumptions*. In the boundary-element method a mass matrix can be constructed which, however, involves selecting an approximating function for the displacements within the bounded medium (Reference [36]). In addition, the mass matrix will not be symmetric.

10.4 ACCURACY

10.4.1 Solid Sphere

The solid sphere of radius r_0, shear modulus G and Poisson ratio ν with a uniform normal displacement on the surface is addressed, which is a one-dimensional problem.

10.4 ACCURACY

The analytical solution of static-stiffness coefficient K^b is derived. The governing equation in the radial displacement u follows from Eq. A.1.5 as

$$u_{,rr} + \frac{2}{r}u_{,r} - \frac{2}{r^2}u = 0 \tag{10.30}$$

Its general solution equals

$$u = c_1 r + c_2 \frac{1}{r^2} \tag{10.31}$$

with the integration constants c_1 and c_2. For the solid sphere, u at the centre $u(r=0)$ must remain finite which yields $c_2 = 0$. For a prescribed displacement at the boundary $u(r = r_0) = u_0$

$$u = \frac{r}{r_0} u_0 \tag{10.32}$$

results. The strains follow from Eq. A.1.2 and the stress σ_r from Eq. A.1.3, leading to the nodal force

$$R = 4\pi r_0^2 \sigma_r(r = r_0) = 8\pi \frac{1+\nu}{1-2\nu} G r_0 u_0 \tag{10.33}$$

The static-stiffness coefficient equals

$$K^b = 8\pi \frac{1+\nu}{1-2\nu} G r_0 \tag{10.34}$$

The coefficients of the consistent infinitesimal finite-element cell equation for the bounded medium which are the same as for the unbounded medium are specified in Eq. 5.204. Substituting the coefficients in Eq. 10.10 with $s = 3$ leads to

$$\frac{1}{8\pi} \frac{1-2\nu}{1-\nu} \frac{1}{Gr_0} (K^b)^2 + \frac{1-5\nu}{1-\nu} K^b - 16\pi \frac{1+\nu}{1-\nu} G r_0 = 0 \tag{10.35}$$

Note that the coefficients in Eq. 10.35 for the solid sphere are the same as those in Eq. 8.37 for the spherical cavity with a sign change in the linear term in K^b. The two solutions are

$$K_1^b = 8\pi \frac{1+\nu}{1-2\nu} G r_0 \tag{10.36a}$$

$$K_2^b = -16\pi G r_0 \tag{10.36b}$$

The positive solution K_1^b represents the exact static-stiffness coefficient of the solid sphere (Eq. 10.34). The negative solution K_2^b, after a sign change, is equal to the exact static-stiffness coefficient K^∞ of the corresponding spherical cavity (Eq. A.1.17), which is also the positive solution of Eq. 10.35 with the sign of the linear term in K^b reversed.

In addition, the static-stiffness coefficient K^b is calculated as a three-dimensional problem for the solid sphere with $\nu = 0.25$ using the spatial discretization shown in Figure 5-7. ε is added to E^2 in Eq. 5.204c with $\varepsilon = 10^{-9} Gr_0$. The analysis with the consistent infinitesimal finite-element cell method leads to $19.67\pi Gr_0$. The analytical solution equals $20\pi Gr_0$.

The displacements and strains in the solid sphere are calculated similarly to these for the spherical cavity in Section 8.3.1. The eigenvalue problem of Eq. 10.15 is written as

$$\begin{bmatrix} -\dfrac{1-5\nu}{2(1-\nu)} & -\dfrac{1}{8\pi}\dfrac{1-2\nu}{1-\nu} \\ -16\pi\dfrac{1+\nu}{1-\nu} & \dfrac{1-5\nu}{2(1-\nu)} \end{bmatrix} \begin{Bmatrix} \Phi_{1i} \\ \Phi_{2i} \end{Bmatrix} = \lambda_i \begin{Bmatrix} \Phi_{1i} \\ \Phi_{2i} \end{Bmatrix} \qquad (10.37)$$

Note that the diagonal elements in Eq. 10.37 have signs opposite to those in Eq. 8.39. The eigenvalues and eigenvectors are equal to

$$\lambda_1 = -\frac{3}{2} \qquad (10.38a)$$

$$\lambda_2 = -\lambda_1 \qquad (10.38b)$$

$$\begin{bmatrix} \Phi_{11} & \Phi_{12} \\ \Phi_{21} & \Phi_{22} \end{bmatrix} = \begin{bmatrix} 1 & 1 \\ 8\pi\dfrac{1+\nu}{1-2\nu} & -16\pi \end{bmatrix} \qquad (10.39)$$

The internal displacements follow from Eq. 10.21 as

$$u = \frac{r}{r_0} u_0 \qquad (10.40)$$

which is the exact solution (Eq. 10.32). The strains are calculated from Eq. 8.36 with $\{B^1\}$ and $\{B^2\}$ in Eq. 8.44 and u in Eq. 10.40 as

$$\begin{Bmatrix} \varepsilon_r \\ \varepsilon_\theta \\ \varepsilon_\varphi \end{Bmatrix} = \frac{1}{r_0} \begin{Bmatrix} 1 \\ 1 \\ 1 \end{Bmatrix} u_0 \qquad (10.41)$$

Again, the exact solution results which is verified by substituting Eq. 10.40 in Eq. A.1.2.

10.4 ACCURACY

The low-frequency limit of the analytical solution of the dynamic-stiffness coefficient $S^b(\omega)$ leads to the mass M^b. The solution of the equation of motion in the frequency domain specified in Eq. A.1.10 for the solid sphere equals

$$u(\omega) = c_1 j_1\left(\frac{\omega r}{c_p}\right) + c_2 y_1\left(\frac{\omega r}{c_p}\right) \tag{10.42}$$

with the first-order spherical Bessel functions of the first kind j_1 and the second kind y_1

$$j_1\left(\frac{\omega r}{c_p}\right) = \left(\frac{c_p}{\omega r}\right)^2\left(\sin\left(\frac{\omega r}{c_p}\right) - \frac{\omega r}{c_p}\cos\left(\frac{\omega r}{c_p}\right)\right) \tag{10.43a}$$

$$y_1\left(\frac{\omega r}{c_p}\right) = -\left(\frac{c_p}{\omega r}\right)^2\left(\cos\left(\frac{\omega r}{c_p}\right) + \frac{\omega r}{c_p}\sin\left(\frac{\omega r}{c_p}\right)\right) \tag{10.43b}$$

and the integration constants c_1 and c_2. For the limit $r \to 0$, $y_1(\omega r/c_p) \to -\infty$ results. As the displacement amplitude at the centre of the solid sphere must remain finite, $c_2 = 0$ follows. For a prescribed $u_0(\omega)$ at $r = r_0$, Eq. 10.42 is written as

$$u(\omega) = \frac{j_1\left(\dfrac{\omega r}{c_p}\right)}{j_1\left(\dfrac{\omega r_0}{c_p}\right)} u_0(\omega) \tag{10.44}$$

The amplitudes of the strains follow from Eq. A.1.2 and those of the stress $\sigma_r(\omega)$ from Eq. A.1.3 leading to the nodal force amplitude

$$R(\omega) = 4\pi r_0^2 \sigma_r(\omega, r = r_0) = S^b(\omega) u_0(\omega) \tag{10.45}$$

where the dynamic-stiffness coefficient equals

$$S^b(a_0) = \frac{8\pi}{1-2\nu} G r_0 \frac{(1-\nu)a_0 j_1(a_0)_{,a_0} + 2\nu j_1(a_0)}{j_1(a_0)} \tag{10.46}$$

with the dimensionless frequency

$$a_0 = \frac{\omega r_0}{c_p} \tag{10.47}$$

Performing a Taylor expansion at $a_0 = 0$ results in

$$S^b(\omega) = 8\pi \frac{1+\nu}{1-2\nu} G r_0 - \omega^2 \frac{4}{5}\pi r_0^3 \rho + O(\omega^4) \tag{10.48}$$

The first term is the static-stiffness coefficient K^b (Eq. 10.34), and the coefficient of $-\omega^2$ in the second term is equal to the mass

$$M^b = \frac{4}{5}\pi r_0^3 \rho \tag{10.49}$$

With the consistent infinitesimal finite-element cell method the mass follows from (Eq. 10.23) with M^0 in Eq. 5.205

$$M^b = \frac{4}{5}\pi r_0^3 \rho \qquad (10.50)$$

which again coincides with the value determined from the asymptotic expansion of the exact solution at low frequency (Eq. 10.49).

10.4.2 Cantilever

As a two-dimensional case a cantilever with one end fixed of length l and height h with modulus of elasticity E and Poisson's ratio $v = 0.2$ in plane stress is examined (Figure 10-2). First, statics is addressed. The ratio $h/l = 1/3$ is selected. The

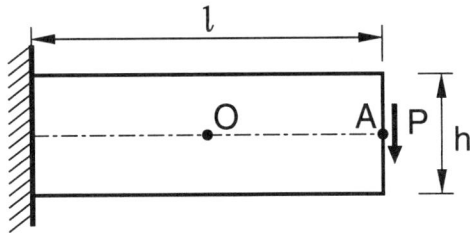

Figure 10-2 Cantilever in plane stress

similarity centre O is chosen at the centre of the cantilever. The discretization of the boundary with n_l and n_h 3-node line elements per side in the horizontal and vertical directions yields $2(n_l + n_h)$ elements. In the static-stiffness matrix $[K^b]$ of order $8(n_l + n_h) \times 8(n_l + n_h)$ the boundary condition at the fixed end is enforced. For the sake of comparison a distributed shear force P varying parabolically over the height is applied at the free end. For various discretizations the vertical displacement u of Point A (Figure 10-2) is listed in Table 10.1, together with the exact solution taken from a figure in Reference [21]. As expected in a procedure based on finite elements, the highly accurate results of the consistent infinitesimal finite-element cell method converge from the stiff side.

Table 10.1 Vertical displacement of cantilever

	$n_l = 2$, $n_h = 1$	$n_l = 3$, $n_h = 1$	$n_l = 4$, $n_h = 1$	$n_l = 4$, $n_h = 2$	$n_l = 6$, $n_h = 2$	EXACT
uE/P	115.2	115.7	115.8	116.0	116.1	≈116.2

10.4 ACCURACY

Second, the mass matrix is determined which permits the natural frequencies to be calculated (Figure 10-2). The following parameters are selected: $l = 24m$, $h = 6m$, Poisson's ratio $= 0.2$, $\sqrt{E/\rho} = 100m/s$ (mass density ρ). In the discretization of the boundary $n_h = 1$ is selected and n_l is varied. The total number of line elements equals $2(n_l + 1)$. In the static-stiffness matrix $[K^b]$ and in the mass matrix $[M^b]$ of orders $8(n_l + 1) \times 8(n_l + 1)$ the boundary condition at the fixed end is enforced yielding $[K]$ and $[M]$ of order $(8n_l + 2) \times (8n_l + 2)$. The natural frequencies ω_i follow from the eigenvalue problem

$$\left|[K] - \omega_i^2[M]\right| = 0 \tag{10.51}$$

The periods $T_i = 2\pi/\omega_i$ of the first five natural frequencies are plotted verses the number of elements in Figure 10-3. Boundary-element results taken from

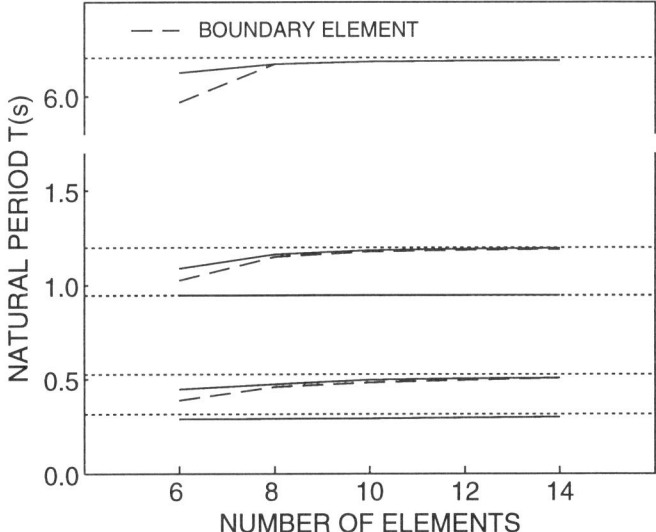

Figure 10-3 Natural periods of cantilever

Reference [36] are also presented where the solution using a very fine finite-element mesh is also specified. This is shown as a dotted line. Convergence of the results of the consistent infinitesimal finite-element cell method from the stiff side is achieved. Even for 6 elements the results are accurate exhibiting smaller deviations than those of the boundary-element solution.

10.4.3 Edge-cracked Plate

The consistent infinitesimal finite-element cell method is applied to linear fracture mechanics where stress singularities must be represented. The in-plane motion of a square isotropic homogeneous plate with a crack in plane stress is addressed (Figure 10-4). The straight crack of length a is perpendicular to the side of length b, as shown in the figure. The applied normal stress σ_0 leads to an opening mode (mode I).

The similarity centre O is chosen at the tip of the crack. The straight crack faces are traction free. As these pass through the similarity centre, no discretization on the crack faces is necessary. At the intersection point of each crack face with the side of the plate, a node is chosen. These two nodes P_1 and P_2 with the same coordinates are independent. The discretization of the sides with 15 3-node line elements is shown in the figure.

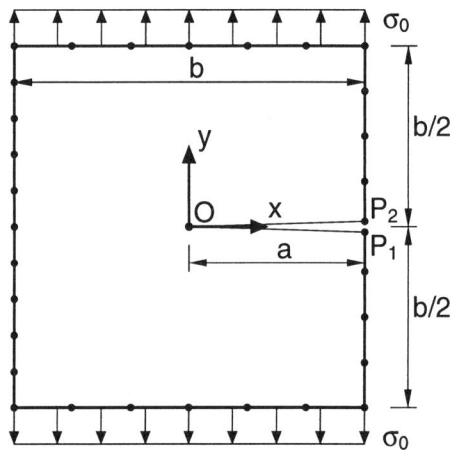

Figure 10-4 Edge-cracked homogeneous plate

The displacement in the interior of the plate is described in Eq. 10.19 with r measured from the crack tip. For $s = 2$, the spatial variation of the displacements in the radial direction is represented by $r^{-\lambda_i}$ with the ith eigenvalue λ_i determined from Eq. 10.15. (Note that in the theory of elasticity a sign change for λ_i applies.)

The first few λ_i are compared. The eigenvalues 0, 0 and -1 correspond to the 2 translational and the rotational rigid-body motions. For $a = 0.5b$ the first 5 of the remaining eigenvalues are -0.49988, -0.50002, -1.00000, -1.49992, -1.50017. Changing a e.g. to $a = 0.3b$ only affects the last digit. The values are very close to those of the eigenvalues of the theory of elasticity $-0.5, -0.5, -1, -1.5, -1.5$. The first 2 eigenvalues result in the singularities of the stresses at the crack tip ($\sim r^{-0.5}$), and the third corresponds to constant stresses. The remaining eigenvalues do not lead to stresses at the crack tip.

In standard finite-element and boundary-element methods special techniques are required to calculate the stress intensity factors. In the consistent infinitesimal finite-

10.4 ACCURACY

Table 10.2 Dimensionless stress intensity factor of edge-cracked homogeneous plate

CRACK LENGTH (a/b)	FINITE-ELEMENT DISCRETIZATION		REFERENCE [11]
	15 LINE ELEMENTS	29 LINE ELEMENTS	
0.2	1.482	1.488	1.488
0.3	1.845	1.849	1.848
0.4	2.326	2.324	2.325
0.5	3.017	3.014	3.010
0.6	4.168	4.157	4.153

element cell method the stress distribution in the radial direction is expressed analytically, and this permits the stress intensity factors to be calculated directly based on their definitions. The stress intensity factor for mode I of the edge-cracked plate (Figure 10-4) equals

$$K_I = \sqrt{2\pi} \lim_{r \to 0} \left(r^{1+\lambda} \sigma_y(r, \theta = 0) \right) \qquad (10.52)$$

with $\lambda = -0.49988$. The expression $r^{1+\lambda}\sigma_y(r, \theta = 0)$ follows from the strains in Eq. 8.36 and the elasticity matrix $[D]$.

The stress intensity factors K_I non-dimensionalized with $\sigma_0\sqrt{\pi a}$ are listed for various crack lengths a/b in Table 10.2. The results for a finer finite-element discretization with 29 3-node line elements, where the first 5 eigenvalues (excluding the 3 corresponding to the rigid-body motion) are -0.50000, -0.50000, -1.00000, -1.50000, -1.50000, are also shown. The analytical solution of Reference [11] is used for comparison. The finer discretization leads to highly accurate results with a deviation of 0.1%. The discretization with 15 line elements, yielding a deviation of 0.4%, is comparable to the accuracy of boundary-element results with 32 3-node boundary elements (see Reference [1]).

As a first step towards analysing inhomogeneous and anisotropic cases of fracture mechanics, where the consistent infinitesimal finite-element cell method is extremely powerful, the in-plane motion of an edge-cracked plate with two zones of different materials is examined (Figure 10-5). The 2 eigenvalues resulting in stress singularities are complex conjugates. Their analytical solution for an infinite plate is specified for plane stress as (Reference [42])

$$\lambda = -0.5 \pm i\frac{1}{2\pi} \ln \frac{\frac{3-v_1}{1+v_1}\frac{G_2}{G_1} + 1}{\frac{3-v_2}{1+v_2} + \frac{G_2}{G_1}} \qquad (10.53)$$

with the shear modulus G and Poisson's ratio v. The subscripts 1 and 2 refer to the two materials.

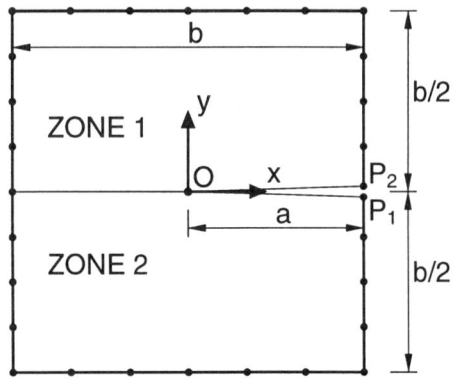

Figure 10-5   Edge-cracked inhomogeneous plate

Table 10.3   Eigenvalues of edge-cracked inhomogeneous plate

	FINITE-ELEMENT DISCRETIZATION		
G_2/G_1	14 LINE ELEMENTS	28 LINE ELEMENTS	Eq. 10.53
1	−0.49993	−0.50000	−0.5
5	−0.49998 ± i0.08679	−0.50000 ± i0.08697	−0.5 ± i0.08699
10	−0.49997 ± i0.10791	−0.50000 ± i0.10814	−0.5 ± i0.10815

The boundary of the plate is discretized with 14 3-node line elements. The similarity centre is selected at the crack tip. As the interface between the two materials passes through the similarity centre, equilibrium and compatibility are satisfied without any discretization. Again, the crack faces are not discretized. The crack length equals $a = 0.5b$. For a constant $v_1 = v_2 = 0.2$, G_2/G_1 is varied. A finer mesh with 28 3-node elements is also processed. The eigenvalues λ are compared with Eq. 10.53 in Table 10.3. Excellent agreement results.

The consistent infinitesimal finite-element cell method does not require any discretization of the straight crack faces and the interfaces between materials, in contrast to the boundary-element procedure. No error from a practical point of view is introduced in the radial direction permitting a rigorous representation of the stress singularities around the crack tip. The stress intensity factor is calculated directly from its definition. Interior cracks with more than one crack tip can also be addressed by division into subdomains containing only one crack tip, where the similarity centre is chosen.

Part II

DAMPING-SOLVENT EXTRACTION FOR DYNAMIC STIFFNESS AND INTERACTION FORCE

What you can see, yet cannot see over, is as good as infinite.

>					Thomas Carlyle 1795–1881

PART II DAMPING-SOLVENT EXTRACTION

As another procedure for modelling the unbounded medium for use in the substructure method of a dynamic unbounded medium-structure-interaction analysis, the so-called damping-solvent extraction method is discussed. It is based on the physical notion that *waves propagating in a damped medium decay*. The damping-solvent extraction method permits an efficient analysis of the dynamic response of an unbounded medium by calculating the part of the unbounded medium adjacent to the structure-medium interface only, in the same way as for a (bounded) structure. It is approximate but easy to implement, as the standard finite-element method for a bounded medium is sufficient for the analysis. In the frequency domain the dynamic stiffness of the unbounded medium is calculated whereby any specific frequency of interest can be evaluated directly. In the time domain a transient can be processed directly leading to the interaction forces of the unbounded medium without first determining the unit-impulse response functions.

Chapter 11 discusses the fundamentals of the damping-solvent extraction method. The effect of damping on the dynamic stiffness is studied, which provides the basis for the extraction of damping.

Chapter 12 addresses the implementation of the damping-solvent extraction method in the frequency domain using linear hysteretic material damping and in the time domain with exterior mass-proportional dashpots. The method is verified analytically using a simple one-dimensional example. High accuracy is demonstrated.

This Part II is based on References [48], [64] and [62].

11

FUNDAMENTALS OF DAMPING-SOLVENT EXTRACTION METHOD

The concept is explained in Section 11.1 without specifying any details. The effect of damping on the dynamic stiffness is examined in Section 11.2, followed by the procedure for extracting the damping in Section 11.3.

11.1 CONCEPT

In the substructure method of dynamic unbounded medium-structure-interaction analysis, the interaction force-displacement relationship in the nodes on the structure-medium interface of the unbounded medium is addressed. In a frequency-domain analysis the dynamic-stiffness matrix is calculated. In a time-domain analysis when addressing a transient excitation directly the time history of the interaction forces is determined from that of the displacements on the structure-medium interface. For both types of analyses the damping-solvent extraction method can be used.

An explanation of the use of a solvent extraction outside computational mechanics is discussed at the beginning of Section 1.5.2. The damping-solvent extraction method consists of three steps.

In the *first step*, a *finite region of the unbounded medium* adjacent to the structure-medium interface, a bounded medium, is modelled with *finite elements* (Figure 11-1), whereby *artificial damping* which is not present in the actual medium *is introduced as a solvent*. Internal damping such as linear hysteretic material damping or exterior damping caused essentially by dashpots can be used. While in a frequency-domain analysis a wide range of choices for damping, including that associated with non-causal behaviour, exists, in a time-domain analysis the introduced damping must be physically realizable. The effect of this damping consists of reducing the amplitudes of the outgoing waves f propagating from the structure-medium interface towards the outer boundary and after reflection, diminishing the amplitudes of the reflected waves g, resulting in negligible amplitudes when reaching the structure-medium interface.

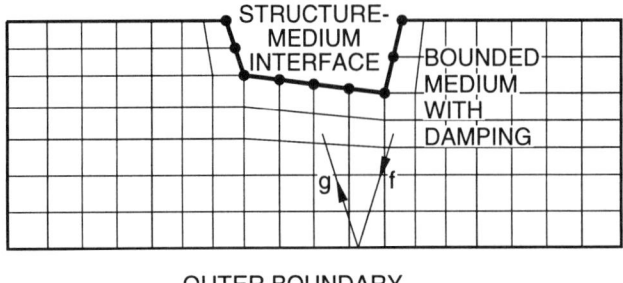

Figure 11-1 Finite-element discretization of finite region of unbounded medium adjacent to structure-medium interface

The *damping acting as a solvent* thus leads to the structure-medium interface's motion depending only on the outgoing waves f. The amplitudes of the reflected waves g can be further reduced by placing viscous dashpots at the outer boundary, instead of fixing it. This allows the dimension of the bounded medium in the radial direction to be reduced. In a frequency-domain analysis, all degrees of freedom of this bounded medium with the exception of those on the structure-medium interface can be eliminated, leading to the dynamic-stiffness matrix of the (artificially) damped bounded medium. As this analysis is performed in the frequency domain, hysteretic damping is straightforwardly introduced based on the correspondence principle. In a time-domain analysis for a specified time history of displacements in the nodes on the structure-medium interface, the interaction forces of the (artificially) damped bounded medium can be calculated. The analysis of this first step is analogous to that performed for a bounded structure.

In the *second step*, only assumptions are formulated and no calculations are performed. The dynamic-stiffness matrix of the *damped unbounded medium* and its first derivative with respect to frequency are assumed to be equal to the corresponding values of the (artificially) *damped bounded medium* determined in the first step with the same introduced artificial damping. For a sufficiently large damping ratio and finite region the reflected waves g at the structure-medium interface will be small. If no reflected waves g existed at the structure-medium interface, the two dynamic-stiffness matrices would be exactly equal.

In the *third step*, the influence of the *introduced artificial damping*, the solvent, on the dynamic-stiffness matrix *is extracted*. This extraction of the damping solvent involves a Taylor expansion of the dynamic-stiffness matrix of the (artificially) damped bounded medium using the constant and linear terms in the dimensionless frequency. In a frequency-domain analysis the dynamic-stiffness matrix of the corresponding undamped unbounded medium for the degrees of freedom in the nodes on the structure-medium interface thus results. This elimination of the damping solvent can be performed for each element of the matrix independently of the others and for each frequency. This operation is very simple. The computational effort of this third step can be neglected when compared with that of the first step. In the derivation for a time-domain analysis, the inverse Fourier transformation is performed. For the exterior mass-proportional dashpot system, it can be shown that the interaction forces follow as a linear combination of two loading cases for the damped bounded medium.

To make the explanation of the damping-extraction method easier to understand, it is assumed that the unbounded medium is undamped. Of course, the procedure also works for a damped unbounded medium, whereby only the introduced artificial damping is extracted in the third step.

11.2 EFFECT OF DAMPING ON DYNAMIC STIFFNESS

The damping-solvent extraction method works with a bounded medium, where artificial damping is introduced in the first step and extracted in the third step. It is thus necessary to investigate the effect of introducing damping on the dynamic stiffness. In a frequency-domain analysis damping is extracted directly in the frequency domain leading to the final dynamic stiffness. In a time-domain analysis the derivation to extract damping is also performed in the frequency domain. Two types of damping are addressed: interior material damping and essentially exterior mass-proportional dashpots. The effect of damping on the dynamic stiffness is demonstrated using the finite-element formulation.

In the frequency domain for an undamped bounded medium the dynamic-stiffness matrix corresponding to all degrees of freedom (superscript t for total) is written as

$$[S^t(\omega)] = [K] - \omega^2[M] \tag{11.1}$$

The dimensionless matrices $[\overline{K}]$ and $[\overline{M}]$ are introduced

$$[K] = Gr_0^{s-2}[\overline{K}] \tag{11.2a}$$

$$[M] = \rho r_0^s [\overline{M}] \tag{11.2b}$$

with the shear modulus G, mass density ρ, length r_0 and spatial dimension s ($=2$ or $=3$). $[\overline{K}]$ does not depend on G nor $[\overline{M}]$ on ρ. Eq. 11.1 is reformulated as

$$[S^t(\omega)] = Gr_0^{s-2}[\overline{K}] - \omega^2 \rho r_0^s [\overline{M}] \tag{11.3}$$

Introducing the dimensionless frequency

$$a_0 = \frac{\omega r_0}{c_s} \tag{11.4}$$

with the shear-wave velocity

$$c_s = \sqrt{\frac{G}{\rho}} \tag{11.5}$$

Eq. 11.3 is written as

$$[S^t(\omega)] = G r_0^{s-2}([\overline{K}] - a_0^2[\overline{M}]) \tag{11.6}$$

After elimination of all degrees of freedom which are not located on the structure-medium interface, the (reduced) dynamic-stiffness matrix on the structure-medium interface will still be proportional to $G r_0^{s-2}$

$$[S(\omega)] = G r_0^{s-2}[\overline{S}(a_0)] \tag{11.7}$$

The dimensionless dynamic-stiffness matrix $[\overline{S}(a_0)]$ depends on a_0; the actual function is determined by $[\overline{K}]$ and $[\overline{M}]$.

Interior material damping is introduced using the correspondence principle by replacing G by G^*. G^* is a complex function which can also depend on ω. Eq. 11.6 is transformed to (subscript ζ for damping)

$$[S_\zeta^t(\omega)] = G^* r_0^{s-2}([\overline{K}] - a_0^{*2}[\overline{M}]) \tag{11.8}$$

where

$$a_0^* = \frac{\omega r_0}{c_s^*} \tag{11.9}$$

with

$$c_s^* = \sqrt{\frac{G^*}{\rho}} \tag{11.10}$$

The dynamic-stiffness matrix on the structure-medium interface (Eq. 11.7) is then written as

$$[S_\zeta(\omega)] = G^* r_0^{s-2}[\overline{S}(a_0^*)] \tag{11.11}$$

Note that $[\overline{S}(a_0^*)]$ in Eq. 11.11 is the same function as in Eq. 11.7, as the same matrices $[\overline{K}]$ and $[\overline{M}]$ are present in Eqs. 11.8 and 11.6. In Eq. 11.11 $[\overline{S}(a_0^*)]$ is evaluated at a_0^* specified in Eq. 11.9 and in Eq. 11.7 $[\overline{S}(a_0)]$ at a_0. Thus, *the dimensionless dynamic-stiffness matrices $[\overline{S}(a_0)]$ and $[\overline{S}(a_0^*)]$ are the same functions evaluated at a_0 for the undamped medium and at a_0^* for the damped medium, respectively.*

11.2 EFFECT OF DAMPING ON DYNAMIC STIFFNESS

Linear hysteretic material damping is used in the frequency-domain analysis in Chapter 12 where

$$G^* = G(1+2i\zeta) \tag{11.12}$$

with the damping ratio ζ.

In the time domain the equations of motion of the undamped bounded medium with all degrees of freedom corresponding to Eq. 11.1 equal

$$[M]\{\ddot{u}^t\} + [K]\{u^t\} = \{P\} \tag{11.13}$$

with the displacement vector $\{u^t\}$ and the load vector $\{P\}$.

Exterior damping can be chosen for a time-domain analysis. In principle, various possibilities exist. However, it is important to select a damping mechanism whose effect can easily be extracted in the time domain in the third step. A formulation which avoids convolution integrals is described.

To provide artificial damping, exterior dashpots are added to the bounded medium. A mass-proportional damping matrix $2\zeta[M]$ is selected with the damping coefficient ζ of dimension c_s/r_0. As will become apparent after transformation to the frequency domain, a concise formulation of the dynamic-stiffness matrix can be obtained by augmenting the static-stiffness matrix by $\zeta^2[M]$. For the damped modified bounded medium, the equations of motion are equal to

$$[M]\{\ddot{u}^t\} + 2\zeta[M]\{\dot{u}^t\} + ([K] + \zeta^2[M])\{u^t\} = \{P\} \tag{11.14}$$

For short, this added system consisting of dashpots and springs is called the *exterior mass-proportional dashpots*.

The dynamic-stiffness matrix of Eq. 11.14 is written in the frequency domain as

$$[S^t_\zeta(\omega)] = [K] + \zeta^2[M] + i\omega 2\zeta[M] - \omega^2[M] = [K] - (\omega - i\zeta)^2[M] \tag{11.15}$$

Using Eq. 11.2, Eq. 11.15 can be reformulated as

$$[S^t_\zeta(\omega)] = Gr_0^{s-2}([\overline{K}] - a_0^{*2}[\overline{M}]) \tag{11.16}$$

where

$$a_0^* = \frac{(\omega - i\zeta)r_0}{c_s} \tag{11.17}$$

Eq. 11.16 for the damped medium is in the same form as that for the undamped medium (see Eq. 11.6). Eliminating all degrees of freedom not located on the structure-medium interface yields

$$[S_\zeta(\omega)] = Gr_0^{s-2}[\overline{S}(a_0^*)] \tag{11.18}$$

Again, the function $[\bar{S}(a_0^*)]$ in Eq. 11.18 is the same as in Eq. 11.7, but evaluated at a_0^* specified in Eq. 11.17. Note that for the artificial damping introduced by the exterior mass-proportional dashpots, the dynamic-stiffness matrix is proportional to G (Eq. 11.18) which is not affected by the artificial damping (in contrast to the internal material damping, Eq. 11.11).

11.3 EXTRACTION OF EFFECT OF DAMPING

In the second step, it is assumed that the dynamic-stiffness matrix of the damped unbounded medium $[S_\zeta^\infty(\omega)]$ is approximately equal to that of the damped bounded medium $[S_\zeta(\omega)]$. The same also applies to their first derivatives with respect to ω

$$[S_\zeta^\infty(\omega)] = [S_\zeta(\omega)] \tag{11.19}$$

$$[S_\zeta^\infty(\omega)]_{,\omega} = [S_\zeta(\omega)]_{,\omega} \tag{11.20}$$

Dividing Eq. 11.19 by $G^* r_0^{s-2}$ with

$$[S_\zeta^\infty(\omega)] = G^* r_0^{s-2} [\bar{S}^\infty(a_0^*)] \tag{11.21}$$

and with Eq. 11.11 yields

$$[\bar{S}^\infty(a_0^*)] = [\bar{S}(a_0^*)] \tag{11.22}$$

Analogously, dividing Eq. 11.20 by $G^* r_0^{s-2}$, substituting Eqs. 11.21 and 11.11 and with $d\omega = da_0^*/a_{0,\omega}^*$ results in

$$[\bar{S}^\infty(a_0^*)]_{,a_0^*} = [\bar{S}(a_0^*)]_{,a_0^*} \tag{11.23}$$

Thus, the assumptions of Eqs. 11.19 and 11.20 lead to the same properties formulated in the dimensionless dynamic-stiffness matrices (denoted by a bar) as a function of a_0^* (Eqs. 11.22 and 11.23).

In the third step, to extract the effect of damping from $[\bar{S}^\infty(a_0^*)]$, $[\bar{S}^\infty(a_0)]$ is evaluated with a_0 corresponding to the same ω as a_0^*. To calculate $[\bar{S}^\infty(a_0)]$ from $[\bar{S}^\infty(a_0^*)]$ the first two terms of a Taylor expansion of $[\bar{S}^\infty(a_0^*)]$ are formulated

$$[\bar{S}^\infty(a_0)] = [\bar{S}^\infty(a_0^*)] + [\bar{S}^\infty(a_0^*)]_{,a_0^*}(a_0 - a_0^*) \tag{11.24}$$

Note that Eq. 11.24 is valid for each coefficient of the matrix. To revert on the right-hand side of Eq. 11.24 to the dynamic-stiffness matrix of the damped bounded medium $[S_\zeta(\omega)]$, Eq. 11.24 is multiplied by $G^* r_0^{s-2}$, and $da_0^* = a_{0,\omega}^* d\omega$ as well as Eqs. 11.22 and 11.23 being substituted. In addition

$$[S^\infty(\omega)] = G r_0^{s-2} [\bar{S}^\infty(a_0)] \tag{11.25}$$

11.3 EXTRACTION OF EFFECT OF DAMPING

is substituted on the left-hand side yielding

$$[S^\infty(\omega)] = \frac{G}{G^*} \left([S_\zeta(\omega)] + [S_\zeta(\omega)]_{,\omega} \frac{a_0 - a_0^*}{a_{0,\omega}^*} \right) \tag{11.26}$$

The dynamic-stiffness matrix of the undamped unbounded medium $[S^\infty(\omega)]$ can thus be calculated from that of the damped bounded medium $[S_\zeta(\omega)]$.

The exterior mass-proportional dashpots can be treated as a special case with $G^* = G$ (Eq. 11.18) and a_0^* defined in Eq. 11.17.

12

IMPLEMENTATION, VERIFICATION AND ACCURACY OF DAMPING-SOLVENT EXTRACTION METHOD

The implementations in the frequency domain using linear hysteretic material damping and in the time domain with exterior mass-proportional dashpots are discussed in Section 12.1 and Section 12.2. A flexibility formulation is outlined in Section 12.3. An analytical verification is presented in Section 12.4. The accuracy of the damping-solvent extraction method is evaluated for the dynamic-stiffness matrix in the frequency domain in Section 12.5 and for a transient in the time domain in Section 12.6.

12.1 IMPLEMENTATION IN FREQUENCY DOMAIN

The damping-solvent extraction method in the frequency domain for calculating the dynamic-stiffness matrix of the unbounded medium is explained for the linear hysteretic material damping which is commonly used in the dynamic analysis of bounded structures.

The three steps of the procedure are described in detail. It is assumed that the reader is familiar with the concept outlined in Section 11.1.

FIRST STEP

To introduce artificial linear hysteretic damping with the damping ratio ζ, the *correspondence principle* is applied. Both Lamé constants λ and G are multiplied by $1+2i\zeta$

$$\lambda^* = \lambda(1+2i\zeta) \qquad (12.1a)$$
$$G^* = G(1+2i\zeta) \qquad (12.1b)$$

This results in the same factor $\sqrt{1+2i\zeta}$ being applied to the shear-wave and dilatational-wave velocities c_s and c_p

$$c_s^* = c_s\sqrt{1+2i\zeta} \qquad (12.2a)$$
$$c_p^* = c_p\sqrt{1+2i\zeta} \qquad (12.2b)$$

Poisson's ratio ν is not modified by this choice.

The terms undamped medium and damped medium refer to the original medium before and after introducing artificial damping, respectively. The original medium could exhibit any type of damping of its own, which is not affected by introducing and extracting the artificial damping.

The finite region of the unbounded medium adjacent to the structure-medium interface, a bounded medium, is discretized with finite elements (Figure 12-1). Only

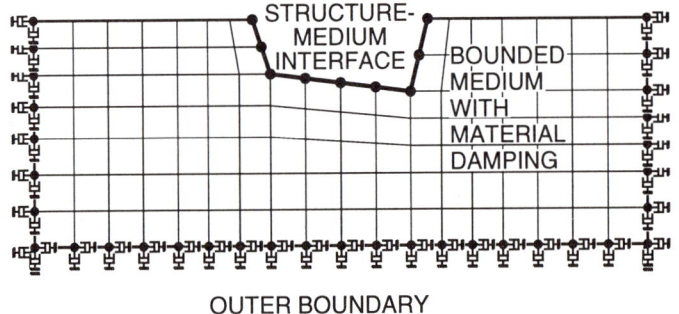

Figure 12-1 Finite-element discretization of finite region of unbounded medium adjacent to structure-medium interface with dashpots on outer boundary

a few rows (e.g. 4 parabolic elements) in the direction from the structure-medium interface to infinity are necessary when the familiar viscous dashpots are used on the outer boundary to further reduce the amplitudes of the reflected waves.

To introduce linear hysteretic damping, the static-stiffness matrix of the finite-element assemblage of the bounded medium is multiplied by $1+2i\zeta$ (Eq. 12.1). The coefficients of the viscous dashpots per unit surface area at the outer boundary are equal to ρc_p^* in the perpendicular direction and to ρc_s^* in the two tangential directions (Eq. 12.2). All other modelling aspects of the first step are the same as encountered in conventional structural dynamics for harmonic excitation and need thus not be discussed.

After eliminating for every frequency all degrees of freedom not located on the structure-medium interface, the *dynamic-stiffness matrix of the damped bounded medium at the structure-medium interface* $[S_\zeta(\omega)]$ results.

12.2 IMPLEMENTATION IN TIME DOMAIN

SECOND STEP

As this step consists of assumptions only and does not involve any calculations, it is not addressed in this section describing the algorithm.

THIRD STEP

The extraction of the artificial damping is described in Eq. 11.26. G^* is specified in Eq. 12.1, and a_0^* follows from Eq. 11.9 with Eq. 12.2a as (characteristic length of structure-medium interface r_0)

$$a_0^* = \frac{\omega r_0}{c_s \sqrt{1+2i\zeta}} \quad (12.3)$$

The derivative of Eq. 12.3 yields

$$a_{0,\omega}^* = \frac{r_0}{c_s \sqrt{1+2i\zeta}} \quad (12.4)$$

As a large hysteretic damping ratio ($0.2 \leq \zeta \leq 0.4$) is chosen, no expansion of $\sqrt{1+2i\zeta}$ is performed. Substituting Eqs. 12.1, 12.3 and 12.4 in Eq. 11.26 yields

$$[S^\infty(\omega)] = \frac{1}{1+2i\zeta}\left([S_\zeta(\omega)] + (\sqrt{1+2i\zeta}-1)\omega[S_\zeta(\omega)]_{,\omega}\right) \quad (12.5)$$

Eq. 12.5 applies for each element of the dynamic-stiffness matrix independently of the others. It can be evaluated separately for each ω.

In an actual calculation, the first step provides each element $S_\zeta(\omega_j)$ of the dynamic-stiffness matrix of the damped bounded medium at discrete frequencies ω_j. The derivative in Eq. 12.5 is e.g. expressed as

$$S_\zeta(\omega)_{,\omega}\Big|_{\omega=\omega_j} = \frac{S_\zeta(\omega_{j+1}) - S_\zeta(\omega_{j-1})}{\omega_{j+1} - \omega_{j-1}} \quad (12.6)$$

The corresponding *dynamic-stiffness coefficient of the undamped unbounded medium* $S^\infty(\omega_j)$ is calculated as

$$S^\infty(\omega_j) = \frac{1}{1+2i\zeta}\left(S_\zeta(\omega_j) + (\sqrt{1+2i\zeta}-1)\omega_j\, S_\zeta(\omega)_{,\omega}\Big|_{\omega=\omega_j}\right) \quad (12.7)$$

As the damping extraction is performed for each frequency independent of the others, the damping ratio ζ can also be varied as a function of frequency.

12.2 IMPLEMENTATION IN TIME DOMAIN

The damping-solvent extraction method in the time domain to calculate the interaction forces of the unbounded medium is derived for the exterior mass-proportional dashpots defined in Section 11.2 (Eq. 11.14). The concept is outlined in Section 11.1.

Instead of determining the unit-impulse response as the inverse Fourier transform of the dynamic stiffness, the interaction forces are calculated directly analysing the damped bounded medium in the time domain for two loading cases.

Eq. 11.26 still applies when the exterior mass-proportional dashpots are used with $G^* = G$ (see Eq. 11.18) and the dimensionless frequency (Eq. 11.17) defined correspondingly

$$a_0^* = \frac{(\omega - i\zeta)r_0}{c_s} \tag{12.8}$$

Substituting in Eq. 11.26 with

$$a_{0,\omega}^* = \frac{r_0}{c_s} \tag{12.9}$$

yields

$$[S^\infty(\omega)] = [S_\zeta(\omega)] + i\zeta[S_\zeta(\omega)]_{,\omega} \tag{12.10}$$

Applying the inverse Fourier transformation to Eq. 12.10 leads to the displacement unit-impulse response matrix

$$[S^\infty(t)] = (1 + \zeta t)[S_\zeta(t)] \tag{12.11}$$

The interaction forces of the undamped unbounded medium (Eq. 1.4) are specified as

$$\{R(t)\} = \int_0^t [S^\infty(t-\tau)]\{u(\tau)\}d\tau \tag{12.12}$$

Substituting Eq. 12.11 in Eq. 12.12 results in

$$\{R(t)\} = (1+\zeta t)\int_0^t [S_\zeta(t-\tau)]\{u(\tau)\}d\tau - \zeta \int_0^t [S_\zeta(t-\tau)]\tau\{u(\tau)\}d\tau \tag{12.13}$$

or

$$\{R(t)\} = (1+\zeta t)\{R_\zeta(t)\} - \zeta\{R_{\zeta r}(t)\} \tag{12.14}$$

with the interaction forces of the damped bounded medium for the (original) loading case $\{u(t)\}$ on the structure-medium interface

$$\{R_\zeta(t)\} = \int_0^t [S_\zeta(t-\tau)]\{u(\tau)\}d\tau \tag{12.15}$$

and with the interaction forces of the damped bounded medium for the loading case (subscript r for ramp function)

$$\{u_r(t)\} = t\{u(t)\} \tag{12.16}$$

on the structure-medium interface

$$\{R_{\zeta r}(t)\} = \int_0^t [S_\zeta(t-\tau)]\{u_r(\tau)\}d\tau \tag{12.17}$$

The three steps of the procedure are formulated in detail as follows

FIRST STEP

$\{R_\zeta(t)\}$ and $\{R_{\zeta r}(t)\}$ are not calculated from Eqs. 12.15 and 12.17 but from solving Eq. 11.14. $\{u(t)\}$ and $\{u_r(t)\}$ (Eq. 12.16) are enforced in the nodes on the structure-medium interface of the damped bounded medium (Figure 12-1). All nodes not located on the structure-medium interface are unloaded. The interaction forces on the structure-medium interface determined by solving Eq. 11.14 are equal to $\{R_\zeta(t)\}$ and $\{R_{\zeta r}(t)\}$.

If viscous dashpots are placed at the outer boundary to further reduce the amplitudes of the reflected waves, springs must also be attached to achieve a consistent formulation for the damped case. This is verified as follows. The dynamic-stiffness coefficients of the dashpots per unit area for the undamped medium equal $i\omega\rho c$ with the mass density ρ and the wave velocity c where $c = c_p$ (dilatational-wave velocity) in the perpendicular direction and $c = c_s$ (shear-wave velocity) in the tangential directions. As explained in connection with Eqs. 11.16 and 11.17, the exterior mass-proportional dashpots correspond to replacing ω by $\omega - i\zeta$. This yields $i\omega\rho c + \zeta\rho c$. The second term represents the coefficients of the springs per unit area, which are also present in the damped system. The dashpots and the springs at the outer boundary result in additional terms in the damping and static-stiffness matrices of Eq. 11.14.

SECOND STEP

As the assumptions of this step are incorporated in the derivation of Eq. 12.14, no additional calculations are necessary.

THIRD STEP

The final interaction forces $\{R(t)\}$ at the structure-medium interface of the undamped unbounded medium are calculated from Eq. 12.14 using $\{R_\zeta(t)\}$ and $\{R_{\zeta r}(t)\}$ determined in the first step. Note that *a transient can be processed directly* without first evaluating unit-impulse response functions and then calculating the subsequent convolution integrals. Only two loading cases are analysed in this simple method. The introduced damping is extracted for each degree of freedom and for each time step independently of the others.

12.3 FLEXIBILITY FORMULATION

When a structure is present and thus $\{u(t)\}$ is not known, standard substructure methods are applicable in a dynamic unbounded medium-structure-interaction analysis. As an alternative to the stiffness procedure described above, a flexibility formulation can be applied. The latter can be attractive if only the unbounded medium is to be analysed with a known transient of interaction forces acting on the (structure-medium) interface.

In the frequency domain the dynamic-flexibility matrix corresponds to the inverse of the dynamic-stiffness matrix

$$[F^\infty(\omega)] = [S^\infty(\omega)]^{-1} \tag{12.18}$$

The displacement-interaction force relationship formulated in amplitudes equals

$$\{u(\omega)\} = [F^\infty(\omega)]\{R(\omega)\} \tag{12.19}$$

In the time domain, the displacement-interaction force relationship is formulated as

$$\{u(t)\} = \int_0^t [F^\infty(t-\tau)]\{R(\tau)\}d\tau \tag{12.20}$$

with the displacement response matrix to a unit impulse of the interaction forces.

The derivation of the flexibility formulation in the time domain is analogous to that of the stiffness formulation described in Sections 11.2, 11.3 and 12.2. Only certain key equations are summarized.

Eq. 11.18 corresponds to

$$[F_\zeta(\omega)] = \frac{1}{Gr_0^{s-2}}[\overline{F}(a_0^*)] \tag{12.21}$$

with the dimensionless dynamic-flexibility matrix $[\overline{F}(a_0^*)]$ being a function of the dimensionless frequency a_0^* (Eq. 11.17) only. The assumptions in Eqs. 11.19 and 11.20 are written as

$$[F_\zeta^\infty(\omega)] = [F_\zeta(\omega)] \tag{12.22}$$

$$[F_\zeta^\infty(\omega)]_{,\omega} = [F_\zeta(\omega)]_{,\omega} \tag{12.23}$$

For the exterior mass-proportional dashpots, the dynamic-flexibility matrix follows as (Eq. 12.10)

$$[F^\infty(\omega)] = [F_\zeta(\omega)] + i\zeta[F_\zeta(\omega)]_{,\omega} \tag{12.24}$$

Eq. 12.14 corresponds to

$$\{u(t)\} = (1+\zeta t)\{u_\zeta(t)\} - \zeta\{u_{\zeta r}(t)\} \tag{12.25}$$

where $\{u_\zeta(t)\}$ and $\{u_{\zeta r}(t)\}$ are the displacements on the structure-medium interface caused by the applied interaction force $\{R(t)\}$ and

$$\{R_r(t)\} = t\{R(t)\} \tag{12.26}$$

respectively.

Thus, in the first step the known interaction forces $\{R(t)\}$ and $\{R_r(t)\}$ (Eq. 12.26) are applied to the damped bounded medium (Figure 12-1) to calculate the displacements $\{u_\zeta(t)\}$ and $\{u_{\zeta r}(t)\}$ at the structure-medium interface by solving Eq. 11.14. In the third step the final displacements $\{u(t)\}$ of the undamped unbounded medium are determined from Eq. 12.25.

12.4 ANALYTICAL VERIFICATION

A simple one-dimensional example is addressed to illustrate the key features of the three steps of the damping-solvent extraction method applying the analytical expressions instead of the results from a finite-element discretization which is used in an actual application. The same example is examined in a parametric study to investigate the influence of the ratio of the introduced artificial damping and the length of the finite region of the unbounded medium (measured from the structure-medium interface to the outer boundary). Also discussed is why, after performing the third step, the dynamic response of the undamped *bounded* medium is not recovered. All studies are performed in the frequency domain using linear hysteretic material damping (Eq. 12.1).

Guidelines on the size of the bounded medium, its finite-element discretization and the damping ratio are formulated.

The semi-infinite rod on an elastic foundation (Figure A-3) discussed in Appendix A.2 is examined. Analytical solutions for the damped case are derived by applying the correspondence principle.

The finite region adjacent to the structure-medium interface, the bounded medium with introduced artificial damping, of the semi-infinite rod on an elastic foundation is shown in Figure 12-2. At the outer boundary a viscous dashpot is introduced to reduce

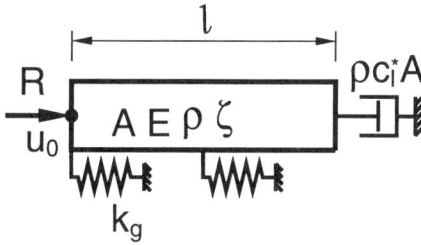

Figure 12-2 Finite region of semi-infinite rod on elastic foundation representing damped bounded medium

the length l. As a damped system is calculated, the dashpot coefficient equals $\rho c_l^* A$ with

$$c_l^* = c_l \sqrt{1 + 2i\zeta} \tag{12.27}$$

The corresponding boundary condition equals

$$N(\omega, x = l) = -\rho c_l^* A i \omega u(\omega, x = l) \tag{12.28}$$

Substituting Eq. A.2.2 yields

$$u_{,x}(\omega, x = l) + \frac{1}{c_l^*} i\omega u(\omega, x = l) = 0 \tag{12.29}$$

The solution of the equation of motion (Eq. A.2.10) for a damped medium is written as

$$u(\omega) = c_1 e^{+i\sqrt{a_0^{*2}-1}\frac{x}{r_0}} + c_2 e^{-i\sqrt{a_0^{*2}-1}\frac{x}{r_0}} \tag{12.30}$$

where

$$a_0^* = \frac{\omega r_0}{c_l^*} = \frac{a_0}{\sqrt{1+2i\zeta}} \tag{12.31}$$

c_1 and c_2 are the amplitudes of the incoming and outgoing waves. At the outer boundary the outgoing wave corresponds to the incident wave and the incoming wave to the reflected wave. Substituting Eq. 12.30 in Eq. 12.29 yields

$$\frac{c_1}{c_2} = \frac{\sqrt{a_0^{*2}-1}-a_0^*}{\sqrt{a_0^{*2}-1}+a_0^*} e^{-2i\sqrt{a_0^{*2}-1}\frac{l}{r_0}} \tag{12.32}$$

The interaction force amplitude (Eq. A.2.12) follows as

$$R(\omega) = -E(1+2i\zeta)Au_0(\omega)_{,x} \tag{12.33}$$

The dynamic-stiffness coefficient is defined by

$$R(\omega) = S_\zeta(\omega)u_0(\omega) \tag{12.34}$$

Substituting Eq. 12.30 yields the dynamic-stiffness coefficient at the beginning of the damped rod with finite length

$$S_\zeta(\omega) = K^\infty(1+2i\zeta)\sqrt{1-a_0^{*2}} \frac{1-\dfrac{c_1}{c_2}}{1+\dfrac{c_1}{c_2}} \tag{12.35}$$

with the static-stiffness coefficient of the undamped semi-infinite rod $K^\infty = \sqrt{EAk_g}$ (Eq. A.2.13). When $c_1/c_2 = 0$, Eq. 12.35 yields the exact result for the damped semi-infinite rod on an elastic foundation

$$S_\zeta^\infty(\omega) = K^\infty(1+2i\zeta)\sqrt{1-a_0^{*2}} \tag{12.36}$$

or

$$S_\zeta^\infty(\omega) = K^\infty(1+2i\zeta)\overline{S}^\infty(a_0^*) \tag{12.37}$$

In this example $K^\infty(1+2i\zeta)$ is used to define $\overline{S}^\infty(a_0^*) = \sqrt{1-a_0^{*2}}$, i.e. K^∞ corresponds to $G^* r_0^{s-2}$ in Eq. 11.21.

12.4 ANALYTICAL VERIFICATION

For the representation of the results, a_0^* of Eq. 12.31 is substituted in Eq. 12.35 which is then decomposed to introduce the dimensionless spring coefficient $k(a_0)$ and damping coefficient $c(a_0)$

$$S_\zeta^\infty(\omega) = K^\infty(k(a_0) + ia_0 c(a_0)) \tag{12.38}$$

For the first step of the damping-solvent extraction method the amplitude ratio c_1/c_2 (Eq. 12.32) is examined. As an example, the bounded medium with the viscous dashpot at the outer boundary has the dimensionless length $l/r_0 = 4$ and the linear hysteretic damping ratio $\zeta = 0.2$. The real and imaginary parts of the amplitude ratio c_1/c_2 are plotted as a dotted-dashed line in Figure 12-3. For comparison, the

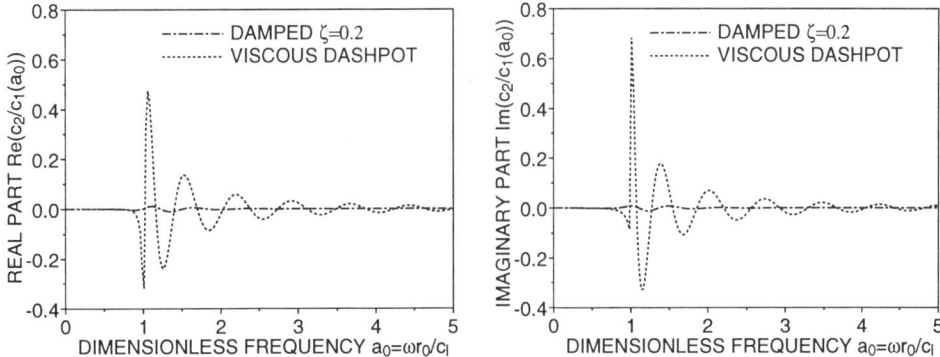

Figure 12-3 Amplitude ratio of reflected wave to incident wave with and without introduced artificial damping after the first step

undamped finite region of the same length with the viscous dashpot at the outer boundary is used to model the semi-infinite rod, as is often done in practice. Its result is plotted as a dotted line in Figure 12-3. Introducing artificial damping in the finite region drastically reduces the reflected wave amplitude c_1.

From c_1/c_2 the corresponding dynamic-stiffness coefficient $S_\zeta(\omega)$ follows from Eq. 12.35. The spring and damping coefficients $k(a_0)$ and $c(a_0)$ follow from Eq. 12.38 and are plotted in Figure 12-4. The exact solution for the undamped semi-infinite rod (Eq. A.2.14) is also shown. As expected, just introducing material damping, and thus performing the first step only, leads to totally unacceptable results (dotted-dashed line). In contrast, after extraction of the material damping as described in Eq. 12.5 of the third step, the dynamic-stiffness coefficient plotted as a solid line is accurate. For comparison, the dynamic-stiffness coefficient of the undamped finite region with a viscous dashpot at the outer boundary, calculated on the basis of Eq. 12.35 with $\zeta = 0$, $a_0^* = a_0$ and the corresponding c_1/c_2, exhibits significant oscillations with large amplitudes around the exact solution (dotted line in Figure 12-4).

Turning to the parametric study using the semi-infinite rod on an elastic foundation, the amplitude ratio c_1/c_2 at the end of the first step (Eq. 12.32) and the dynamic-

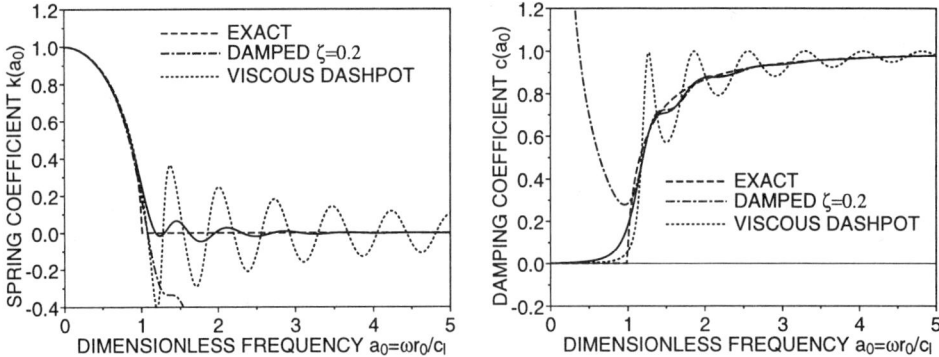

Figure 12-4 Dynamic-stiffness coefficient after extracting introduced artificial damping in third step as well as without and with introduced artificial damping after first step

stiffness coefficient after extraction of the material damping are calculated varying the dimensionless length l/r_0 of the bounded medium as a function of a_0. The material damping ratio is equal to $\zeta = 0.2$ in Figure 12-5. As expected, the larger the length l/r_0 is, the smaller the peaks of the amplitude ratio c_1/c_2 are, tending towards zero (Figure 12-5a) and thus the more accurate the dynamic-stiffness coefficients are (Figure 12-5b). However, if no artificial damping is introduced, increasing l/r_0 will not necessarily improve the accuracy of the dynamic-stiffness coefficients, as is demonstrated in Reference [58] on page 140.

Although for a larger damping ratio ζ, the reflected wave amplitude c_1 becomes smaller, the error introduced in the damping extraction of the third step becomes larger in the region near the cut-off frequency $a_0 = 1$. The derivative $S_\zeta(\omega),_\omega$ close to the cut-off frequency in Eq. 12.5 becomes very large. The effect of the damping ratio is demonstrated for $l/r_0 = 4$ in Figure 12-6 using $\zeta = 0.2$, 0.4 and 0.6. For this very stringent test with a cut-off frequency, the optimum ζ is thus quite small (0.2). It is worth mentioning that for a sufficiently small artificial damping ratio (but not zero), the result converges as the length increases. This is illustrated in Figure 12-7 with $\zeta = 0.05$ ($l/r_0 = 5$, 10 and 15).

In the damping-solvent extraction method, two terms in the Taylor expansion of the dimensionless dynamic-stiffness matrix (up to the first-derivative term) in Eq. 11.24 are retained. The question arises of what the effect of the higher-order terms might be.

Again, the semi-infinite rod on an elastic foundation (Figure A-3) with an analytical solution is examined. For the damped bounded medium the outer boundary is fixed, which provides a clearer physical insight as only one type of damping, the introduced and extracted artificial material damping, is present

$$u(\omega, x = l) = 0 \tag{12.39}$$

For the investigation it is sufficient to address the dimensionless dynamic-stiffness

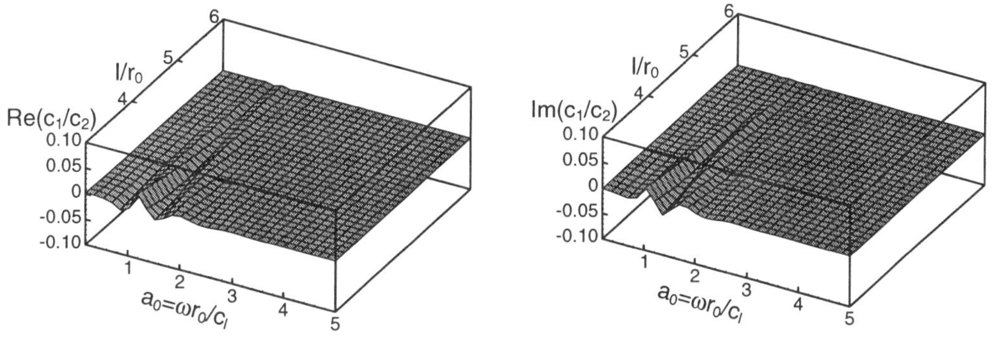

12.4 ANALYTICAL VERIFICATION

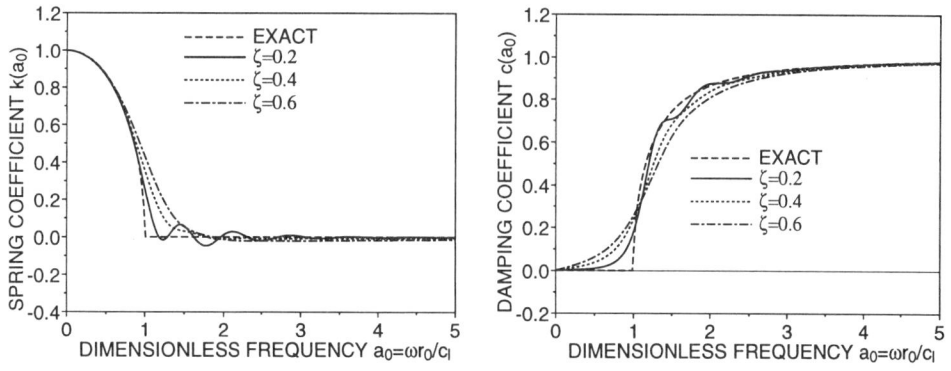

(a) amplitude ratio after first step

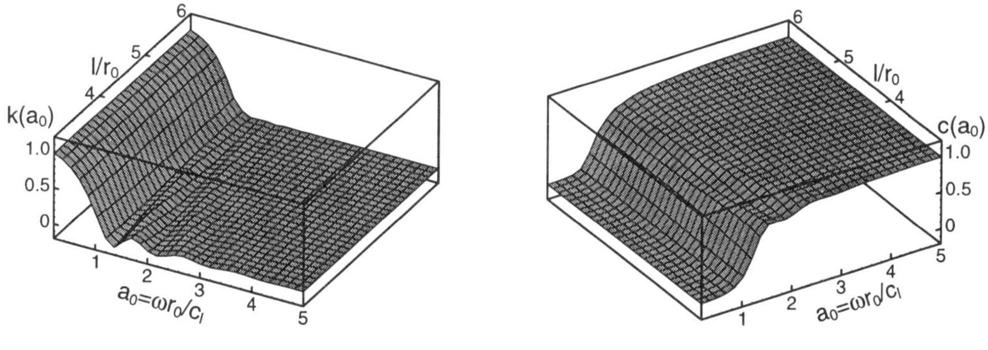

(b) dynamic-stiffenss coefficient after third step

Figure 12-5 Parametric study of length of bounded medium with fixed introduced artificial damping ratio

Figure 12-6 Effect of introduced artificial damping on dynamic-stiffness coefficient near cut-off frequency

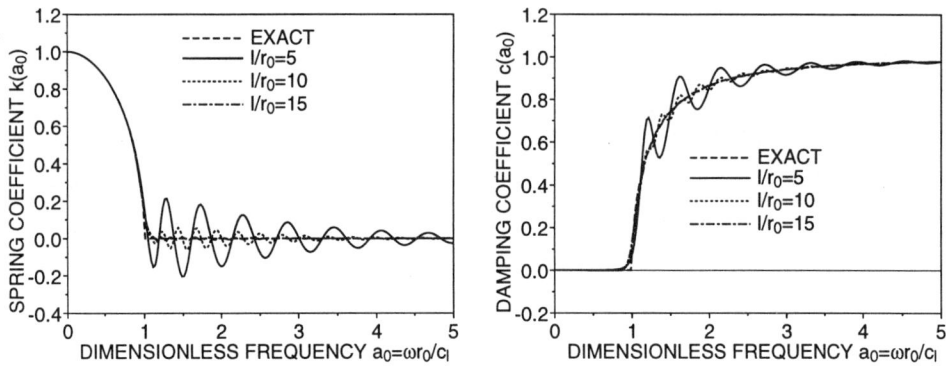

Figure 12-7 Effect of length of bounded medium on dynamic-stiffness coefficient near cut-off frequency with small introduced artificial damping ratio

coefficient of the undamped unbounded medium $\bar{S}^\infty(a_0)$ (which is the same function as that of the damped unbounded medium $\bar{S}^\infty(a_0^*)$)

$$\bar{S}^\infty(a_0) = \frac{S^\infty(\omega)}{K^\infty} = \sqrt{1 - a_0^2} \tag{12.40}$$

$$\bar{S}^\infty(a_0^*) = \frac{S_\zeta^\infty(\omega)}{K^\infty(1 + 2i\zeta)} = \sqrt{1 - a_0^{*2}} \tag{12.41}$$

and the dimensionless dynamic-stiffness coefficient of the damped bounded medium $\bar{S}(a_0^*)$ (which is again the same function as that of the undamped bounded medium $\bar{S}(a_0)$)

$$\bar{S}(a_0^*) = \frac{S_\zeta(\omega)}{K^\infty(1 + 2i\zeta)} = \sqrt{1 - a_0^{*2}} \frac{1 + e^{-2\sqrt{1-a_0^{*2}}\frac{l}{r_0}}}{1 - e^{-2\sqrt{1-a_0^{*2}}\frac{l}{r_0}}} \tag{12.42}$$

$$\bar{S}(a_0) = \frac{S(\omega)}{K^\infty} = \sqrt{1 - a_0^2} \frac{1 + e^{-2\sqrt{1-a_0^2}\frac{l}{r_0}}}{1 - e^{-2\sqrt{1-a_0^2}\frac{l}{r_0}}} \tag{12.43}$$

Eq. 12.40 is identical to Eq. A.2.14. Eq. 12.41 follows from Eq. 12.36. Substituting Eq. 12.30 in Eq. 12.39 leads to c_1/c_2 which is substituted in Eq. 12.35 resulting in Eq. 12.42. Eq. 12.43 corresponds to the undamped case. Note that the dimensionless dynamic-stiffness coefficients of the undamped and damped media are the same function but evaluated at different values, i.e. at a_0 and a_0^* corresponding to the same ω.

12.4 ANALYTICAL VERIFICATION

For the presentation of the results of the *dimensionless* dynamic-stiffness coefficients the dimensionless spring and damping coefficients $k(a_0)$ and $c(a_0)$ are defined directly as

$$\overline{S}^\infty(a_0^*) = k(a_0) + ia_0 c(a_0) \tag{12.44}$$

$$\overline{S}(a_0^*) = k(a_0) + ia_0 c(a_0) \tag{12.45}$$

This decomposition for the dimensionless dynamic-stiffness coefficient differs by the factor $1 + 2i\zeta$ from the standard definition used for the dynamic-stiffness coefficient with a dimension in Eq. 12.38 for the damped case.

The unbounded medium is addressed. To calculate the corresponding $\overline{S}^\infty(a_0)$ from $\overline{S}^\infty(a_0^*)$, the following Taylor expansion at a_0^* is formulated

$$\overline{S}^\infty(a_0) = \overline{S}^\infty(a_0^*) + \frac{d\overline{S}^\infty(a_0^*)}{da_0^*}(a_0 - a_0^*) + \frac{1}{2}\frac{d^2\overline{S}^\infty(a_0^*)}{da_0^{*2}}(a_0 - a_0^*)^2$$
$$+ \ldots + \frac{1}{n!}\frac{d^n\overline{S}^\infty(a_0^*)}{da_0^{*n}}(a_0 - a_0^*)^n + \ldots \tag{12.46}$$

A damping ratio $\zeta = 0.2$ is selected. Considering one term only corresponds to the damped case $\overline{S}^\infty(a_0^*)$ (dashed line in Figure 12-8) with unacceptable deviations.

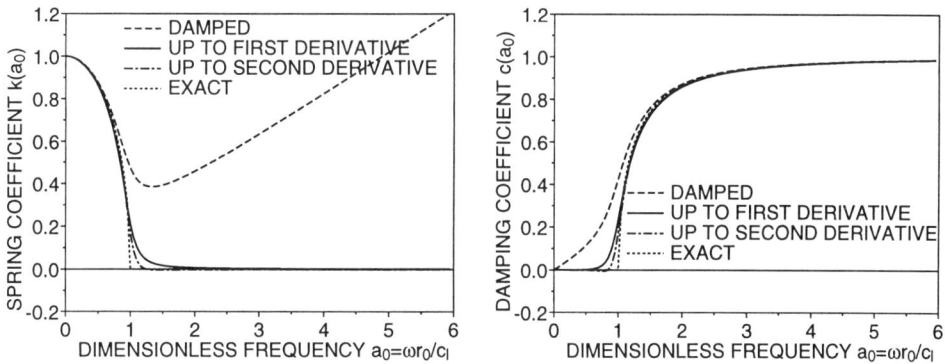

Figure 12-8 Dynamic-stiffness coefficient of undamped unbounded medium calculated from that of damped unbounded medium

An accurate undamped $\overline{S}^\infty(a_0)$ is obtained using the Taylor expansion up to the first-derivative term (solid line). Including the second-derivative term improves the accuracy slightly close to the cut-off frequency (dashed-dotted line). As the dynamic-stiffness matrix of an *unbounded* medium is a smooth function of a_0, the contribution of the second- and higher-derivative terms is negligible. It is thus demonstrated that by applying Eq. 12.46 the dynamic-stiffness coefficient of the undamped unbounded medium can be determined from that of the damped unbounded medium. The two terms (up to the first-derivative term) are sufficient (and necessary) for the selected damping. All terms on the right-hand side can be calculated analytically using Eq. 12.41. However, at the cut-off frequency $a_0 = 1$ where all derivatives for the undamped case are infinite, the deviations from the exact solution (dotted line) are the largest.

Replacing the damped unbounded medium by the damped bounded medium on the right-hand side of Eq. 12.46 yields

$$\overline{S}^{\infty}(a_0) = \overline{S}(a_0^*) + \frac{d\overline{S}(a_0^*)}{da_0^*}(a_0 - a_0^*) + \frac{1}{2}\frac{d^2\overline{S}(a_0^*)}{da_0^{*2}}(a_0 - a_0^*)^2$$

$$+ \ldots + \frac{1}{n!}\frac{d^n\overline{S}(a_0^*)}{da_0^{*n}}(a_0 - a_0^*)^n + \ldots \quad (12.47)$$

All terms on the right-hand side can be evaluated analytically using Eq. 12.42. A term-by-term comparison is performed for the first three terms of Eqs. 12.46 and 12.47 in Figure 12-9 choosing $l/r_0 = 5$ and $\zeta = 0.2$. The solid and dashed lines correspond to the results of the damped finite rod and of the damped semi-infinite rod, respectively. The deviations are caused by the wave reflected at the fixed boundary present only in the finite rod. The higher the derivative in a term is, the larger the deviation is, and thus the larger the contribution of the reflected wave. As already established in connection with Figure 12-8, the two terms up to the first-derivative term are sufficient to determine the dynamic-stiffness coefficient of the undamped unbounded medium from that of the damped bounded medium. The damping ratio and the length of the bounded medium can be selected large enough to reduce the reflected wave, resulting in small deviations in the first-derivative term. As in an unbounded medium no reflected waves occur, the first two terms up to the first-derivative term of the bounded medium are used in the damping-solvent extraction method (assumptions of the second step). Stated differently, this procedure filters the contribution of the reflected waves which affect the higher-derivative terms to a larger extent that the first two terms. The resulting high accuracy of the damping-solvent extraction method is confirmed for the semi-infinite rod on an elastic foundation using $l/r_0 = 4$ and $\zeta = 0.2$ in Figure 12-4.

For illustration, the influence of the high-derivative terms is investigated further. For $l/r_0 = 5$ and $\zeta = 0.2$ and including the first 6 terms (up to the fifth-derivative term) on the right-hand side of Eq. 12.47, the dimensionless spring coefficient is plotted as a solid line in Figure 12-10 together with the analytical spring coefficient of the corresponding undamped finite rod presented as a dashed line (Eq. 12.43). For comparison the result is also shown as a dotted line using only the first two terms. As expected from Figure 12-9, including higher-derivative terms does indeed recover the behaviour of the undamped bounded medium exhibiting resonances.

12.4 ANALYTICAL VERIFICATION

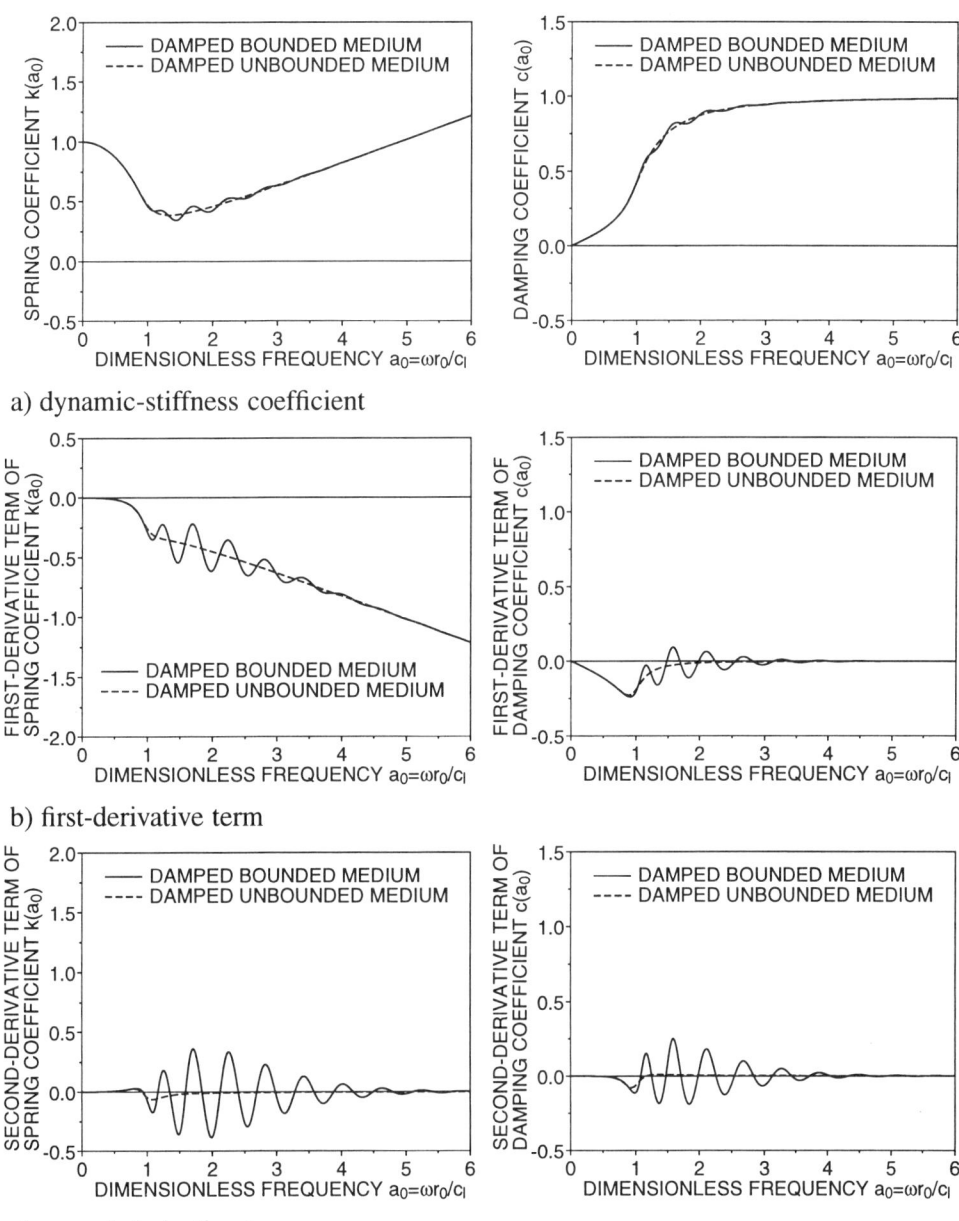

a) dynamic-stiffness coefficient

b) first-derivative term

c) second-derivative term

Figure 12-9 First three terms of Taylor expansion

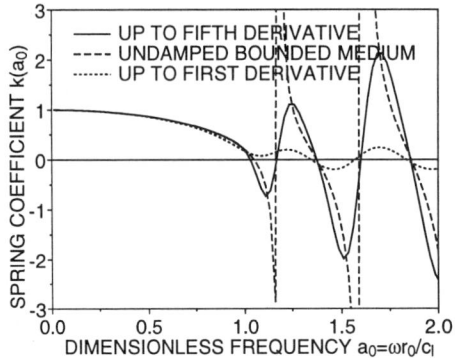

Figure 12-10 Dimensionless spring coefficient calculated from dynamic-stiffness coefficient of damped bounded medium using various numbers of terms in Taylor expansion compared with that of undamped bounded medium

The decay of the wave amplitude in the first step is governed by the triple product of the frequency ω, the damping ratio ζ and the distance from the structure-medium interface to the outer boundary of the bounded medium. For a given ω, the same decay can be achieved by a large distance with a small ζ or by a small distance and a large ζ. In the third step, the smaller ζ is, the more accurate the proposed simple damping extraction is.

The finite element mesh can be constructed as follows. The static case determines the distance from the structure-medium interface to the outer boundary. The highest frequency whose wave length must at least be represented by 6 nodes leads to the discretization into nodes.

A linear hysteretic damping ratio $0.2 < \zeta < 0.4$ leads to accurate results. A smaller damping ratio $0.05 < \zeta < 0.1$ should be used near the cut-off frequencies.

12.5 ACCURACY IN FREQUENCY DOMAIN

12.5.1 Out-of-plane Motion of Semi-infinite Layer of Constant Depth

The out-of-plane motion of a semi-infinite layer of constant depth (Figure A-7) described in Appendix A.3.1 is analysed. The analytical expression of the dynamic-stiffness matrix is specified in Eq. A.3.28.

The finite region of the semi-infinite layer of length l with viscous dashpots, whose coefficients per unit length equal ρc_s, located on the outer boundary is shown in Figure 12-11. The dynamic-stiffness matrix is calculated with the damping-solvent extraction method for $l/d = 3$ and $\zeta = 0.2$. Over the depth of the layer 11 nodes with linear shape functions and in the horizontal direction 31 nodes with quadratic shape functions are selected. In Figures 12-12a and 12-12b. the coefficients of the

12.5 ACCURACY IN FREQUENCY DOMAIN

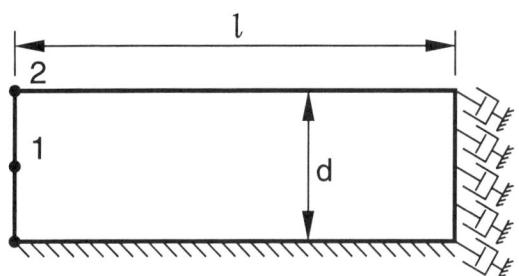

Figure 12-11 Finite region of semi-infinite layer of constant depth

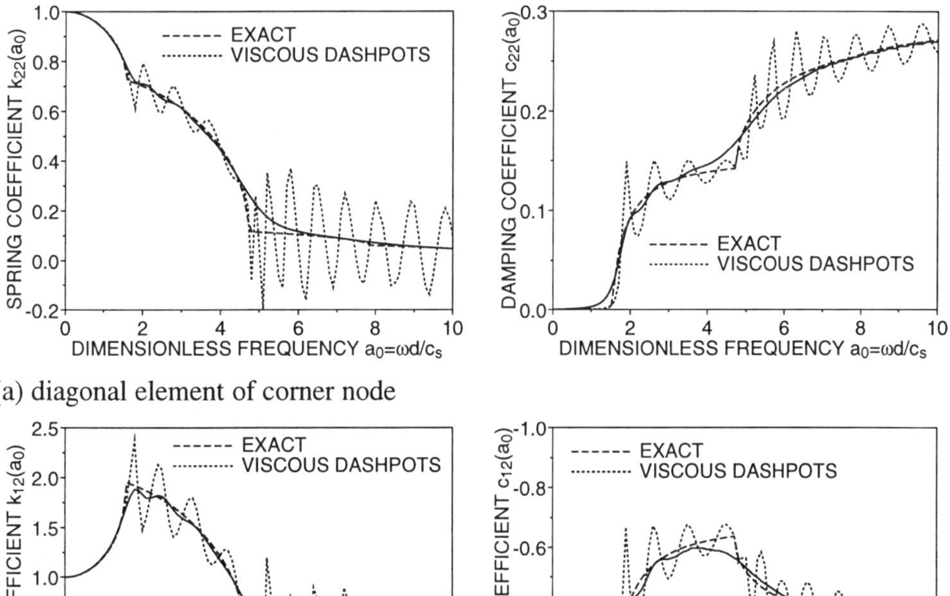

(a) diagonal element of corner node

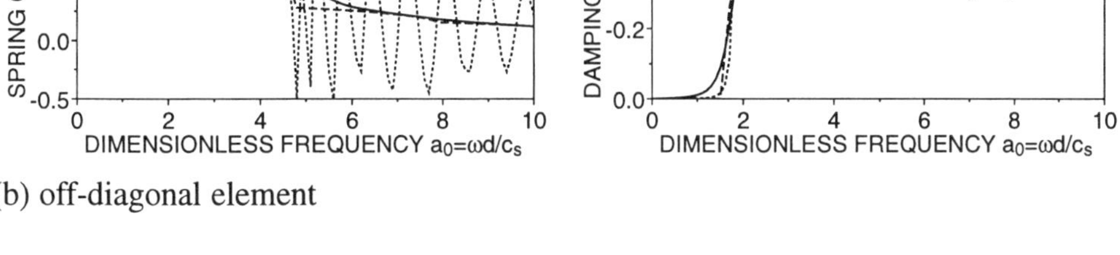

(b) off-diagonal element

Figure 12-12 Dynamic-stiffness matrix of semi-infinite layer

corner node 2 $S_{22}^{\infty}(a_0)$ and of the coupling term $S_{12}^{\infty}(a_0)$ (normalized with their static-stiffness coefficients) are compared to the exact solution (Eq. A.3.28). The agreement is good; in particular the behaviour near the cut-off frequency (at $a_0 = \pi/2$) of the first mode, which has a small participation factor, is well represented. It is important to stress that the damping-solvent extraction method also results in accurate off-diagonal terms (Figure 12-12b). For the sake of comparison, the undamped finite region of the same length with the viscous dashpots at the outer boundary is also evaluated. This corresponds to the direct method of analysis based on the viscous dashpots serving as a transmitting boundary. As can be seen from the corresponding dynamic-stiffness coefficients shown as a dotted line, large deviations exist.

The largest deviation of the solution of the damping-solvent extraction method occurs close to the cut-off frequency (at $a_0 = 3\pi/2$) of the second mode which for the prescribed quadratic displacement on the structure-medium interface exhibits a large participation factor. As demonstrated in connection with Figure 12-6, a smaller damping ratio improves the accuracy close to the cut-off frequencies. This is confirmed for $S_{22}^{\infty}(a_0)$ in Figure 12-13, where $\zeta = 0.05$ for $4 \leq a_0 \leq 5.5$ and the same $\zeta = 0.2$ is kept for a_0 outside this region.

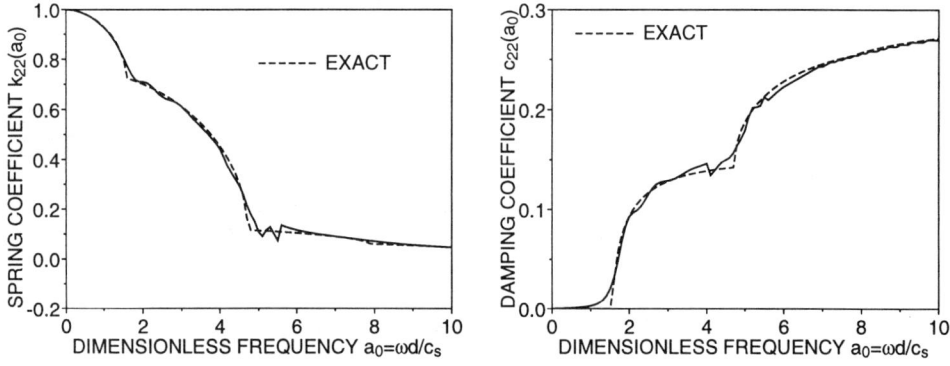

Figure 12-13 Diagonal element of corner node in dynamic-stiffness matrix of semi-infinite layer

12.5.2 In-plane Motion of Semi-infinite Wedge

The in-plane motion of a semi-infinite wedge with an opening angle $\alpha = 30°$ and Poisson's ratio $\nu = 0.25$ discussed in Appendix A.4.2 (Figure A-13) is addressed. The dynamic-stiffness coefficient corresponding to the horizontal in-plane motion $u_0(\theta)$ prescribed as a linear function in the circumferential direction on the arc and zero vertical motion is calculated. As no analytical solution exists, the result determined by the consistent infinitesimal finite-element cell method in the frequency domain

12.5 ACCURACY IN FREQUENCY DOMAIN

(Sections 7.1 and 7.2) is used for comparison.

The mesh of the bounded medium is constructed as follows. Each finite element with a linear shape function in the circumferential direction and a quadratic shape function in the radial direction has 6 nodes. 6 nodes with non-prescribed displacements and 1 fixed node are selected in the circumferential direction and 9 nodes are "proportionally spaced" in the radial direction. For a specific frequency, the ratio γ of the radial coordinates of consecutive nodes is constant (see Figure 12-14). The length of the finite element in the radial direction depends on the frequency.

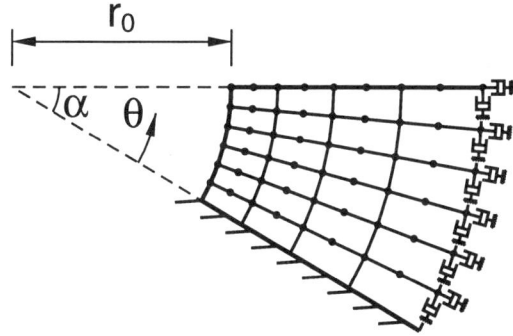

Figure 12-14 Finite-element discretization of finite region of semi-infinite wedge

For the static case $\gamma = 1.3$ is chosen and then reduced to guarantee 6 points per wavelength for the outermost finite element in the dynamic case. The introduced artificial material damping ratio is selected as $\zeta = 0.3$. On the outer boundary viscous dashpots with the coefficients per unit length equal to ρc_p^* in the radial direction and ρc_s^* in the circumferential direction are placed. The dimensionless frequency is defined as $a_0 = \omega r_0 / c_s$. The spring and damping coefficients $k(a_0)$ and $c(a_0)$ of the dynamic-stiffness coefficient $S^\infty(a_0)$, non-dimensionalized with the static value $K^\infty = 0.759G$ ($G=$ shear modulus) as in Eq. A.0.3, agree well with the solution of the consistent infinitesimal finite-element cell method (Figure 12-15). This is remarkable, as only 4 rows of finite elements are selected in the radial direction. In this example, for the sake of illustration, as already mentioned, the mesh is adapted to the frequency being analysed to allow calculation of the dynamic-stiffness coefficient for high frequencies with only 4 rows of finite elements. If the response is of interest in the intermediate frequency range only ($a_0 < 2$), a finite-element mesh which is independent of frequency can of course be applied, without a substantial increase of the number of finite elements.

250   IMPLEMENTATION, VERIFICATION AND ACCURACY

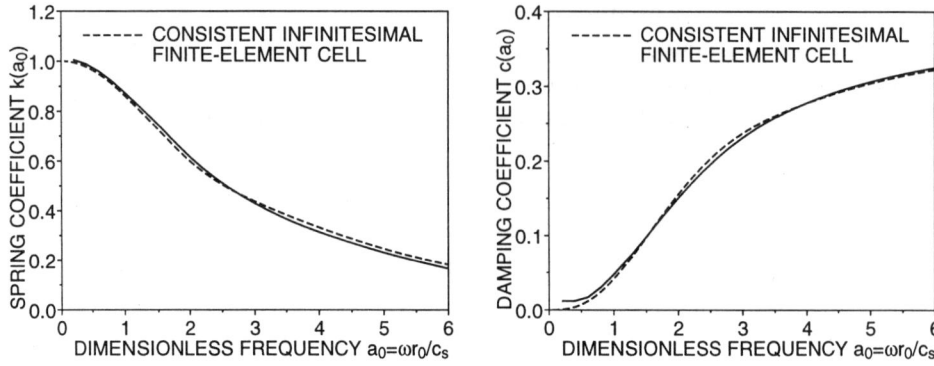

Figure 12-15 Dynamic-stiffness coefficient of in-plane motion of semi-infinite wedge

12.5.3 Strip Foundation with Rectangular Cross-section Embedded in Half-plane

The in-plane motion of a strip foundation with a rectangular cross-section embedded in a half-plane is addressed, which is described in Appendix A.6 (Figure A-16). The homogeneous case $G_1 = G_2 = G_3 = G$ is examined with Poisson's ratio $\nu = 0.25$. The mesh of finite elements with a linear shape function parallel to the structure-medium interface and a quadratic shape function in the radial direction is shown in Figure 12-16. Note that only 4 rows of finite elements are used in the radial

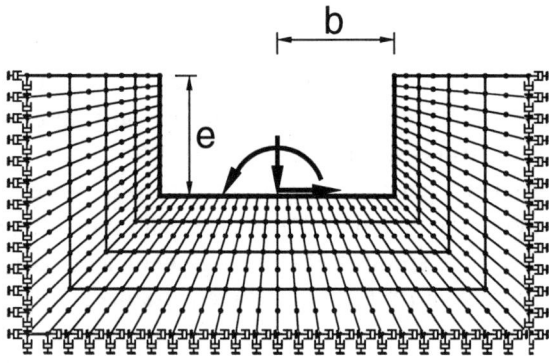

Figure 12-16 Finite-element discretization of finite region of unbounded medium for strip foundation embedded in half-plane in frequency-domain analysis

direction starting with the aspect ratio $\gamma = 1.5$ for the static case. Viscous dashpots with the coefficients per unit length ρc_p^* in the perpendicular direction and ρc_s^* in the tangential direction on the outer boundary are introduced. The dimensionless frequency is defined as $a_0 = \omega b/c_s$. This is a stringent test, as $k(a_0)$ and $c(a_0)$ for the two translational degrees of freedom start for $a_0 = 0$ at zero and infinity, respectively.

For $a_0 < 2$ a boundary-element solution is available in Reference [56] for the dynamic-stiffness matrix. To be able to compare the results also for higher frequencies, the values of the consistent infinitesimal finite-element cell method of Sections 7.1 and 7.2 are also shown in Figure 12-17. The coefficients of the dynamic-stiffness matrix are non-dimensionalized for the translational degrees of freedom with the shear modulus G and for the rocking degree of freedom with Gb^2. The applied linear hysteretic damping ratio of the solvent equals $\zeta = 0.4$. The results of the damping-solvent extraction method are highly accurate (Figure 12-17). The results using only the viscous dashpots (and no damping solvent) as a transmitting boundary deviate significantly, although for this analysis the number of rows of finite elements in the radial direction is selected as 12, leading to a finite-element mesh which is 3 times as large.

12.6 ACCURACY IN TIME DOMAIN

The same example of the strip foundation embedded in a half-plane analysed in the frequency domain in Section 12.5.3 is investigated in the time domain.

The finite-element discretization of the bounded medium of length l analysed in the first step is shown in Figure 12-18. 4-node isoparametric finite elements are used. On the structure-medium interface the lengths of the elements equal $b/5$. On the free surface of the half-plane the length is not changed. The number of elements in the radial direction is varied parametrically. 5, 10 and 15 finite elements are chosen, resulting in $l = b$, $l = 2b$ and $l = 3b$, respectively. Viscous dashpots with the coefficients per unit length ρc_p in the perpendicular direction and ρc_s in the tangential direction leading also to springs with the coefficients $\zeta \rho c_p$ and $\zeta \rho c_s$ are introduced on the outer boundary (see first step in Section 11.2).

It is appropriate to define a dimensionless artificial damping factor $\bar{\zeta} = (b/c_s)\zeta$. The decay of a wave propagating in a medium with the exterior mass-proportional dashpots is evaluated approximately. For a typical shear wave, the phase angle equals $e^{-i\omega r/c_s}$ (coordinate r). Replacing ω by $\omega - i\zeta$ (analogous to Eq. 11.17) with $\zeta = (c_s/b)\bar{\zeta}$ yields $e^{-\bar{\zeta} r/b} e^{-i\omega r/c_s}$. The additional decay caused by the artificial damping is described by $e^{-\bar{\zeta} r/b}$. For the propagation distance from the structure-medium interface to the outer boundary and back again ($r = 2l$), the decay factor equals $(1 - e^{-2\bar{\zeta} l/b})$. As the product of $\bar{\zeta}$ and l/b appears, increasing the dimensionless length of the bounded medium l/b permits a decrease of the dimensionless artificial damping factor $\bar{\zeta}$ to achieve the same decay factor.

For the parametric study, $\bar{\zeta} l/b$ is selected as 0.5, 1 and 2, yielding the decay factors 0.63, 0.86 and 0.98, respectively. For a given l/b, the $\bar{\zeta}$ can be calculated for each

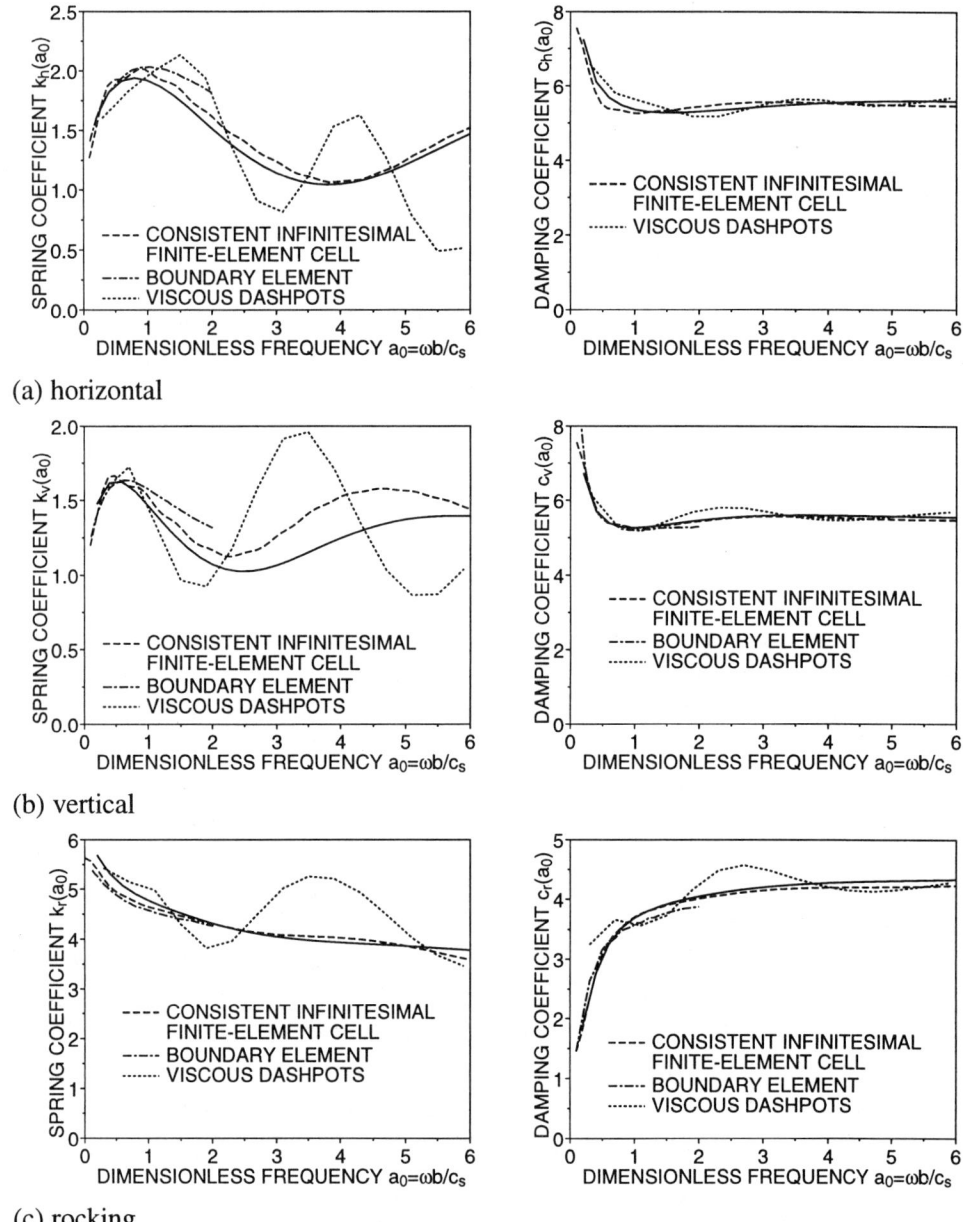

(a) horizontal

(b) vertical

(c) rocking

Figure 12-17 Dynamic-stiffness coefficients of strip foundation embedded in half-plane

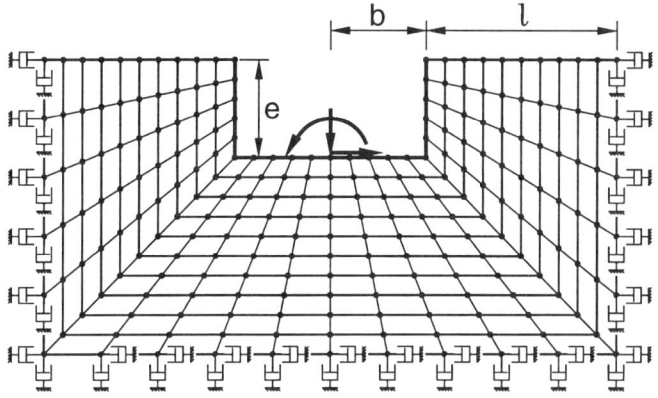

Figure 12-18 Finite-element discretization of finite region of unbounded medium for strip foundation embedded in half-plane in time-domain analysis

decay factor.

The transient excitation consists of a prescribed horizontal displacement at the centre of the rigid base (with zero values for the vertical and rocking motions)

$$u(t) = \frac{u_0}{2}(1 - \cos 2\pi \frac{t}{T}) \qquad 0 \le \frac{t}{T} \le 2$$

$$= 0 \qquad \frac{t}{T} > 2 \qquad (12.48)$$

with the period $T = 8b/c_s$.

The interaction forces $R_\zeta(t)$ and $R_{\zeta r}(t)$ of the bounded damped medium to the two loading cases (Eqs. 12.48 and 12.16) are calculated on the basis of the Newmark method with $\gamma = 1/2$ and $\beta = 1/4$ and with $\Delta t = 0.1b/c_s$. The horizontal interaction force $R(t)$ then follows from Eq. 12.14.

The dimensionless horizontal interaction force $R(\bar{t})$ is plotted versus the dimensionless time $\bar{t} = tc_s/b$ for $l/b = 1, = 2$ and $= 3$ in Figures 12-19a, 12-19b and 12-19c. The results are shown for the three values of $\bar{\zeta}l/b$ described above, which determine $\bar{\zeta}$ shown in the figures. For comparison, the results for an extended mesh, which is selected so large that the reflected wave does not reach the structure-medium interface in the time of interest, are also presented. The solution using viscous dashpots on the outer boundary as a transmitting boundary, i.e. without any artificial damping, is also specified. In all cases the damping-solvent extraction method leads to significantly more accurate results than the viscous dashpots do. For the damping-solvent extraction method for each l/b, $\bar{\zeta}l/b$ in the range from 1 to 2 yields the most accurate solution. For each product $\bar{\zeta}l/b$, the smaller l/b is, the larger the deviation is, as $\bar{\zeta}$ is larger, resulting in a loss of accuracy when extracting damping in the third step.

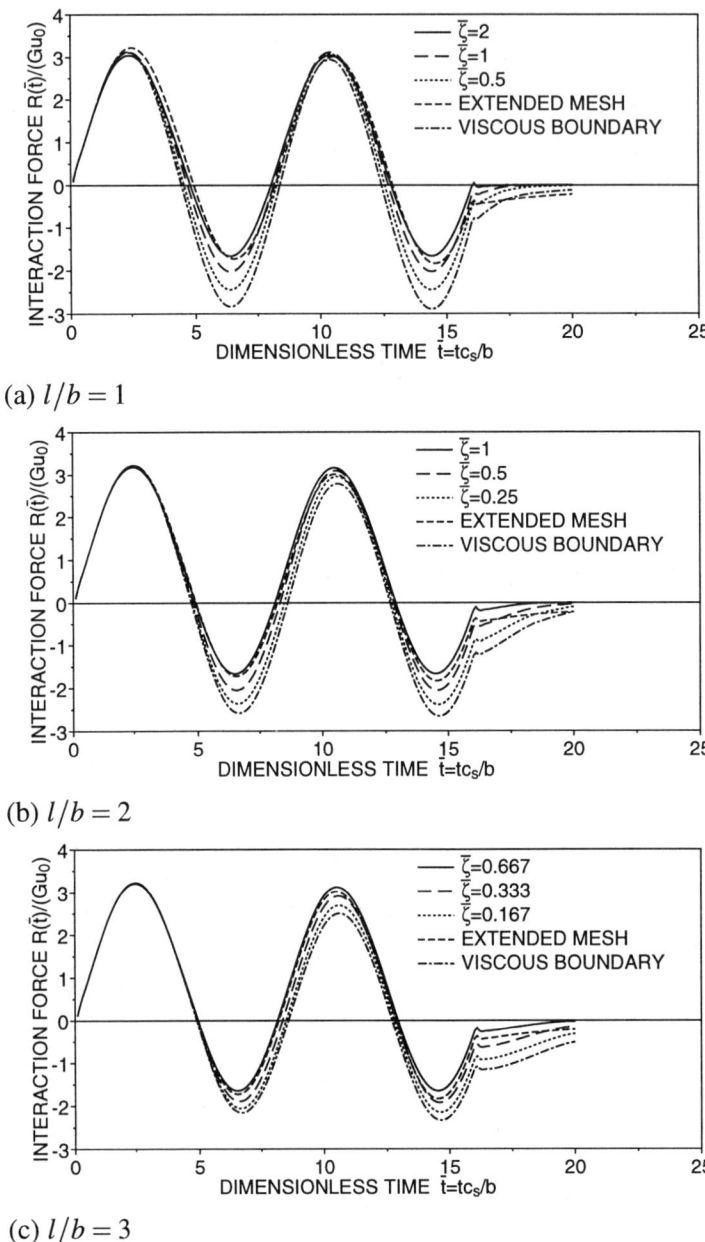

(a) $l/b = 1$

(b) $l/b = 2$

(c) $l/b = 3$

Figure 12-19 Time history of horizontal interaction force varying the dimension of bounded medium l/b and dimensionless artificial damping factor $\bar{\zeta}$

12.6 ACCURACY IN TIME DOMAIN

It is, of course, important to extract the introduced artificial damping in the third step. If this is not done, disastrous results are obtained. This is demonstrated in Figure 12-20, where for $l/b = 2$ and $\bar{\zeta}l/b = 2$, i.e. $\bar{\zeta} = 1$, the bounded damped system

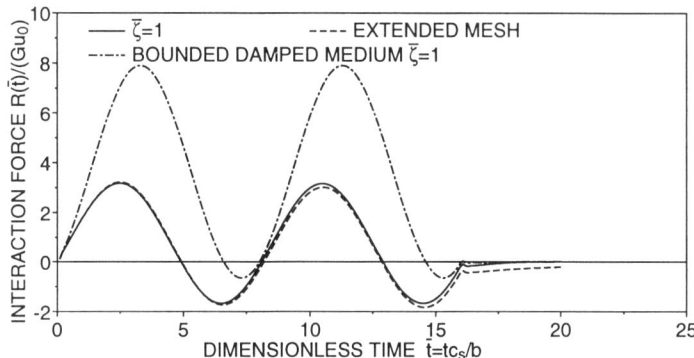

Figure 12-20 Time history of horizontal interaction force of bounded damped medium with dimensionless length of bounded medium $l/b = 2$ and dimensionless artificial damping factor $\bar{\zeta} = 1$

is analysed for the prescribed displacement yielding $R_\zeta(\bar{t})$ (dashed-dotted line)

The accuracy is also evaluated for the vertical and rocking degrees of freedom with the same prescribed time history of the motion as in Eq. 12.48. As an example, for $l/b = 2$ and $\bar{\zeta}l/b = 2$ (which results in $\bar{\zeta} = 1$) the vertical interaction force $R_v(\bar{t})$ caused by a vertical displacement $u_v(t)$ is shown in Figure 12-21a and the interaction moment $M(\bar{t})$ caused by rocking $\theta(t)$ in Figure 12-21b. Again, the damping-solvent extraction method yields higher accuracy than using viscous dashpots only.

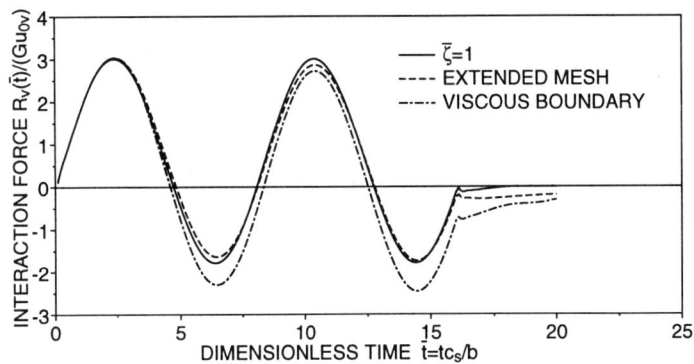

(a) vertical

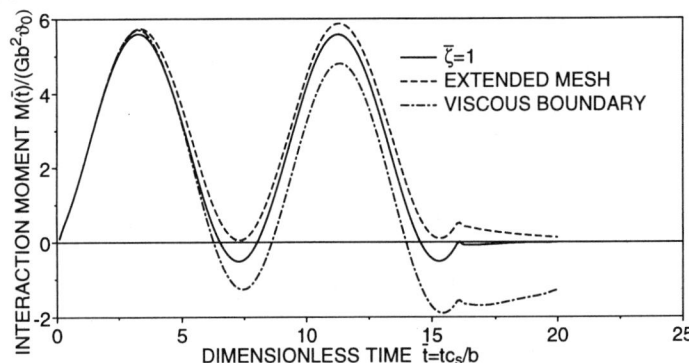

(b) rocking

Figure 12-21 Time history of interaction force with dimensionless length of bounded medium $l/b = 2$ and dimensionless artificial damping factor $\bar{\zeta} = 1$

Part III

DOUBLY-ASYMPTOTIC MULTI-DIRECTIONAL TRANSMITTING BOUNDARY

On ne peut pas partir de l'infini, on peut y aller.

(One cannot come from infinity, but one can go towards it.)

Jules Lachelier 1832–1918

PART III TRANSMITTING BOUNDARY

As an efficient procedure to model approximately the unbounded medium for use in the direct method of a dynamic unbounded medium-structure-interaction analysis, the so-called doubly-asymptotic multi-directional transmitting boundary is developed. A transient excitation can be processed directly in the time domain. The method combines the advantages of the doubly-asymptotic and multi-directional transmitting boundaries.

Chapter 13 discusses the concept and the numerical implementation. To construct the doubly-asymptotic multi-directional transmitting boundary, the multi-directional outward plane wave boundary condition is formulated for the interaction forces (and not for the displacements), where the contributions of the low- and high-frequency limits covered rigorously by the doubly-asymptotic boundary are subtracted beforehand.

Chapter 14 demonstrates that the accuracy of the doubly-asymptotic multi-directional transmitting boundary is higher than that of other transmitting boundaries with the same finite-element discretization. A parametric study of the accuracy permits a decision flowchart for the modelling procedure to be established.

This Part III is based on Reference [67].

13

CONCEPT AND NUMERICAL IMPLEMENTATION OF DOUBLY-ASYMPTOTIC MULTI-DIRECTIONAL TRANSMITTING BOUNDARY

The doubly-asymptotic transmitting boundary and the multi-directional transmitting boundary are outlined in Section 13.1, followed by the concept of the combined formulation. The numerical implementation of the doubly-asymptotic multi-directional transmitting boundary is addressed in Section 13.2.

13.1 CONCEPT

In the direct method of dynamic unbounded medium-structure-interaction analysis the unbounded medium is discretized with e.g. finite elements up to the artificial boundary (Figure 1-2b). To represent that part of the unbounded medium up to infinity which is not explicitly modelled, a boundary condition must be introduced on the artificial boundary. To formulate the interaction force-displacement relationship on the boundary, several schemes called transmitting boundaries have been developed resulting in efficient highly absorbing approximations which are mentioned in Section 1.4.2. The transmitting boundaries are local in time. The novel *doubly-asymptotic multi-directional transmitting boundary*, which will be discussed in detail, combines the advantages of the doubly-asymptotic [54] and multi-directional [19] formulations. It is rigorous for the low-frequency limit (statics) and the high-frequency limit in the wave-propagation direction perpendicular to the artificial boundary and at all the preselected angles. It is highly accurate for plane waves at intermediate frequencies and at other angles. Owing to the fully-coupled static-stiffness matrix the doubly-asymptotic multi-directional transmitting boundary is global in space. Using an approximate banded static-stiffness matrix yields a transmitting boundary which is also local in space.

To gain physical insight, the two-dimensional scalar wave equation is addressed. It governs for instance the out-of-plane (anti-plane) motion of elasto-dynamics with shear waves. The wave equation equals

$$u_{,xx}+u_{,yy}-\frac{1}{c_s^2}\ddot{u}=0 \tag{13.1}$$

with the displacement u and the shear-wave velocity $c_s = \sqrt{G/\rho}$ (shear modulus G, mass density ρ). This equation is e.g. derived in Appendix A.3.1. For plane waves propagating outwards at the angle of incidence α measured from the normal to the boundary (Figure 13-1)

$$u = f(x\cos\alpha + y\sin\alpha - c_s t) \tag{13.2}$$

satisfies Eq. 13.1. The artificial boundary coincides with the y-axis. Eq. 13.2 also

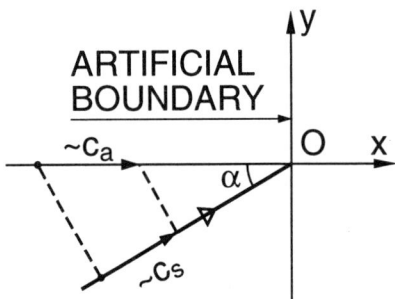

Figure 13-1 Inclined plane wave at artficial boundary with apparent velocity in perpendicular direction

satisfies the *outward plane wave boundary condition*

$$\frac{c_s}{\cos\alpha}u_{,x}+\dot{u}=0 \tag{13.3}$$

with the apparent velocity

$$c_a = \frac{c_s}{\cos\alpha} \tag{13.4}$$

in the direction perpendicular to the boundary.

A physical interpretation of Eq. 13.3 is obtained by multiplication with ρc_s, yielding

$$Gu_{,x}+\rho c_s \cos\alpha\, \dot{u}=0 \tag{13.5}$$

13.1 CONCEPT

With the shear stress acting on the artificial boundary

$$\tau_{zx} = G u_{,x} \tag{13.6}$$

Eq. 13.5 formulated as

$$\tau_{zx} + \rho c_s \cos \alpha \dot{u} = 0 \tag{13.7}$$

represents distributed viscous dashpots with the coefficient per unit length $\rho c_s \cos \alpha$. Note that the dashpot coefficient depends on the angle of incidence α. Eq. 13.7 is in the same form as Eq. 13.2 derived at the wave front formulating the law of conservation of momentum (see also Section 1.2).

A plane wave propagating with $-\alpha$ also satisfies Eq. 13.3. As the angles of incidence are, in general, unknown, Eq. 13.3 is written for distinct values α_i ($i = 1, 2, \ldots, m$). To absorb plane waves propagating at angles $\pm \alpha_i$, the *multi-directional* boundary condition is formulated as a product of differential operators

$$\left[\prod_{i=1}^{m} \left(\frac{c_s}{\cos \alpha_i} \frac{\partial}{\partial x} + \frac{\partial}{\partial t} \right) \right] u = 0 \tag{13.8}$$

For $m = 1$, Eq. 13.8 yields Eq. 13.3 for $\alpha_1 = \alpha$ with the change in nomenclature $u_{,x} = \partial u/\partial x$ and $\dot{u} = \partial u/\partial t$. Waves propagating outwardly at α_i are absorbed completely and those at other angles are to a large extent.

Note that the multi-directional boundary condition is formulated at a specific location at a specific time and is thus local in space and time.

The paraxial boundary [12] is derived setting $\alpha_1 = \alpha_2 = 0$. Eq. 13.8 yields

$$u_{,xx} + 2c_s \dot{u}_{,x} + \ddot{u} = 0 \tag{13.9}$$

Solving Eq. 13.1 for $u_{,xx}$ and substituting in Eq. 13.9 leads to

$$-\frac{1}{2} u_{,yy} + \frac{1}{c_s} \dot{u}_{,x} + \frac{1}{c_s^2} \ddot{u} = 0 \tag{13.10}$$

which corresponds to the paraxial boundary.

The differential operator applies to any wave field. Instead of u, Eq. 13.8 can also be formulated for forces and this is done in the doubly-asymptotic multi-directional transmitting boundary.

For the implementation of the doubly-asymptotic multi-directional transmitting boundary, the medium between the structure-medium interface and the artificial boundary is discretized with finite elements. The mesh adjacent to the artificial boundary coinciding (locally) with the y-axis is shown in Figure 13-2 with the x-axis in the perpendicular direction. To construct the multi-directional transmitting boundary, Eq. 13.8 formulated at the boundary is discretized in space and time.

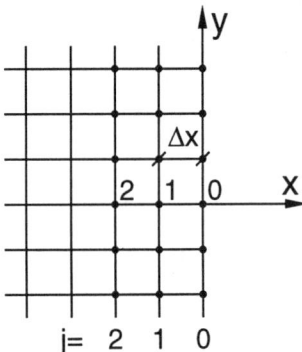

Figure 13-2 Finite-element discretization adjacent to artficial boundary with rows of nodes parallel to artificial boundary forming boundaries

This is discussed in the implementation of the doubly-asymptotic multi-directional transmitting boundary in Section 13.2. It leads to an explicit expression for the displacement on the boundary at the current time station which is a linear function of the displacements at the nodes nearby in the perpendicular direction at a few recent time stations. The discretized formulation is thus also local in space and time.

The multi-directional transmitting boundary is based on propagation of plane waves and thus does not cover statics and evanescent waves such as exponentially decaying displacement patterns below the cut-off frequency which do not propagate.

Statics, the low-frequency limit, is modelled exactly in the *doubly-asymptotic transmitting boundary* which also absorbs completely the high-frequency limit of waves propagating perpendicularly to the artificial boundary.

The low-frequency limit is represented by the static-stiffness matrix $[K^\infty]$ (superscript ∞ for unbounded medium) which couples all degrees of freedom on the artificial boundary. The high-frequency limit is modelled by the viscous diagonal dashpot matrix $[C_\infty]$ (subscript ∞ for $\omega \to \infty$) with the diagonal element for each degree of freedom equal to the product of ρc and the area (perpendicular direction $c = c_p$, tangential directions $c = c_s$). Adding the contributions of the two limits, the interaction force-displacement relationship on the artificial boundary equals

$$\{R\} = [K^\infty]\{u\} + [C_\infty]\{\dot{u}\} \tag{13.11}$$

with the interaction forces $\{R\}$ and displacements $\{u\}$. This boundary condition is straightforwardly assembled with the discretized equation of motion of the finite-element mesh of the medium enclosed by the artificial boundary. The discretized formulation of the doubly-asymptotic transmitting boundary is global in space due to $[K^\infty]$ and local in time. The accuracy can be limited as the response at intermediate frequencies is not well represented by Eq. 13.11.

It is desirable to combine the advantages of the multi-directional and the doubly-asymptotic transmitting boundaries. A straightforward implementation using displacements does not seem to be feasible. A simple formulation to construct the doubly-asymptotic multi-directional transmitting boundary proceeds as follows. *The contributions of the low- and high-frequency limits are first subtracted from the interaction forces. The local multi-directional outward plane wave boundary condition is then formulated explicitly for the remaining interaction forces. The implementation in the finite-element method is straightforward.*

13.2 NUMERICAL IMPLEMENTATION

The doubly-asymptotic multi-directional transmitting boundary is constructed as follows. The medium enclosed by the artificial boundary is discretized with finite elements (Figure 13-2). The actual artificial boundary with its nodes coinciding in the figure with the y-axis is denoted by the index $j = 0$. The row of nodes denoted with the index $j = 1$, shown in the figure at a constant distance Δx for simplicity, can be regarded as another artificial boundary, i.e. the actual artificial boundary has been moved conceptually inwards deleting one row of finite elements. This concept can be repeated defining an additional artificial boundary with the index $j = 2$. The vector $\{u_j\}$ denotes the displacement in all nodes on the jth row away from the actual artificial boundary ($j = 0, 1, 2$) and $\{R_j\}$ the corresponding total interaction forces under the assumption that the actual artificial boundary is conceptually moved inwards to the jth row.

The conditions on the artificial boundary at the jth row at time station n ($t = n\Delta t$) are formulated as

$$\{R_j\}_n = [K_j^\infty]\{u_j\}_n + [C_{\infty j}]\{\dot{u}_j\}_n + \{Q_j\}_n \qquad (j = 0, 1, 2) \qquad (13.12)$$

The first two terms on the right-hand side correspond to the interaction forces of the doubly-asymptotic boundary condition (Eq. 13.11) written for the artificial boundary at the jth row. The third term $\{Q_j\}_n$ satisfies in its continuous form with the function Q the multi-directional boundary condition (Eq. 13.8)

$$\left[\prod_{i=1}^{m}\left(\frac{c_s}{\cos\alpha_i}\frac{\partial}{\partial x} + \frac{\partial}{\partial t}\right)\right] Q = 0 \qquad (13.13)$$

The discretized $\{Q_j\}_n$ represents the "remaining" interaction forces. The doubly-asymptotic multi-directional transmitting boundary is described by Eqs. 13.12 and 13.13.

The implementation proceeds as follows. The algorithm is discussed for the nth time step with all variables (displacements, velocities, accelerations and interaction forces) known up to and including the time station $n-1$. For practical applications, $m=2$ in Eq. 13.13 is sufficient, yielding

$$\frac{c_s}{\cos\alpha_1}\frac{c_s}{\cos\alpha_2}Q_{,xx}+\left(\frac{c_s}{\cos\alpha_1}+\frac{c_s}{\cos\alpha_2}\right)\dot{Q}_{,x}+\ddot{Q}=0 \qquad (13.14)$$

To derive an explicit expression for an element Q_{0n} of $\{Q_0\}_n$ at the actual artificial boundary $j=0$, Eq. 13.14 is discretized on the actual artificial boundary at time station $n-2$ using forward difference formulae

$$(Q_{0,xx})_{n-2} = \frac{Q_{0n-2}-2Q_{1n-2}+Q_{2n-2}}{\Delta x^2} \qquad (13.15a)$$

$$(\dot{Q}_{0,x})_{n-2} = \frac{Q_{0n-1}-Q_{0n-2}-Q_{1n-1}+Q_{1n-2}}{\Delta x \Delta t} \qquad (13.15b)$$

$$(\ddot{Q}_0)_{n-2} = \frac{Q_{0n}-2Q_{0n-1}+Q_{0n-2}}{\Delta t^2} \qquad (13.15c)$$

with the finite-element length Δx. This results in

$$Q_{0n} = (2-a_1)Q_{0n-1}+(-1+a_1-a_2)Q_{0n-2}+a_1Q_{1n-1}$$
$$+(-a_1+2a_2)Q_{1n-2}-a_2Q_{2n-2} \qquad (13.16)$$

where

$$a_1 = \frac{\Delta t}{\Delta x}\left(\frac{c_s}{\cos\alpha_1}+\frac{c_s}{\cos\alpha_2}\right) \qquad (13.17a)$$

$$a_2 = \frac{\Delta t^2}{\Delta x^2}\frac{c_s}{\cos\alpha_1}\frac{c_s}{\cos\alpha_2} \qquad (13.17b)$$

As the right-hand side of Eq. 13.16 is known, each element Q_{0n} of $\{Q_0\}_n$ is calculated independently of the others. $\{Q_0\}_n$ thus represents the "remaining" interaction forces at time station n which are predicted from $\{Q_j\}$ ($j=0,1,2$) at previous time steps using the multi-directional boundary condition. The condition on the actual artificial boundary at time station n follows from Eq. 13.12 with $j=0$

$$\{R_0\}_n = [K_0^\infty]\{u_0\}_n+[C_{\infty 0}]\{\dot{u}_0\}_n+\{Q_0\}_n \qquad (13.18)$$

Eq. 13.18 is incorporated straightforwardly in a finite-element formulation by assemblage with the discretized equation of motion of the medium enclosed by the actual artificial boundary. Its solution yields the displacements, velocities and accelerations at time station n. This permits the calculation of the nodal forces of the finite elements using their static-stiffness and mass matrices. Conceptually moving the artificial boundary to rows $j=1$ and $=2$ by deleting one and two rows of finite elements, respectively, the interaction forces $\{R_j\}_n$ ($j=1,2$) can be determined from

13.2 NUMERICAL IMPLEMENTATION

the nodal forces using equilibrium. An equation for the "remaining" interaction forces is formulated using Eq. 13.12

$$\{Q_j\}_n = \{R_j\}_n - [K_j^\infty]\{u_j\}_n - [C_{\infty j}]\{\dot{u}_j\}_n \tag{13.19}$$

$\{Q_j\}_n$ ($j = 1, 2$) are used in later time steps. This completes the calculation for the nth time step.

The physical insight of the outward plane wave boundary condition for the in-plane motion of elasto-dynamics becomes blurred, as the displacements of the dilatational and shear waves cannot be separated. The extension to the in-plane case can be performed as follows. The directions perpendicular and tangential to the artificial boundary are treated independently. For each of these directions a dominant wave velocity is assumed; e.g. the dilatational-wave velocity c_p in the perpendicular direction and the shear-wave velocity c_s in the tangential. In addition, for $m = 2$, two angles are selected. Eq. 13.13 can then be formulated for each direction. The generalization of the doubly-asymptotic multi-directional transmitting boundary to the three-dimensional case of elasto-dynamics is feasible.

The implementation of the doubly-asymptotic part of Eq. 13.13 is implicit and that of the multi-directional part explicit. Stability for the implementation of Eq. 13.14 is guaranteed when the apparent velocities c_{ai} ($i = 1, 2$) in the perpendicular direction are smaller than $\Delta x/\Delta t$. This is satisfied by choosing a sufficiently small Δt.

It is worth mentioning that the extrapolation boundary [27] follows as a special case of the multi-directional boundary. Selecting e.g. $c_a = c_s/\cos\alpha_1 = c_s/\cos\alpha_2$ and for $\Delta x = c_a \Delta t$ leads to $a_1 = 2$ and $a_2 = 1$ (Eq. 13.17). u replacing Q (Eq. 13.16) results in

$$u_{0n} = 2u_{1n-1} - u_{2n-2} \tag{13.20}$$

which corresponds to linear extrapolation.

14

ACCURACY AND MODELLING PROCEDURE OF DOUBLY-ASYMPTOTIC MULTI-DIRECTIONAL TRANSMITTING BOUNDARY

Using examples of increasing complexity it is demonstrated in Section 14.1 that the doubly-asymptotic multi-directional transmitting boundary is far superior to any other transmitting boundary. A decision flowchart for choosing the optimum location of the artificial boundary with the corresponding spatial coupling is presented in Section 14.2.

14.1 PERFORMANCE AND COMPARISON WITH OTHER TRANSMITTING BOUNDARIES

In contrast to the similarity-based formulation of Part I where in the limit of infinitesimal finite-element size the exact solution is obtained, the transmitting boundaries are approximations. Their accuracy depends on the type of transmitting boundary applied as well as on the location of the artificial boundary. In addition, the dynamic characteristics of the problem to be analysed such as the presence of propagating and evanescent waves are of paramount importance. The accuracy of the doubly-asymptotic multi-directional transmitting boundary and of other transmitting boundaries is evaluated by comparing with the results of an extended mesh. As the error due to the finite-element discretization is also present in the extended mesh, the differences are caused by the inherent error of the transmitting boundary.

14.1.1 Semi-infinite Rod on Elastic Foundation

The semi-infinite rod on an elastic foundation described in Appendix A.2 (Figure A-3) is examined for the rounded triangular displacement pulse of Eq. A.2.24

272   ACCURACY AND MODELLING PROCEDURE

enforced on the structure-medium interface. The resulting interaction force is plotted in Figure A-6.

This one-dimensional example is a stringent test as the waves are dispersive. Varying apparent wave velocities thus exist. In addition, a cut-off frequency occurs below which no waves propagate, i.e. evanescent waves are present.

Between the beginning of the rod, where the structure-medium interface is located, and the artificial boundary, 5 one-dimensional finite elements of equal length $\Delta l = 0.1 r_0$ are selected (Figure 14-1). A linear shape function for the displacement is

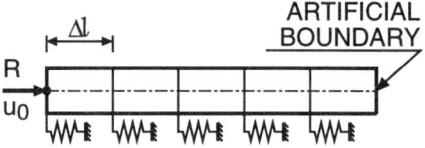

Figure 14-1 Finite-element mesh up to artificial boundary of semi-infinite rod on elastic foundation

assumed. The Newmark algorithm with $\beta = 0.25$, $\gamma = 0.5$ and $\Delta t = 0.05 r_0/c_l$ is used. For comparison, an extended mesh with the same finite-element length and time step is analysed, for which no reflected waves arrive at the beginning of the rod during the calculation.

The interaction force at the beginning of the rod $R(\bar{t})$, non-dimensionalized with $K^\infty u_0$ (static-stiffness coefficient $K^\infty = \sqrt{EAk_g}$, Eq. A.2.13) is plotted as a function of $\bar{t} = tc_l/r_0$ in Figure 14-2a. The doubly-asymptotic multi-directional transmitting

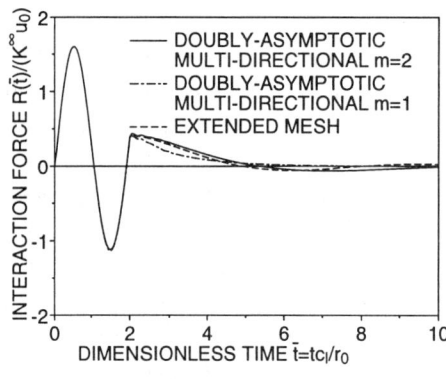

(a) doubly-asymptotic multi-directional transmitting boundary

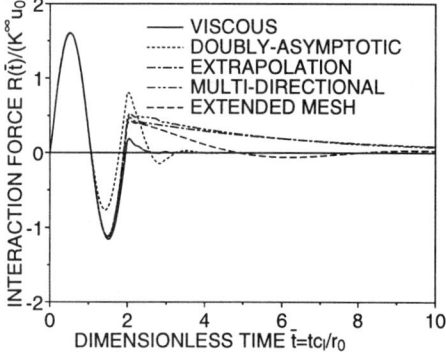

(b) various other transmitting boundaries

Figure 14-2 Semi-infinite rod on elastic foundation discretized with 5 finite elements

14.1 PERFORMANCE AND COMPARISON

boundary with $m = 2$ and $c_{a1}/c_l = 1$, $c_{a2}/c_l = \sqrt{2}$ leads to excellent accuracy for all times. For comparison, the result is also specified for $m = 1$ and $c_{a1}/c_l = 1$ which reduces the accuracy. For the same spatial and temporal discretizations the interaction force is also plotted in Figure 14-2b for the viscous boundary [32], the doubly-asymptotic boundary of first order [54], the extrapolation boundary [27] (with linear extrapolation and $c_a/c_l = 1$) and the multi-directional boundary [19] (with $m = 2$ and $c_{a1}/c_l = 1$, $c_{a2}/c_l = \sqrt{2}$). The performance of all these boundaries is much poorer than the doubly-asymptotic multi-directional transmitting boundary with $m = 2$.

To permit a comparison also with the results of the superposition boundary [46], the paraxial boundary [12] and the extrapolation boundary [27] with different parameters, as specified in Chapter 3 of Reference [58], the number of finite elements is increased to 10 without any other modification. The result for the viscous boundary in this study is the most accurate and is reproduced in Figure 14-3. However, the multi-directional boundary, which is not addressed in Reference [58], performs better but the best result is obtained with the doubly-asymptotic multi-directional transmitting boundary.

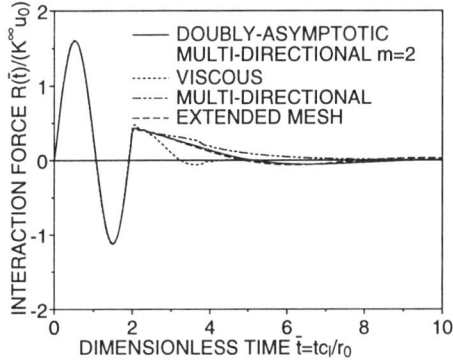

Figure 14-3 Semi-infinite rod on elastic foundation discretized with 10 finite elements

14.1.2 In-plane Motion of Semi-infinite Layer of Constant Depth

As a two-dimensional example, the in-plane motion of a semi-infinite layer of constant depth fixed at its base with Poisson's ratio $\nu = 0.25$ as discussed in Appendix A.3.2 (Figure A-10) is addressed. On the vertical structure-medium interface a horizontal displacement varying linearly $u_0(y,t)$ is applied

$$u_0(y,t) = \frac{y}{d} u_0(t) \tag{14.1}$$

where $u_0(t)$ is specified in Eq. A.2.24 ($t_0 = 2d/c_s$). The equivalent interaction force $R(t)$ is determined by integrating the nodal forces of the finite elements with the displacement as a weighting function over the structure-medium interface, which is equivalent to the procedure described in Eq. A.0.2.

The discretization up to the artificial boundary is presented in Figure 14-4 consisting

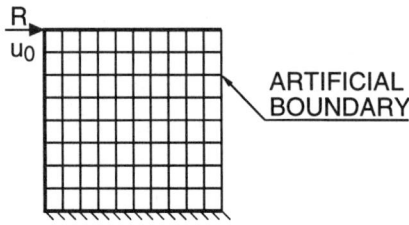

Figure 14-4 Finite-element mesh up to artificial boundary of semi-infinite layer of constant depth

of 8 finite elements with a linear displacement shape function of length $0.125d$ in the vertical direction and 10 finite elements of length $0.1d$ in the horizontal direction. The Newmark algorithm with $\beta = 0.25$, $\gamma = 0.5$ and $\Delta t = 0.05 d/c_s$ is applied.

In the doubly-asymptotic multi-directional transmitting boundary with $m = 2$ the same apparent wave velocities $c_{a1}/c_s = \sqrt{2}$, $c_{a2}/c_p = \sqrt{2}$ are used in the two directions perpendicular and tangential to the artificial boundary. The fully-coupled static-stiffness matrix $[K^\infty]$ is determined by the consistent infinitesimal finite-element cell method of Chapter 8. The interaction force $R(\bar{t})$ non-dimensionalized with $K^\infty u_0$ (static-stiffness coefficient K^∞) is plotted as a function of the dimensionless time $\bar{t} = t c_s/d$. The accuracy of the doubly-asymptotic multi-directional transmitting boundary (Figure 14-5a) is good for this finite-element mesh and is much better than that of all the other boundaries (viscous, doubly-asymptotic, extrapolation (with linear extrapolation and $c_a/c_p = 1$), multi-dimensional boundary ($m = 2$ and $c_{a1}/c_s = \sqrt{2}$, $c_{a2}/c_p = \sqrt{2}$)) presented in Figure 14-5b.

14.1.3 In-plane Motion of Semi-infinite Wedge

As an example where a curved artificial boundary is introduced, the in-plane motion of a semi-infinite wedge described in Appendix A.4.2 (Figure A-13) is examined for the homogeneous case ($G_1 = G_2 = G_3 = G$). On the arc forming the structure-medium interface a horizontal displacement $u_0(\theta, t)$ is prescribed as a linear function of θ

$$u_0(\theta, t) = \frac{\theta}{\alpha} u_0(t) \tag{14.2}$$

where $u_0(t)$ is specified in Eq. A.2.24 ($t_0 = 1.5 r_0/c_s$). The vertical displacement vanishes. The equivalent interaction force $R(t)$ determined by integrating the nodal forces of the finite elements with the displacement as a weighting function over

14.1 PERFORMANCE AND COMPARISON

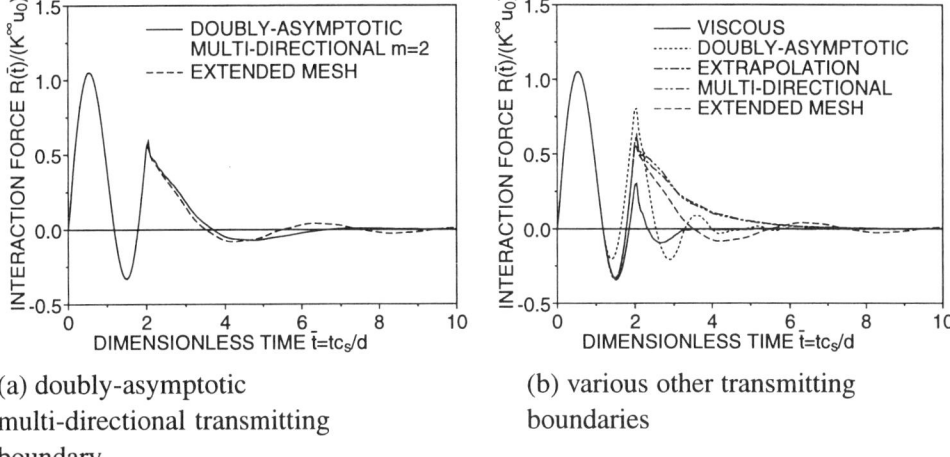

(a) doubly-asymptotic multi-directional transmitting boundary

(b) various other transmitting boundaries

Figure 14-5 Semi-infinite layer of constant depth

the structure-medium interface, which is equivalent to the procedure described in Eq. A.0.2, is addressed.

The discretization shown in Figure 14-6 consists of 8 finite elements in the circumferential direction and 10 finite elements of length $0.05r_0$ in the radial direction. The Newmark algorithm with $\beta = 0.25$, $\gamma = 0.5$ and $\Delta t = 0.025 r_0/c_s$ is used. The doubly-asymptotic multi-directional transmitting boundary is constructed exactly as described in Section 14.1.2. The interaction force $R(\bar{t})$ non-dimensionalized with $K_\infty u_0$ is plotted as a function of the dimensionless time $\bar{t} = tc_s/r_0$. The accuracy is excellent (Figure 14-7a), in contrast to those of the other transmitting boundaries (Figure 14-7b).

To construct a *local* doubly-asymptotic multi-directional transmitting boundary, a banded static-stiffness matrix is obtained for instance by setting the coefficients in the node and the adjacent nodes equal to those of the fully-coupled matrix and setting the

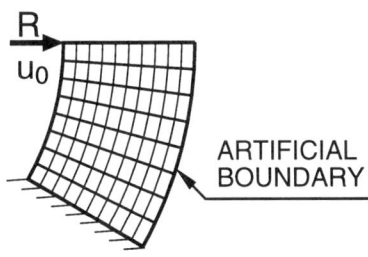

Figure 14-6 Finite-element mesh up to artificial boundary of semi-infinite wedge

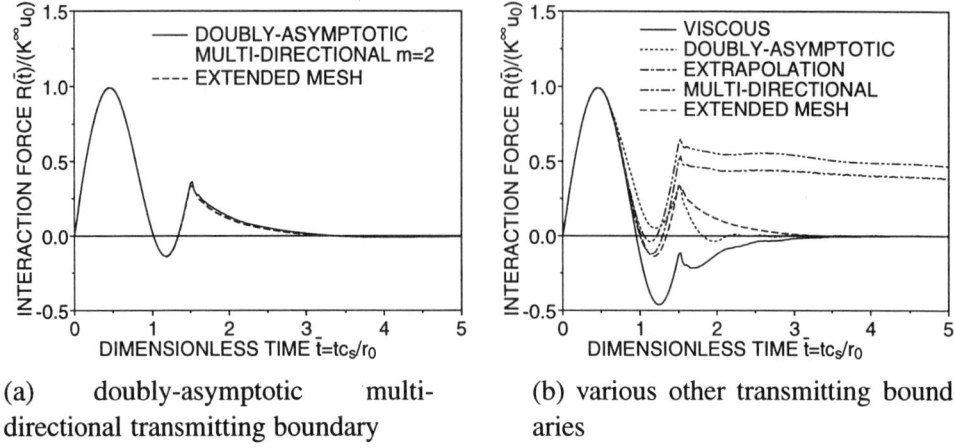

(a) doubly-asymptotic multi-directional transmitting boundary

(b) various other transmitting boundaries

Figure 14-7 Semi-infinite wedge discretized with 10 elements in radial direction

other coefficients equal to zero. The accuracy deteriorates significantly (Figure 14-8). A partially global doubly-asymptotic multi-directional transmitting boundary where the coefficients of the static-stiffness matrix in the node and in the 2 adjacent nodes on each side are kept results in intermediate accuracy. This example demonstrates that the accuracy is governed by the representation of the low-frequency and thus the static behaviour. To increase the accuracy in the local doubly-asymptotic multi-directional transmitting boundary, the number of finite elements in the radial direction is increased from 10 to 20 without any other modification. Using the banded static-stiffness matrix leads to a certain decrease in accuracy (Figure 14-9), which is, however, much smaller than for the smaller mesh (Figure 14-8). This local doubly-asymptotic multi-directional transmitting boundary is much more accurate than the multi-directional boundary.

14.2 DECISION FLOWCHART OF MODELLING PROCEDURE

When applying the direct method of dynamic unbounded medium-structure-interaction analysis, the location of the artificial boundary has to be selected. In addition, for the doubly-asymptotic multi-directional transmitting boundary, the fully-coupled static-stiffness matrix leading to a spatially global formulation or some banded matrix yielding a spatially local procedure can be chosen. Guidelines covering these two aspects and the relationship of the direct method to the substructure method are addressed in the following.

In the substructure method (see Section 1.3, Figure 1-2a) the rigorous interaction force-displacement relationship in the nodes on the generalized structure-medium

14.2 DECISION FLOWCHART OF MODELLING PROCEDURE

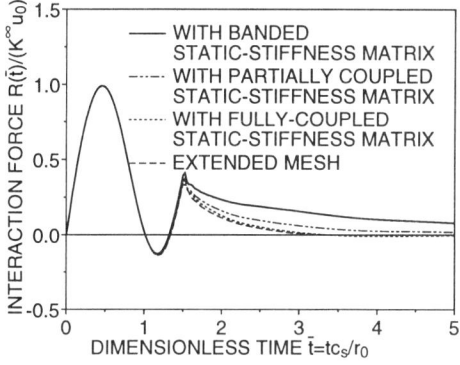

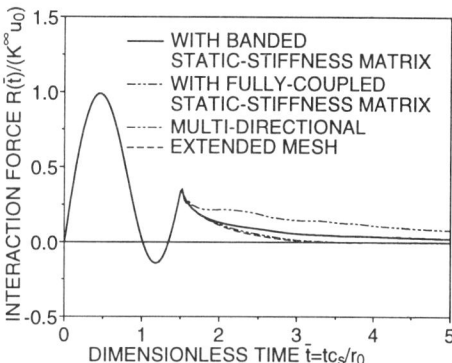

Figure 14-8 Semi-infinite wedge discretized with 10 elements in radial direction and with banded and partially coupled static-stiffness matrices used in local and partially global doubly-asymptotic multi-directional transmitting boundaries, respectively

Figure 14-9 Semi-infinite wedge discretized with 20 elements in radial direction and with banded static-stiffness matrix used in local doubly-asymptotic multi-directional transmitting boundary

interface (Figure 1-1) of the unbounded medium is required, which is global in space and time.

In the direct method (see Section 1.3, Figure 1-2b) the approximate boundary condition formulated on the artificial boundary must permit an adequate representation of wave propagation and static stiffness on the structure-medium interface for accurate results within the structure. As demonstrated in the examples of Section 14.1 the modelling of static and low-frequency behaviours requires a larger mesh than that of the intermediate and high-frequency responses. The accuracy and mesh size are thus governed in dynamic unbounded medium-structure-interaction analysis quite surprisingly by the representation of the low-frequency and static behaviours.

The doubly-asymptotic multi-directional transmitting boundary is addressed in the following. To represent accurately wave propagation in the intermediate and high-frequency ranges at the structure-medium interface, the artificial boundary, where the outward plane wave boundary condition is formulated, cannot be placed too close. The corresponding criterion is based on numerical experience. Using the fully-coupled static-stiffness matrix representing the low-frequency and static behaviours then results in good accuracy in this global formulation. If the artificial boundary must be placed at a large distance, e.g. owing to irregularities in the medium adjacent to the structure-medium interface, the fully-coupled static-stiffness matrix on the artificial boundary can be localized without reducing significantly the accuracy of the static stiffness at the structure-medium interface. In this case, the banded static-stiffness matrix on the artificial boundary can be used in the doubly-asymptotic multi-directional transmitting boundary which thus becomes spatially local.

The optimum location of the interaction horizon with the corresponding spatial coupling of the boundary condition depends on the particularities of the problem, the software available and on personal preference. A decision flowchart is presented in Figure 14-10 with the boundaries mentioned shown in Figures 1-2a and 1-2b.

If it is unrealistic to enforce the rigorous boundary condition on the generalized structure-medium interface (Figure 1-1) because e.g. too many degrees of freedom are present, the direct method with the doubly-asymptotic multi-directional transmitting boundary is used. First, the artificial boundary is placed 2 rows outwards from the generalized structure-medium interface for the sake of implementation. Then the distance from the structure-medium interface to the artificial boundary, which depends on the size of the irregular bounded medium, is examined. If it is large enough to model accurately the intermediate and high-frequency wave propagations, only the static and low-frequency behaviours at the structure-medium interface are of concern. A sequence of increasing the accuracy of the static-stiffness matrix on the artificial boundary starting with vanishing values up to the fully-coupled matrix is then examined addressing its accuracy on the structure-medium interface. This results in various implementations of the doubly-asymptotic multi-directional transmitting boundary with different spatial couplings. If the required accuracy is not reached with an acceptable spatial coupling, the artificial boundary is moved outwards.

14.2 DECISION FLOWCHART OF MODELLING PROCEDURE

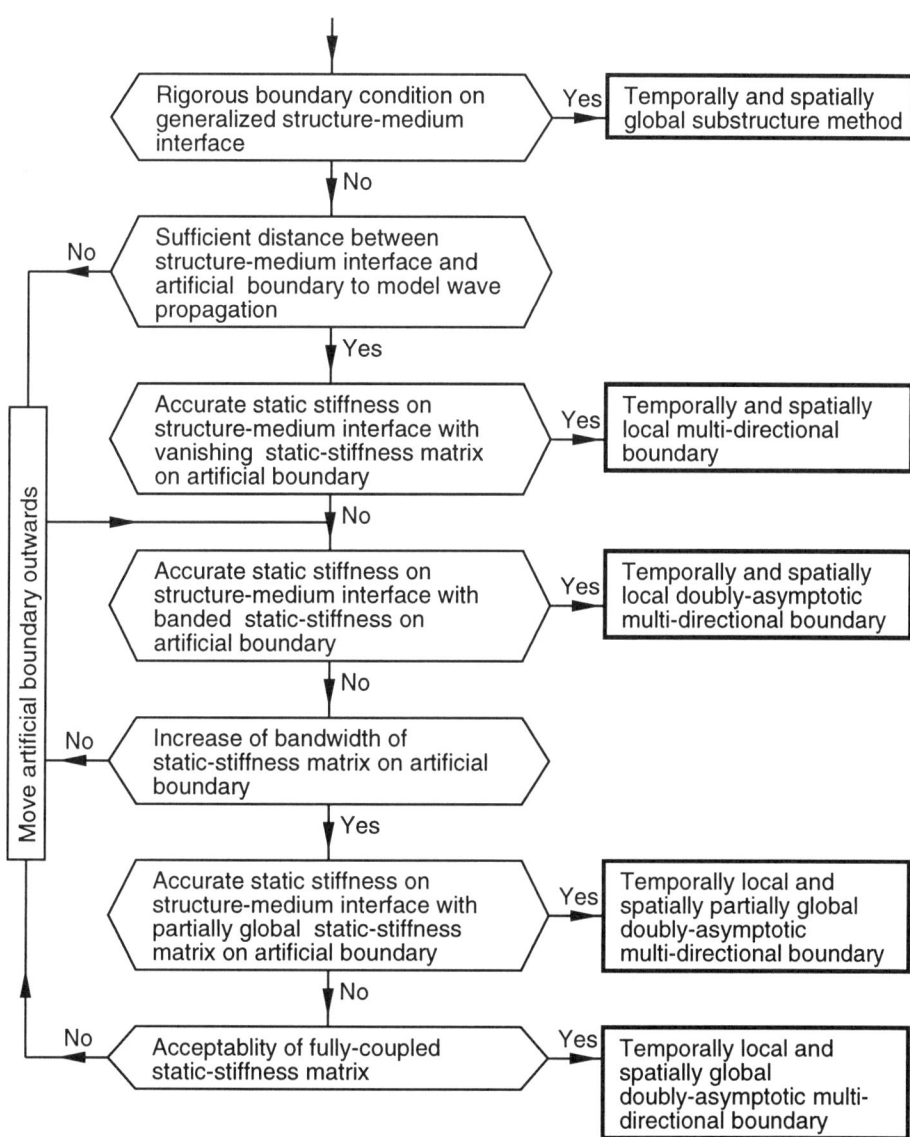

Figure 14-10 Decision flowchart of location of interaction horizon with corresponding spatial coupling of boundary condition

Appendix A

BENCHMARK EXAMPLES

Before discussing the examples, common points are addressed.

The dynamic properties of the unbounded medium are described by the interaction force-motion relationship. In two- and three-dimensional cases the structure-medium interface is discretized spatially introducing a certain number of degrees of freedom. The interaction force-motion relationship is represented in the time domain by the unit-impulse response matrix and in the frequency domain by the dynamic-stiffness matrix. To ease the presentation of the results, equivalent coefficients corresponding to spatial motion patterns are calculated from the matrix. This then permits a direct comparison with the results reported in the literature.

The acceleration unit-impulse response matrix $[M^\infty(t)]$ is used for illustration. The selected spatial motion pattern on the structure-medium interface which does not vary with time is denoted as $\{\phi\}$. From virtual work considerations as applied in the finite-element method the equivalent coefficient is calculated as

$$M^\infty(t) = \{\phi\}^T [M^\infty(t)] \{\phi\} \tag{A.0.1}$$

For the equivalent dynamic-stiffness coefficient

$$S^\infty(\omega) = \{\phi\}^T [S^\infty(\omega)] \{\phi\} \tag{A.0.2}$$

applies.

It is customary to normalize the dynamic-stiffness coefficient with the static-stiffness coefficient K^∞ and to introduce the dimensionless spring coefficient $k(a_0)$ and damping coefficient $c(a_0)$ as

$$S^\infty(a_0) = K^\infty (k(a_0) + ia_0 c(a_0)) \tag{A.0.3}$$

with the dimensionless frequency

$$a_0 = \frac{\omega r_0}{c_s} \tag{A.0.4}$$

r_0 is the characteristic length of the structure-medium interface and c_s the shear-wave velocity.

As in the time domain analytical results and even boundary-element solutions are rare, a transformation of the unit-impulse response coefficient to the frequency domain is sometimes performed to enable a comparison with the corresponding dynamic-stiffness coefficient reported in the literature (see also Section 2.2, Eqs. 2.23 and 2.25).

As a generally applicable procedure an extended mesh can be analysed to generate results in the time domain for comparison. The extended mesh adjacent to the structure-medium interface must be so large that the influence of its outer boundary on the results at the structure-medium interface is negligible or for wave propagation the waves reflected at the exterior boundary have not reached the structure-medium interface during the time of analysis. The response at the structure-medium interface of the bounded medium, the extended mesh, is identical to that of the unbounded medium based on the uniqueness theorem discussed in Section 1.2. This is verified by evaluating the domain of influence first at the structure-medium interface and second at the outer boundary. As the computational effort for analysing the extended mesh increases rapidly as time proceeds, the results are only calculated up to a limited time.

Another possibility for generating results for comparison in the frequency domain is based on dynamic condensation and similarity. The method is described in depth in Section 7.3. Its concept can be summarized as follows. If the dynamic-stiffness matrix at the exterior boundary characterized by r_e is known, the dynamic-stiffness matrix can be calculated at the interior boundary with r_i by discretizing the region between the two boundaries with finite elements and applying standard assemblage. For a fixed frequency starting at an exterior boundary with a very large characteristic length, this procedure can be applied repeatedly, permitting the dynamic-stiffness matrices corresponding to the boundaries of smaller characteristic lengths down to an arbitrarily small value to be determined. The decrease in r corresponds to a decrease in a_0. The dynamic-stiffness matrix is thus known from this arbitrarily small value up to the a_0 corresponding to the original exterior boundary. The variation of a_0 can also be interpreted as varying ω for a fixed r. The starting dynamic-stiffness coefficient for a very large a_0 is selected as the dashpot matrix $[C_\infty]$ of Eq. 2.7 of the singular part. It can be determined straightforwardly using the law of conservation of momentum. As demonstrated in Section 7.3 the error of the dynamic-stiffness matrix decreases as dynamic condensation proceeds.

A.1 SPHERICAL CAVITY EMBEDDED IN FULL-SPACE

As a very simple one-dimensional wave propagation problem with an analytical solution readily available, the spherical cavity embedded in a full-space with a uniform normal displacement enforced on its wall, the structure-medium interface, is addressed (Figure A-1). The radius of the spherical cavity is r_0, and the material properties of the full-space are defined by the shear modulus G, Poisson's ratio ν

A.1 SPHERICAL CAVITY EMBEDDED IN FULL-SPACE

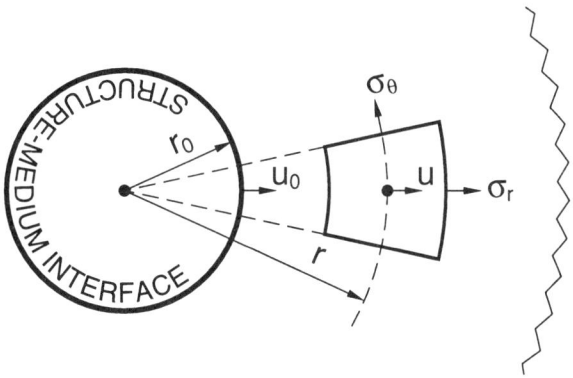

Figure A-1 Spherical cavity embedded in full-space with symmetric waves (section)

and mass density ρ. The prescribed displacement $u_0 = u_0(t)$ generates a symmetric dilatational wave with the wave velocity

$$c_p = \sqrt{\frac{G}{\rho}\frac{2(1-\nu)}{1-2\nu}} \qquad (A.1.1)$$

For spherical symmetry, all variables depend on the radial coordinate r only. The only displacement component, which arises in the radial direction, is denoted as u. The radial strain ε_r and the circumferential strains ε_θ, ε_φ are specified as

$$\varepsilon_r = u_{,r} \qquad (A.1.2a)$$

$$\varepsilon_\theta = \varepsilon_\varphi = \frac{1}{r}u \qquad (A.1.2b)$$

The stresses σ_r, σ_θ and σ_φ in the corresponding directions follow from Hooke's law

$$\begin{Bmatrix} \sigma_r \\ \sigma_\theta \\ \sigma_\varphi \end{Bmatrix} = \frac{2G}{1-2\nu} \begin{bmatrix} 1-\nu & \nu & \nu \\ \nu & 1-\nu & \nu \\ \nu & \nu & 1-\nu \end{bmatrix} \begin{Bmatrix} \varepsilon_r \\ \varepsilon_\theta \\ \varepsilon_\varphi \end{Bmatrix} \qquad (A.1.3)$$

The dynamic equilibrium equation is formulated as

$$\sigma_{r,r} + \frac{(2\sigma_r - \sigma_\theta - \sigma_\varphi)}{r} - \rho \ddot{u} = 0 \qquad (A.1.4)$$

Substituting Eq. A.1.2 in Eq. A.1.3 and then substituting the stresses in Eq. A.1.4 leads to the equation of motion

$$u_{,rr} + \frac{2}{r}u_{,r} - \frac{2}{r^2}u - \frac{1}{c_p^2}\ddot{u} = 0 \qquad (A.1.5)$$

The boundary condition at the wall is formulated as

$$u(r = r_0) = u_0 \qquad (A.1.6)$$

The initial conditions

$$u(t = 0) = 0 \qquad (A.1.7a)$$
$$\dot{u}(t = 0) = 0 \qquad (A.1.7b)$$

specify that the spherical cavity is initially at rest.

First, the displacement dynamic-stiffness coefficient in the frequency domain $S^\infty(\omega)$ is discussed, which relates the displacement amplitude at the wall ($r = r_0$) $u_0(\omega)$ to the corresponding interaction force amplitude

$$R(\omega) = -4\pi r_0^2 \sigma_r(\omega, r = r_0) \qquad (A.1.8)$$

as

$$R(\omega) = S^\infty(\omega) u_0(\omega) \qquad (A.1.9)$$

The superscript ∞ denotes the unbounded medium. Often, the word displacement is omitted in the term displacement dynamic-stiffness coefficient. The equation of motion (Eq. A.1.5) in the frequency domain equals

$$u(\omega)_{,rr} + \frac{2}{r} u(\omega)_{,r} - \frac{2}{r^2} u(\omega) + \frac{\omega^2}{c_p^2} u(\omega) = 0 \qquad (A.1.10)$$

The displacement amplitude can be expressed as a function of the potential $\phi(\omega)$

$$u(\omega) = \phi(\omega)_{,r} \qquad (A.1.11)$$

Substituting Eq. A.1.11 in Eq. A.1.10 leads to the wave equation expressed in the potential. This equation is satisfied if the following equation applies for the product $r\phi(\omega)$

$$(r\phi(\omega))_{,rr} - \frac{\omega^2}{c_p^2} r\phi(\omega) = 0 \qquad (A.1.12)$$

The solution of this one-dimensional wave equation is

$$\phi(\omega) = \frac{c_1}{r} e^{+i\frac{\omega}{c_p}r} + \frac{c_2}{r} e^{-i\frac{\omega}{c_p}r} \qquad (A.1.13)$$

with the two integration constants c_1 and c_2. The displacement amplitude follows from Eq. A.1.11 as

$$u(\omega) = \frac{c_1}{r^2}\left(-1 + i\frac{\omega}{c_p}r\right)e^{+i\frac{\omega}{c_p}r} + \frac{c_2}{r^2}\left(-1 - i\frac{\omega}{c_p}r\right)e^{-i\frac{\omega}{c_p}r} \qquad (A.1.14)$$

A.1 SPHERICAL CAVITY EMBEDDED IN FULL-SPACE

To interpret Eq. A.1.14 physically, recall that the displacement equals the product of the corresponding amplitude and the factor $e^{+i\omega t}$. As the expression $e^{+i\omega(t-r/c_p)}$ describes a wave propagating in the positive radial direction with the velocity c_p, the second term in Eq. A.1.14 corresponds to an outgoing spherical wave. Analogously, the first is associated with an incoming wave. As only outgoing waves will exist, this radiation condition sets $c_1 = 0$. For the prescribed displacement amplitude $u_0(\omega)$

$$u(\omega) = u_0(\omega) \left(\frac{r_0}{r}\right)^2 \left(\frac{c_p + i\omega r}{c_p + i\omega r_0}\right) e^{-i\frac{\omega}{c_p}(r-r_0)} \quad (A.1.15)$$

follows. The interaction force amplitude $R(\omega)$ is calculated from Eqs. A.1.8, A.1.3, A.1.2 and A.1.15. The dynamic-stiffness coefficient defined in Eq. A.1.9 then follows as

$$S^\infty(\omega) = 16\pi G r_0 \left(1 + \frac{1-\nu}{2(1-2\nu)} \frac{1}{c_p} \frac{(i\omega r_0)^2}{c_p + i\omega r_0}\right) \quad (A.1.16)$$

The static-stiffness coefficient is calculated as

$$K^\infty = 16\pi G r_0 \quad (A.1.17)$$

Normalizing $S^\infty(\omega)$ as

$$S^\infty(\omega) = K^\infty (k(a_0) + i a_0 c(a_0)) \quad (A.1.18)$$

defines the dimensionless spring and damping coefficients as

$$k(a_0) = 1 - \frac{1-\nu}{2(1-2\nu)} \frac{a_0^2}{1+a_0^2} \quad (A.1.19a)$$

$$c(a_0) = \frac{1-\nu}{2(1-2\nu)} \frac{a_0^2}{1+a_0^2} \quad (A.1.19b)$$

with the dimensionless frequency

$$a_0 = \frac{\omega r_0}{c_p} \quad (A.1.20)$$

Note that for simplicity a_0 is defined using $c_p = \sqrt{2(1-\nu)/(1-2\nu)}\, c_s$ and not c_s.

In order to determine the Fourier transform, $S^\infty(\omega)$ is decomposed into the singular part $S_s^\infty(\omega)$, which is equal to its asymptotic value for $\omega \to \infty$ and the remaining regular part $S_r^\infty(\omega)$

$$S^\infty(\omega) = S_s^\infty(\omega) + S_r^\infty(\omega) \quad (A.1.21)$$

where

$$S_s^\infty(\omega) = i\omega C_\infty + K_\infty \tag{A.1.22}$$

The subscript ∞ denotes the limit $\omega \to \infty$. For the spherical cavity, Eq. A.1.16 leads to

$$C_\infty = 4\pi r_0^2 \rho c_p \tag{A.1.23a}$$

$$K_\infty = 8\pi \frac{1-3\nu}{1-2\nu} G r_0 \tag{A.1.23b}$$

$$S_r^\infty(\omega) = K^\infty \frac{1-\nu}{2(1-2\nu)} \frac{c_p}{c_p + i\omega r_0} \tag{A.1.23c}$$

The displacement unit-impulse response function in the time domain is equal to the inverse Fourier transform of the dynamic-stiffness coefficient in the frequency domain

$$S^\infty(t) = \frac{1}{2\pi} \int_{-\infty}^{+\infty} S^\infty(\omega) e^{+i\omega t} d\omega \tag{A.1.24}$$

which, when applied to Eqs. A.1.21 and A.1.22, yields

$$S^\infty(t) = S_s^\infty(t) + S_r^\infty(t) \tag{A.1.25}$$

where

$$S_s^\infty(t) = C_\infty \dot{\delta}(t) + K_\infty \delta(t) \tag{A.1.26}$$

For the spherical cavity, using Eq. A.1.23c

$$S_r^\infty(t) = K^\infty \frac{c_p}{r_0} \frac{1-\nu}{2(1-2\nu)} e^{-\frac{c_p}{r_0} t} H(t) \tag{A.1.27}$$

results. $\delta(t)$ represents the Dirac-delta function, $H(t)$ the Heaviside-step function ($H(t<0) = 0$, $H(t \geq 0) = 1$). Thus the displacement unit-impulse response function is written as

$$S^\infty(t) = C_\infty \dot{\delta}(t) + K_\infty \delta(t) + K^\infty \frac{c_p}{r_0} \frac{1-\nu}{2(1-2\nu)} e^{-\frac{c_p}{r_0} t} H(t) \tag{A.1.28}$$

Analogously to the dimensionless frequency a_0, a dimensionless time $\bar{t}$ is defined

$$\bar{t} = t \frac{c_p}{r_0} \tag{A.1.29}$$

Using

$$\delta(t) = \frac{c_p}{r_0} \delta(\bar{t}) \tag{A.1.30a}$$

$$\dot{\delta}(t) = \left(\frac{c_p}{r_0}\right)^2 \delta(\bar{t})_{,\bar{t}} \tag{A.1.30b}$$

A.1 SPHERICAL CAVITY EMBEDDED IN FULL-SPACE

Eq. A.1.28 is formulated as

$$S^\infty(\bar{t}) = C_\infty \left(\frac{c_p}{r_0}\right)^2 \delta(\bar{t})_{,\bar{t}} + K_\infty \frac{c_p}{r_0}\delta(\bar{t}) + K^\infty \frac{c_p}{r_0}\frac{1-\nu}{2(1-2\nu)}e^{-\bar{t}}H(\bar{t}) \quad \text{(A.1.31)}$$

The regular part of the displacement unit-impulse response function $S_r^\infty(\bar{t})$, non-dimensionalized with $K^\infty c_p/r_0$, is plotted for Poisson's ratio $\nu = 0.25$ as a function of $\bar{t}$ in Figure A-2a.

Second, the velocity dynamic-stiffness coefficient $V^\infty(\omega)$ is introduced as

$$R(\omega) = V^\infty(\omega)\dot{u}_0(\omega) \quad \text{(A.1.32)}$$

where $\dot{u}_0(\omega)$ represents the velocity amplitude. With

$$\dot{u}_0(\omega) = i\omega u_0(\omega) \quad \text{(A.1.33)}$$

and comparing Eqs. A.1.9 and A.1.32 yields the relationship between the displacement and velocity dynamic-stiffness coefficients

$$V^\infty(\omega) = \frac{S^\infty(\omega)}{i\omega} \quad \text{(A.1.34)}$$

For the spherical cavity using Eq. A.1.16

$$V^\infty(\omega) = K^\infty \left(\frac{1}{i\omega} + \frac{1-\nu}{2(1-2\nu)}\frac{r_0}{c_p}\frac{i\omega r_0}{c_p + i\omega r_0}\right) \quad \text{(A.1.35)}$$

results. Again, a decomposition into the singular and regular parts is performed

$$V^\infty(\omega) = V_s^\infty(\omega) + V_r^\infty(\omega) \quad \text{(A.1.36)}$$

where

$$V_s^\infty(\omega) = C_\infty \quad \text{(A.1.37)}$$

For the spherical cavity

$$V_r^\infty(\omega) = K^\infty \left(\frac{1}{i\omega} - \frac{1-\nu}{2(1-2\nu)}\frac{r_0}{c_p + i\omega r_0}\right) \quad \text{(A.1.38)}$$

follows from Eq. A.1.35.

The velocity unit-impulse response function, again calculated as the inverse Fourier transform of $V^\infty(\omega)$, and applying Eq. A.1.36, is equal to

$$V^\infty(t) = V_s^\infty(t) + V_r^\infty(t) \quad \text{(A.1.39)}$$

where from Eq. A.1.37

$$V_s^\infty(t) = C_\infty \delta(t) \tag{A.1.40}$$

For the spherical cavity using Eq. A.1.38,

$$V_r^\infty(t) = K^\infty \left(1 - \frac{1-\nu}{2(1-2\nu)} e^{-\frac{c_p}{r_0}t}\right) H(t) \tag{A.1.41}$$

results. Thus, the velocity unit-impulse response function equals

$$V^\infty(t) = C_\infty \delta(t) + K^\infty \left(1 - \frac{1-\nu}{2(1-2\nu)} e^{-\frac{c_p}{r_0}t}\right) H(t) \tag{A.1.42}$$

or

$$V^\infty(\bar{t}) = C_\infty \frac{c_p}{r_0} \delta(\bar{t}) + K^\infty \left(1 - \frac{1-\nu}{2(1-2\nu)} e^{-\bar{t}}\right) H(\bar{t}) \tag{A.1.43}$$

The regular part of the velocity unit-impulse response function $V_r^\infty(\bar{t})$ non-dimensionalized with K^∞ is plotted for Poisson's ratio $\nu = 0.25$ as a function of $\bar{t}$ in Figure A-2b. Note that $V_r^\infty(\bar{t} = 0)$ is equal to the spring coefficient for $\omega \to \infty$, K_∞ (Eq. A.1.23b) and that $V_r^\infty(\bar{t} \to \infty)$ equals the static-stiffness coefficient K^∞ (Eq. A.1.17).

Third, the acceleration dynamic-stiffness coefficient $M^\infty(\omega)$ is defined as

$$R(\omega) = M^\infty(\omega) \ddot{u}_0(\omega) \tag{A.1.44}$$

where $\ddot{u}_0(\omega)$ denotes the acceleration amplitude. With

$$\ddot{u}_0(\omega) = (i\omega)^2 u_0(\omega) \tag{A.1.45}$$

and matching Eqs. A.1.9 and A.1.44 results in the relationship between the displacement and acceleration dynamic-stiffness coefficients

$$M^\infty(\omega) = \frac{S^\infty(\omega)}{(i\omega)^2} \tag{A.1.46}$$

For the spherical cavity applying Eq. A.1.16

$$M^\infty(\omega) = K^\infty \left(\frac{1}{(i\omega)^2} + \frac{1-\nu}{2(1-2\nu)} \frac{r_0^2}{c_p(c_p + i\omega r_0)}\right) \tag{A.1.47}$$

is derived. The singular part vanishes.

The acceleration unit-impulse response function is again determined as the inverse Fourier transform of $M^\infty(\omega)$. For the spherical cavity using Eq. A.1.47

$$M^\infty(t) = K^\infty \left(t + \frac{r_0}{c_p} \frac{1-\nu}{2(1-2\nu)} e^{-\frac{c_p}{r_0}t}\right) H(t) \tag{A.1.48}$$

A.2 SEMI-INFINITE ROD ON ELASTIC FOUNDATION

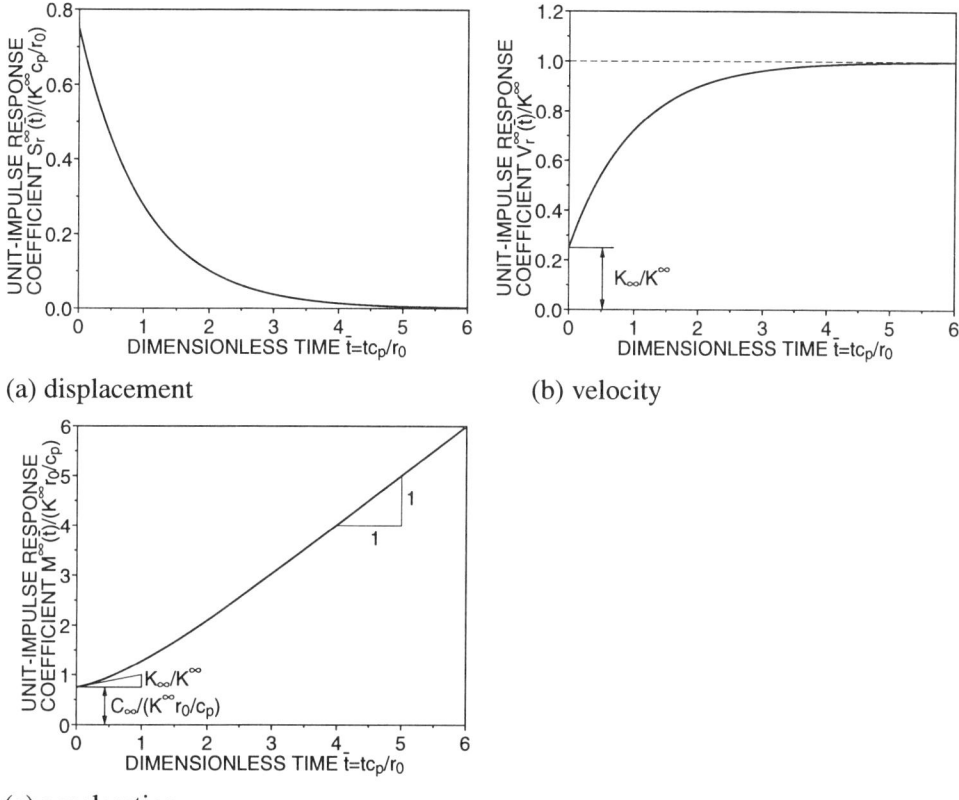

(a) displacement

(b) velocity

(c) acceleration

Figure A-2 Unit-impulse response functions of spherical cavity

follows, or in dimensionless time

$$M^\infty(\bar{t}) = K^\infty \frac{r_0}{c_p}\left(\bar{t} + \frac{1-\nu}{2(1-2\nu)}e^{-\bar{t}}\right)H(\bar{t}) \tag{A.1.49}$$

The acceleration unit-impulse response coefficient $M^\infty(\bar{t})$ non-dimensionalized with $K^\infty r_0/c_p$ is plotted for $\nu = 0.25$ as a function of $\bar{t}$ in Figure A-2c. Note that at $\bar{t} = 0$, $M^\infty(\bar{t} = 0)$ is equal to the damping coefficient for $\omega \to \infty$, C_∞, and that the tangent equals the spring coefficient for $\omega \to \infty$, K_∞. For $\bar{t} \to \infty$, the tangent equals the static-stiffness coefficient K^∞.

A.2 SEMI-INFINITE ROD ON ELASTIC FOUNDATION

As another one-dimensional wave propagation problem with an analytical solution available, the semi-infinite rod on an elastic foundation is addressed (Figure A-3a).

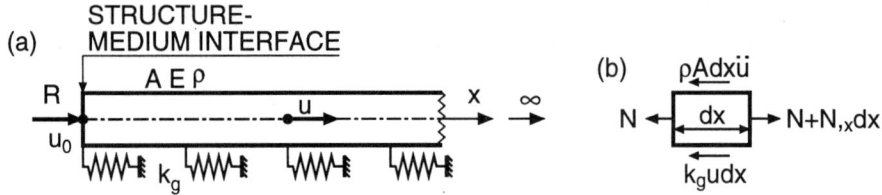

Figure A-3 (a) Semi-infinite rod on elastic foundation; (b) Equilibrium of infinitesimal element

As will become apparent, this is a dispersive system, i.e. the phase velocity is a function of frequency. In addition, a so-called cut-off frequency occurs below which no waves propagate, i.e. evanescent waves are present. The semi-infinite rod on elastic foundation is thus a stringent test. The significance of this example is much greater than expected, as the semi-infinite rod on an elastic foundation represents exactly a single mode of an infinite modal expansion of the out-of-plane response of a layer fixed at its base (Appendix A.3.1).

The area of the undamped rod is denoted by A, the modulus of elasticity by E, the mass density by ρ, and the static spring stiffness per unit length of the elastic foundation by k_g. N represents the normal force and u the axial displacement. Formulating equilibrium (Figure A-3b)

$$N_{,x}\,dx - k_g u\,dx - A\rho\ddot{u}\,dx = 0 \qquad (A.2.1)$$

and substituting the force-displacement relationship

$$N = EAu_{,x} \qquad (A.2.2)$$

leads to the equation of motion

$$u_{,xx} - \frac{1}{r_0^2}u - \frac{\ddot{u}}{c_l^2} = 0 \qquad (A.2.3)$$

with the characteristic length

$$r_0 = \sqrt{\frac{EA}{k_g}} \qquad (A.2.4)$$

and the rod velocity

$$c_l = \sqrt{\frac{E}{\rho}} \qquad (A.2.5)$$

A.2 SEMI-INFINITE ROD ON ELASTIC FOUNDATION

First, the displacement dynamic-stiffness coefficient in the frequency domain $S^\infty(\omega)$, relating the displacement amplitude $u_0(\omega)$ to the interaction force amplitude $R(\omega)$

$$R(\omega) = S^\infty(\omega) u_0(\omega) \tag{A.2.6}$$

is derived. The equation of motion (Eq. A.2.3) in the frequency domain equals

$$u(\omega)_{,xx} - \frac{1}{r_0^2} u(\omega) + \frac{\omega^2}{c_l^2} u(\omega) = 0 \tag{A.2.7}$$

Introducing the dimensionless frequency

$$a_0 = \frac{\omega r_0}{c_l} \tag{A.2.8}$$

Eq. A.2.7 is formulated as

$$r_0^2 u(a_0)_{,xx} + (a_0^2 - 1) u(a_0) = 0 \tag{A.2.9}$$

The solution is given by

$$u(a_0) = c_1 e^{+i\sqrt{a_0^2-1}\frac{x}{r_0}} + c_2 e^{-i\sqrt{a_0^2-1}\frac{x}{r_0}} \tag{A.2.10}$$

When determining the dynamic-stiffness coefficient at $x = 0$ in the semi-infinite system, only outgoing waves exist. As discussed in connection with Eq. A.1.14, the first term on the right-hand side corresponds for $a_0 > 1$ to an incoming wave which violates the radiation condition. Setting $c_1 = 0$ and enforcing $u(x = 0, a_0) = u_0(a_0)$ leads to

$$u(a_0) = u_0(a_0) e^{-i\sqrt{a_0^2-1}\frac{x}{r_0}} \tag{A.2.11}$$

Substituting Eq. A.2.11 in

$$R(a_0) = -EA u_0(a_0)_{,x} \tag{A.2.12}$$

and with the static-stiffness coefficient

$$K^\infty = \sqrt{EA k_g} \tag{A.2.13}$$

the dynamic-stiffness coefficient (Eq. A.2.6) is formulated as

$$S^\infty(a_0) = K^\infty \sqrt{1 - a_0^2} \tag{A.2.14}$$

The dimensionless spring and damping coefficients are defined as

$$S^\infty(a_0) = K^\infty(k(a_0) + ia_0 c(a_0)) \tag{A.2.15}$$

with

$$\text{for } a_0 \leq 1 \quad k(a_0) = \sqrt{1 - a_0^2} \tag{A.2.16a}$$
$$c(a_0) = 0 \tag{A.2.16b}$$
$$\text{for } a_0 > 1 \quad k(a_0) = 0 \tag{A.2.16c}$$
$$c(a_0) = \sqrt{1 - \frac{1}{a_0^2}} \tag{A.2.16d}$$

Below the cut-off frequency $a_0 = 1$, the damping coefficient $c(a_0)$ representing radiation damping vanishes. The dynamic-stiffness coefficient is plotted in Figure A-4.

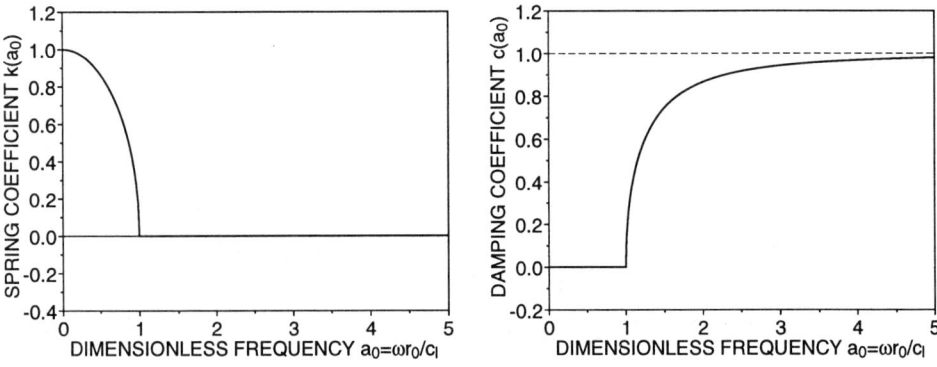

Figure A-4 Dynamic-stiffness coefficient of semi-infinite rod on elastic foundation

$S^\infty(a_0)$ is decomposed into the singular part $S_s^\infty(a_0)$, which is equal to its asymptotic value for $a_0 \to \infty$, and the remaining regular part $S_r^\infty(a_0)$

$$S^\infty(a_0) = S_s^\infty(a_0) + S_r^\infty(a_0) \tag{A.2.17}$$

where

$$S_s^\infty(a_0) = K^\infty i a_0 \tag{A.2.18a}$$
$$S_r^\infty(a_0) = K^\infty(\sqrt{1 - a_0^2} - ia_0) \tag{A.2.18b}$$

A.2 SEMI-INFINITE ROD ON ELASTIC FOUNDATION

The displacement unit-impulse response function in the time domain is equal to the inverse Fourier transform of the dynamic-stiffness coefficient in the frequency domain (Eq. 2.4)

$$S^\infty(t) = S_s^\infty(t) + S_r^\infty(t) \tag{A.2.19}$$

where

$$S_s^\infty(t) = K^\infty \frac{r_0}{c_l} \dot{\delta}(t) \tag{A.2.20a}$$

$$S_r^\infty(t) = K^\infty \frac{1}{t} J_1(\frac{c_l}{r_0} t) \tag{A.2.20b}$$

with the Bessel function of the first kind and of the first order J_1. The Fourier transform of Eq. A.2.18b is specified in Reference [10] (No. 556.1).

The acceleration unit-impulse response function follows from Eq. 2.20 as

$$M^\infty(t) = K^\infty \frac{r_0}{c_l} H(t) + K^\infty \int_0^t \int_0^\tau \frac{1}{\tau'} J_1(\frac{c_l}{r_0}\tau') d\tau' d\tau \tag{A.2.21}$$

A dimensionless time $\bar{t}$ is defined as

$$\bar{t} = \frac{c_l}{r_0} t \tag{A.2.22}$$

Eq. A.2.21 is formulated as

$$M^\infty(\bar{t}) = K^\infty \frac{r_0}{c_l} \left(H(\bar{t}) + \int_0^{\bar{t}} \int_0^\tau \frac{1}{\tau'} J(\tau') d\tau' d\tau \right) \tag{A.2.23}$$

The acceleration unit-impulse response function $M^\infty(\bar{t})$ non-dimensionalized with $\rho c_l A$ is plotted as a function of $\bar{t}$ in Figure A-5.

As an application using the unit-impulse response coefficient, the interaction force $R(t)$ at the end of the rod caused by a rounded triangular displacement pulse is determined

$$u_0(t) = \frac{u_0}{2}\left(1 - \cos(2\pi \frac{t}{t_0})\right) \qquad 0 \le t \le t_0 = 2\frac{r_0}{c_l}$$

$$= 0 \qquad t > t_0 \tag{A.2.24}$$

The interaction force-acceleration relationship (Eq. 2.11) is written as

$$R(t) = \int_0^t M^\infty(t-\tau) \ddot{u}(\tau) d\tau \tag{A.2.25}$$

The accelerations follow from Eq. A.2.24. Eq. A.2.25 is evaluated numerically. The interaction force $R(\bar{t})$ non-dimensionalized with $K^\infty u_0$ is plotted as a function of $\bar{t}$ in Figure A-6. Note that due to the regular part of the displacement unit-impulse response function (Eq. A.2.20b), $R(\bar{t})$ does not vanish for $\bar{t} > t_0 c_l/r_0 = 2$ although the end of the rod does not move. This represents the lingering response (Eq. 2.10).

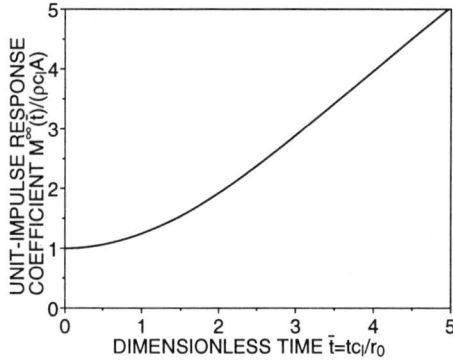
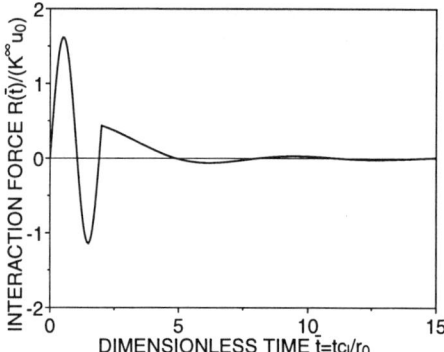

Figure A-5 Acceleration unit-impulse response function of semi-infinite rod on elastic foundation

Figure A-6 Interaction force caused by rounded triangular displacement pulse applied to semi-infinite rod on elastic foundation

A.3 SEMI-INFINITE LAYER OF CONSTANT DEPTH

A.3.1 Out-of-plane Motion

As a two-dimensional wave propagation problem with an analytical solution available, the out-of-plane (anti-plane) motion of a semi-infinite layer of constant depth d is addressed (Figure A-7). The characteristic length d does not change.

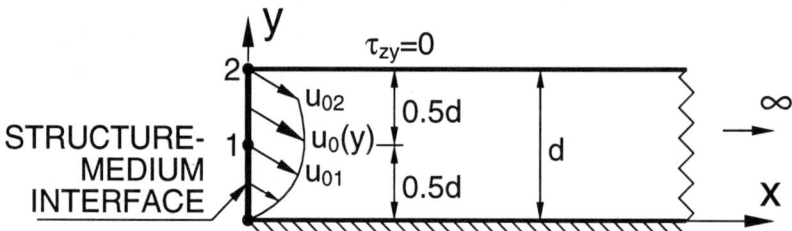

Figure A-7 Out-of-plane motion of semi-infinite layer with prescribed displacement

One of the boundaries extending to infinity in the radial direction is fixed and the other is a free surface. The structure-medium interface coincides with the y-axis. The material behaviour of the homogeneous layer is characterized by the shear modulus G and the mass density ρ. The dynamic-stiffness matrix $[S^\infty(a_0)]$ at the structure-medium interface corresponding to a parabolic variation of the displacement $u_0(y)$ is calculated. As the displacement is zero at the base, only two nodes with the displacements u_{01} and u_{02} assembled in the vector $\{u_0\}$ are introduced. The

A.3 SEMI-INFINITE LAYER OF CONSTANT DEPTH

displacement on the structure-medium interface is formulated as

$$u_0(y) = [N(y)]\{u_0\} \tag{A.3.1}$$

where the shape function equals

$$[N(y)] = \left[\frac{y}{d}\left(-4\frac{y}{d}+4\right) \quad \frac{y}{d}\left(2\frac{y}{d}-1\right) \right] \tag{A.3.2}$$

As will become apparent, this system has a cut-off frequency at the layer's fundamental natural frequency. At this frequency and at each of the higher natural frequencies, a Love mode starts with a phase velocity of infinity which decreases for higher frequencies, converging to the shear-wave velocity for infinite frequency. This case, which often occurs in practice, is thus a very stringent test.

The scalar wave equation for the out-of-plane (anti-plane) motion $u = u(x,y,t)$ is derived. Only shear strains γ_{zx} and γ_{zy} in the z-direction are present

$$\gamma_{zx} = u_{,x} \tag{A.3.3a}$$
$$\gamma_{zy} = u_{,y} \tag{A.3.3b}$$

The shear stresses follow from the stress-strain relationship

$$\tau_{zx} = G\gamma_{zx} \tag{A.3.4a}$$
$$\tau_{zy} = G\gamma_{zy} \tag{A.3.4b}$$

Formulating the equilibrium equation

$$\tau_{zx,x} + \tau_{zy,y} - \rho\ddot{u} = 0 \tag{A.3.5}$$

and substituting Eqs. A.3.3 and A.3.4 results in the wave equation

$$u_{,xx} + u_{,yy} - \frac{1}{c_s^2}\ddot{u} = 0 \tag{A.3.6}$$

with the shear-wave velocity

$$c_s = \sqrt{\frac{G}{\rho}} \tag{A.3.7}$$

The conditions on the free and fixed boundaries equal

$$\tau_{zy}(y = d) = 0 \tag{A.3.8a}$$
$$u(y = 0) = 0 \tag{A.3.8b}$$

A solution is derived in the frequency domain. The corresponding wave equation in the displacement amplitude $u(\omega)$ (Eq. A.3.6) equals

$$u(\omega)_{,xx} + u(\omega)_{,yy} + \frac{\omega^2}{c_s^2} u(\omega) = 0 \qquad (A.3.9)$$

The method of separation of variables is applied to Eq. A.3.9. $u(\omega)$ is written as the product of a function $X(x)$, which is independent of y, and a function $Y(y)$, which is independent of x

$$u(\omega) = X(x)Y(y) \qquad (A.3.10)$$

Substituting Eq. A.3.10 in Eq. A.3.9 yields after division by $X(x)Y(y)/d^2$

$$d^2 \frac{X(x)_{,xx}}{X(x)} + a_0^2 = -d^2 \frac{Y(y)_{,yy}}{Y(y)} \qquad (A.3.11)$$

with the dimensionless frequency

$$a_0 = \frac{\omega d}{c_s} \qquad (A.3.12)$$

The left-hand side of Eq. A.3.11 is a function of x only and the right-hand side of y only. To be able to satisfy Eq. A.3.11, the two sides must be equal to the same constant, denoted as λ^2. Setting

$$d^2 \frac{Y(y)_{,yy}}{Y(y)} = -\lambda^2 \qquad (A.3.13)$$

leads to the following two ordinary differential equations

$$Y(y)_{,yy} + \left(\frac{\lambda}{d}\right)^2 Y(y) = 0 \qquad (A.3.14a)$$

$$X(x)_{,xx} + \frac{1}{d^2}(a_0^2 - \lambda^2) X(x) = 0 \qquad (A.3.14b)$$

Note that Eq. A.3.14b is in the same form as Eq. A.2.9 of the semi-infinite rod on elastic foundation. Eq. A.3.14a is solved, enforcing the boundary conditions of Eq. A.3.8

$$Y(d)_{,y} = 0 \qquad (A.3.15a)$$
$$Y(0) = 0 \qquad (A.3.15b)$$

The general solution of Eq. A.3.14a equals

$$Y(y) = c_1 \cos\left(\lambda \frac{y}{d}\right) + c_2 \sin\left(\lambda \frac{y}{d}\right) \qquad (A.3.16)$$

A.3 SEMI-INFINITE LAYER OF CONSTANT DEPTH

with the integration constants c_1 and c_2. Eq. A.3.15 results in

$$\cos \lambda = 0 \tag{A.3.17}$$

The eigenvalues λ follow as

$$\lambda_i = \frac{(2i+1)\pi}{2} \quad (i = 0, 1, \ldots) \tag{A.3.18}$$

and the eigenfunctions as

$$Y_i(y) = \sin\left(\lambda_i \frac{y}{d}\right) \tag{A.3.19}$$

The general solution of Eq. A.3.14b for λ_i is

$$X_i(x) = c_{1i} e^{+i\sqrt{a_0^2 - \lambda_i^2}\,\frac{x}{d}} + c_{2i} e^{-i\sqrt{a_0^2 - \lambda_i^2}\,\frac{x}{d}} \tag{A.3.20}$$

with the integration constants c_{1i} and c_{2i}. Following the argument presented in connection with Eq. A.1.14, $c_{1i}=0$ is obtained. This leads to

$$X_i(x) = c_i e^{-i\sqrt{a_0^2 - \lambda_i^2}\,\frac{x}{d}} \tag{A.3.21}$$

where the subscript 2 in the integration constant is omitted.

Using Eqs. A.3.21 and A.3.19 together with Eq. A.3.10 permits the displacement amplitude to be expressed as

$$u(\omega) = \sum_{i=0}^{\infty} c_i e^{-i\sqrt{a_0^2 - \lambda_i^2}\,\frac{x}{d}} \sin\left(\lambda_i \frac{y}{d}\right) \tag{A.3.22}$$

The integration constants c_i, which are actually the participation factors of the ith mode, are determined by enforcing the boundary condition specified in Eqs. A.3.1 and A.3.2. After substitution on the left-hand side of Eq. A.3.22 and setting $x = 0$ on the right-hand side yields

$$\sum_{i=0}^{\infty} c_i \sin\left(\lambda_i \frac{y}{d}\right) = [N(y)]\{u_0(\omega)\} \tag{A.3.23}$$

Multiplying Eq. A.3.23 by $\sin(\lambda_i y/d)$ and integrating from 0 to d yields

$$c_i = \frac{2}{\lambda_i^2}\left[(-1)^{i+1}4 + \frac{8}{\lambda_i}(-1)^i 3 - \frac{4}{\lambda_i}\right]\{u_0(\omega)\} \tag{A.3.24}$$

The corresponding amplitude of the shear stress $\tau_{zx}(\omega)$ (Eq. A.3.4a) follows as

$$\tau_{zx}(\omega) = -\frac{G}{d}\sum_{i=0}^{\infty} c_i i \sqrt{a_0^2 - \lambda_i^2}\, e^{-i\sqrt{a_0^2 - \lambda_i^2}\,\frac{x}{d}} \sin\left(\lambda_i \frac{y}{d}\right) \tag{A.3.25}$$

Integrating the surface traction, which is equal to the negative value of $\tau_{zx}(\omega)$ at $x=0$, with the shape function $[N(y)]$, results in the amplitudes of the interaction forces $\{R(\omega)\}$

$$\{R(\omega)\} = -\int_0^d [N(y)]^T \tau_{zx}(\omega, x=0) dy \qquad (A.3.26)$$

Substituting Eqs. A.3.25 and A.3.24 in Eq. A.3.26 leads to

$$\{R(\omega)\} = [S^\infty(a_0)]\{u_0(\omega)\} \qquad (A.3.27)$$

with the dynamic-stiffness matrix

$$[S^\infty(a_0)] = 32G \sum_{i=0}^\infty \frac{1}{(2i+1)^4 \pi^4} i \sqrt{a_0^2 - \frac{(2i+1)^2 \pi^2}{4}}$$

$$\begin{bmatrix} \left(-4 + \dfrac{(-1)^i 16}{(2i+1)\pi}\right)^2 & -12 + \dfrac{(-1)^i 80}{(2i+1)\pi} - \dfrac{128}{(2i+1)^2 \pi^2} \\ -12 + \dfrac{(-1)^i 80}{(2i+1)\pi} - \dfrac{128}{(2i+1)^2 \pi^2} & \left(3 - \dfrac{(-1)^i 8}{(2i+1)\pi}\right)^2 \end{bmatrix} \qquad (A.3.28)$$

The static-stiffness matrix follows from Eq. A.3.28 for $a_0 = 0$

$$[K^\infty] = \begin{bmatrix} 1.337 & -0.241 \\ -0.241 & 0.449 \end{bmatrix} G \qquad (A.3.29)$$

Normalizing $S_{ij}^\infty(a_0)$ with K_{ij}^∞ and then decomposing the coefficient into its real and imaginary parts defines the spring coefficient $k_{ij}(a_0)$ and the damping coefficient $c_{ij}(a_0)$

$$S_{ij}^\infty(a_0) = K_{ij}^\infty(k_{ij}(a_0) + ia_0 c_{ij}(a_0)) \qquad (A.3.30)$$

These dimensionless spring and damping coefficients are plotted versus a_0 in Figure A-8. Below the cut-off frequency $a_0 = \pi/2$, which is equal to the fundamental frequency of the layer fixed at its base ($= \pi c_s/(2d) d/c_s$), $c_{11}(a_0) = c_{12}(a_0) = c_{22}(a_0)$ are equal to zero. For $a_0 \geq (2i+1)\pi/2$, the square root $\sqrt{a_0^2 - (2i+1)^2 \pi^2/4}$ in Eq. A.3.28 becomes real, leading to radiation damping. At each natural frequency of the layer an additional square root becomes real. For $a_0 \to \infty$, all square roots are real and the spring coefficients vanish. The dynamic-stiffness matrix equals

$$[S^\infty(a_0 \to \infty)] = i\omega[C_\infty] = i\omega \frac{d\rho c_s}{15} \begin{bmatrix} 8 & 1 \\ 1 & 2 \end{bmatrix} \qquad (A.3.31)$$

A.3 SEMI-INFINITE LAYER OF CONSTANT DEPTH

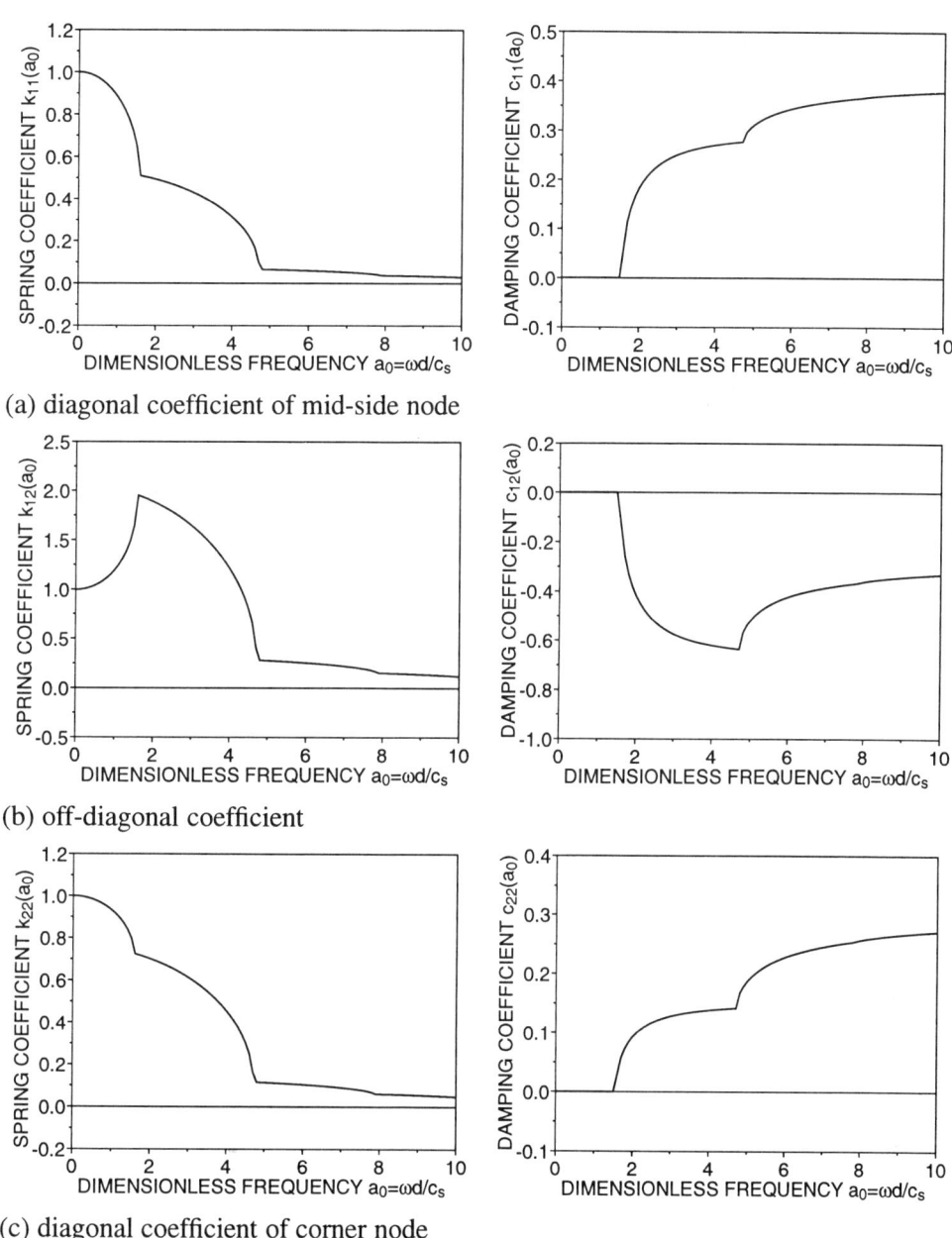

(a) diagonal coefficient of mid-side node

(b) off-diagonal coefficient

(c) diagonal coefficient of corner node

Figure A-8 Dynamic-stiffness matrix in frequency domain of out-of-plane motion of semi-infinite layer of constant depth discretized with quadratic finite element

yielding

$$[C_\infty] = \frac{d\rho c_s}{15} \begin{bmatrix} 8 & 1 \\ 1 & 2 \end{bmatrix} \quad \text{(A.3.32)}$$

This corresponds to concentrated dashpots, which can also be determined by lumping the evenly distributed dashpots with a coefficient $\rho c_s dy$

$$[C_\infty] = \rho c_s \int_0^d [N(y)]^T [N(y)] dy \quad \text{(A.3.33)}$$

For $a_0 \to \infty$, the waves propagate perpendicularly to the structure-medium interface. This wave propagation in the direction of the positive x-axis is thus one-dimensional and a dashpot with the coefficient $\rho c_s dy$ is the exact representation of the unbounded medium.

$[S^\infty(a_0)]$ is decomposed into the singular part $[S_s^\infty(a_0)]$, which is equal to the asymptotic value for $a_0 \to \infty$ specified in Eq. A.3.31, and the remaining regular part $[S_r^\infty(a_0)]$

$$[S_s^\infty(a_0)] = [S^\infty(a_0 \to \infty)] = i\omega[C_\infty] \quad \text{(A.3.34a)}$$
$$[S_r^\infty(a_0)] = [S^\infty(a_0)] - [S_s^\infty(a_0)] \quad \text{(A.3.34b)}$$

The displacement unit-impulse response matrix is equal to the inverse Fourier transform of the dynamic-stiffness matrix (Eq. 2.4)

$$[S^\infty(t)] = [S_s^\infty(t)] + [S_r^\infty(t)] \quad \text{(A.3.35)}$$

where

$$[S_s^\infty(t)] = [C_\infty]\dot{\delta}(t) \quad \text{(A.3.36a)}$$

$$[S_r^\infty(t)] = 16G \sum_{i=0}^{\infty} \frac{1}{(2i+1)^3\pi^3} \frac{1}{t} J_1\left(\frac{(2i+1)\pi}{2} \frac{c_s}{d} t\right)$$

$$\begin{bmatrix} \left(-4 + \frac{(-1)^i 16}{(2i+1)\pi}\right)^2 & -12 + \frac{(-1)^i 80}{(2i+1)\pi} - \frac{128}{(2i+1)^2\pi^2} \\ -12 + \frac{(-1)^i 80}{(2i+1)\pi} - \frac{128}{(2i+1)^2\pi^2} & \left(3 - \frac{(-1)^i 8}{(2i+1)\pi}\right)^2 \end{bmatrix}$$
(A.3.36b)

with the Bessel function of the first kind and of the first order J_1. The Fourier transform of $\sqrt{\text{const}^2 - \omega^2} - i\omega$ is specified in Reference [10] (No. 556.1).

The acceleration unit-impulse response matrix follows from Eq. 2.20 as

$$[M^\infty(t)] = [C_\infty]H(t) + \int_0^t \int_0^\tau [S_r^\infty(\tau')] d\tau' d\tau \quad \text{(A.3.37)}$$

The acceleration unit-impulse response coefficients $M_{ij}^\infty(t)$ non-dimensionalized with $\rho c_s d$ are plotted as a function of the dimensionless time $\bar{t} = tc_s/d$ in Figure A-9.

A.3 SEMI-INFINITE LAYER OF CONSTANT DEPTH

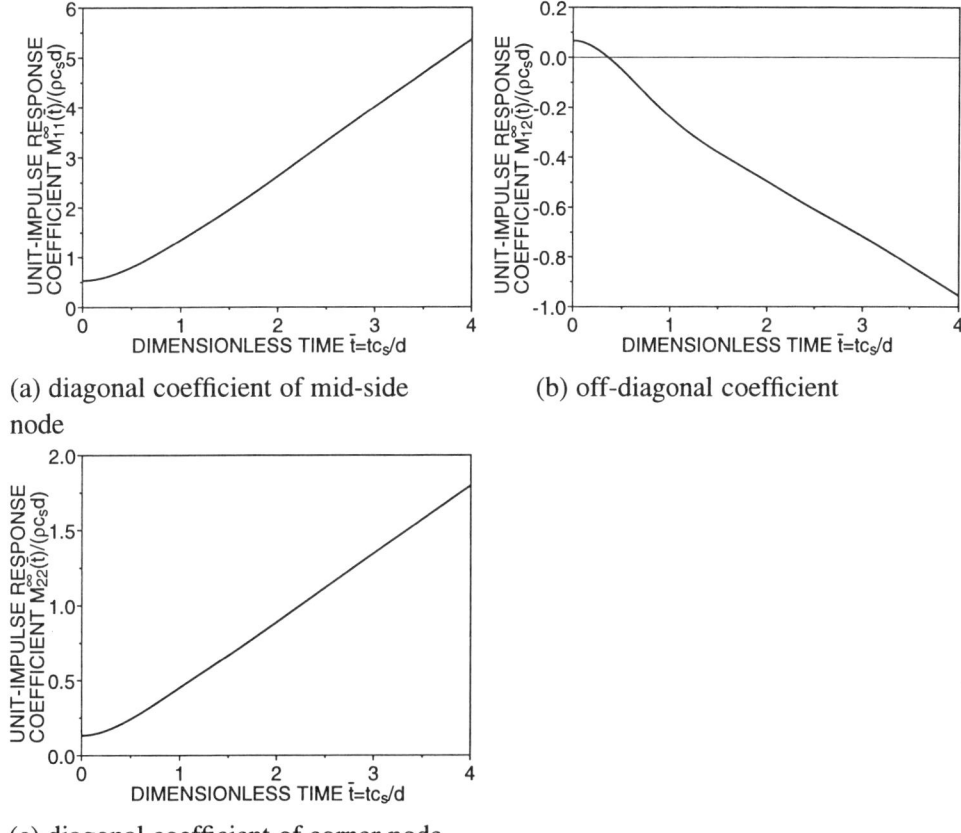

(a) diagonal coefficient of mid-side node

(b) off-diagonal coefficient

(c) diagonal coefficient of corner node

Figure A-9 Acceleration unit-impulse response matrix of out-of-plane motion of semi-infinite layer of constant depth discretized with quadratic finite element

A.3.2 In-plane Motion

As a two-dimensional wave propagation problem (for which the characteristic length is constant), the in-plane motion of a semi-infinite layer of constant depth d is addressed (Figure A-10). One of the boundaries extending to infinity is fixed, and the other is a free surface. The structure-medium interface coincides with the y-axis. The material behaviour is characterized by the shear modulus G, Poisson's ratio ν and the mass density ρ. On the structure-medium interface a linear function in the y-direction of the motion in the x-direction and zero motion in the y-direction are prescribed.

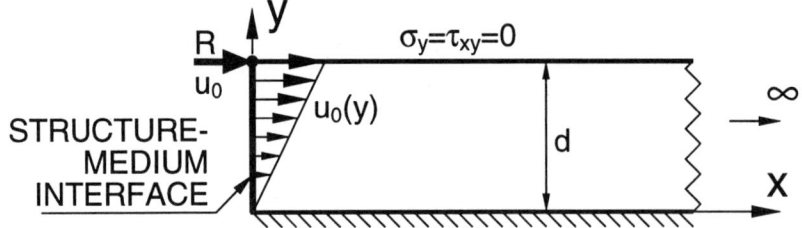

Figure A-10 In-plane motion of semi-infinite layer with prescribed horizontal displacement

A.4 SEMI-INFINITE WEDGE

A.4.1 Out-of-plane Motion

As a two-dimensional wave propagation problem with an analytical solution available, the out-of-plane (anti-plane) motion governed by the scalar wave equation is addressed. A semi-infinite wedge with an opening angle α and with a free and a fixed boundary extending to infinity in the radial direction is investigated (Figure A-11). The material constants of the semi-infinite wedge are the shear modulus G and the

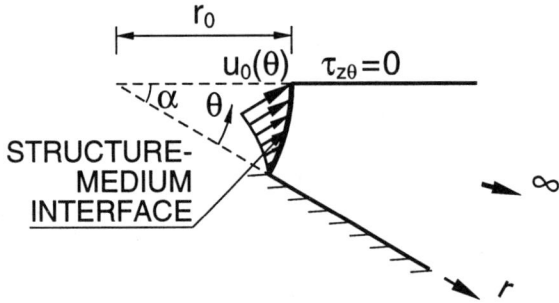

Figure A-11 Out-of-plane motion of semi-infinite wedge with prescribed displacement

mass density ρ. The structure-medium interface coincides with the arc determined by the radius r_0, where the motion as a linear function of the angle is prescribed.

The scalar wave equation in polar coordinates r, θ for the out-of-plane motion $u = u(r,\theta,t)$ is derived. Only shear strains γ_{zr} and $\gamma_{z\theta}$ occurring in the z-direction normal to the plane are present

$$\gamma_{zr} = u_{,r} \qquad \text{(A.4.1a)}$$

$$\gamma_{z\theta} = \frac{1}{r} u_{,\theta} \qquad \text{(A.4.1b)}$$

A.4 SEMI-INFINITE WEDGE

The shear stresses follow from the stress-strain relationship

$$\tau_{zr} = G\gamma_{zr} \tag{A.4.2a}$$
$$\tau_{z\theta} = G\gamma_{z\theta} \tag{A.4.2b}$$

Formulating the equilibrium equation

$$\tau_{zr,r} + \frac{1}{r}\tau_{z\theta,\theta} + \frac{1}{r}\tau_{zr} - \rho\ddot{u} = 0 \tag{A.4.3}$$

and substituting Eqs. A.4.1 and A.4.2 results in the wave equation

$$u_{,rr} + \frac{1}{r}u_{,r} + \frac{1}{r^2}u_{,\theta\theta} - \frac{1}{c_s^2}\ddot{u} = 0 \tag{A.4.4}$$

with the shear-wave velocity

$$c_s = \sqrt{\frac{G}{\rho}} \tag{A.4.5}$$

The conditions on the free and fixed boundaries equal

$$\tau_{z\theta}(\theta = \alpha) = 0 \tag{A.4.6a}$$
$$u(\theta = 0) = 0 \tag{A.4.6b}$$

On the structure-medium interface the boundary condition is written as

$$u(r = r_0) = u_0(\theta) = \frac{\theta}{\alpha}u_0 \tag{A.4.7}$$

A solution is derived in the frequency domain. The corresponding wave equation in the displacement amplitude $u(\omega)$ (Eq. A.4.4) equals

$$u(\omega)_{,rr} + \frac{1}{r}u(\omega)_{,r} + \frac{1}{r^2}u(\omega)_{,\theta\theta} + \frac{\omega^2}{c_s^2}u(\omega) = 0 \tag{A.4.8}$$

The method of separation of variables is applied to Eq. A.4.8. $u(\omega)$ is written as the product of a function $R(r)$, which is independent of θ, and a function $\Theta(\theta)$, which is independent of r

$$u(\omega) = R(r)\Theta(\theta) \tag{A.4.9}$$

Substituting Eq. A.4.9 in Eq. A.4.8 yields after division by $R(r)\Theta(\theta)/r^2$

$$\frac{r^2 R(r)_{,rr} + rR(r)_{,r}}{R(r)} + \left(\frac{\omega r}{c_s}\right)^2 = -\frac{\Theta(\theta)_{,\theta\theta}}{\Theta(\theta)} \tag{A.4.10}$$

The left-hand side is a function of r only and the right-hand side of θ only. To be able to satisfy Eq. A.4.10, the two sides must be equal to the same constant, denoted as λ^2. Setting

$$\frac{\Theta(\theta)_{,\theta\theta}}{\Theta(\theta)} = -\lambda^2 \qquad (A.4.11)$$

leads to the following two ordinary differential equations

$$\Theta(\theta)_{,\theta\theta} + \lambda^2 \Theta(\theta) = 0 \qquad (A.4.12a)$$

$$r^2 R(r)_{,rr} + r R(r)_{,r} + \left(\left(\frac{\omega r}{c_s}\right)^2 - \lambda^2\right) R(r) = 0 \qquad (A.4.12b)$$

The general solution of Eq. A.4.12a equals

$$\Theta(\theta) = c_1 \cos\lambda\theta + c_2 \sin\lambda\theta \qquad (A.4.13)$$

with the integration constants c_1 and c_2. Enforcing the boundary conditions of Eq. A.4.6

$$\Theta(\alpha)_{,\theta} = 0 \qquad (A.4.14a)$$
$$\Theta(0) = 0 \qquad (A.4.14b)$$

results in

$$\cos\lambda\alpha = 0 \qquad (A.4.15)$$

The eigenvalues λ follow as

$$\lambda_i = \frac{(2i+1)\pi}{2\alpha} \qquad (i = 0, 1, \dots) \qquad (A.4.16)$$

and the eigenfunctions as

$$\Theta_i(\theta) = \sin\lambda_i\theta \qquad (A.4.17)$$

The general solution of the Bessel equation Eq. A.4.12b for λ_i is

$$R_i(r) = c_{1i} H^{(1)}_{\lambda_i}\left(\frac{\omega r}{c_s}\right) + c_{2i} H^{(2)}_{\lambda_i}\left(\frac{\omega r}{c_s}\right) \qquad (A.4.18)$$

with the first and second kind Hankel functions of order λ_i and the integration constants c_{1i}, c_{2i}. To introduce the radiation condition, the asymptotic behaviour for $r \to \infty$ is examined

$$H^{(1)}_{\lambda_i}\left(\frac{\omega r}{c_s}\right) \approx \sqrt{\frac{2 c_s}{\pi \omega r}} e^{+i\left(\frac{\omega r}{c_s} - \frac{\lambda_i \pi}{2} - \frac{\pi}{4}\right)} \qquad (A.4.19a)$$

$$H^{(2)}_{\lambda_i}\left(\frac{\omega r}{c_s}\right) \approx \sqrt{\frac{2 c_s}{\pi \omega r}} e^{-i\left(\frac{\omega r}{c_s} - \frac{\lambda_i \pi}{2} - \frac{\pi}{4}\right)} \qquad (A.4.19b)$$

A.4 SEMI-INFINITE WEDGE

Following the argument presented in connection with Eq. A.1.14, $c_{1i}=0$ is obtained. This leads to

$$R_i(r) = c_i H^{(2)}_{\lambda_i}\left(\frac{\omega r}{c_s}\right) \tag{A.4.20}$$

where the subscript 2 in the integration constant is omitted.

Using Eqs. A.4.20 and A.4.17 together with Eq. A.4.9 permits the displacement amplitude to be expressed as

$$u(\omega) = \sum_{i=0}^{\infty} c_i H^{(2)}_{\lambda_i}\left(\frac{\omega r}{c_s}\right) \sin \lambda_i \theta \tag{A.4.21}$$

The integration constants c_i are determined by enforcing the boundary condition Eq. A.4.7

$$\sum_{i=0}^{\infty} c_i H^{(2)}_{\lambda_i}(a_0) \sin \lambda_i \theta = \frac{\theta}{\alpha} u_0(\omega) \tag{A.4.22}$$

with the dimensionless frequency

$$a_0 = \frac{\omega r_0}{c_s} \tag{A.4.23}$$

Multiplying Eq. A.4.22 by $\sin \lambda_i \theta$ and integrating from 0 to α yields

$$c_i = \frac{(-1)^i 2}{\alpha^2 \lambda_i^2 H^{(2)}_{\lambda_i}(a_0)} u_0(\omega) \tag{A.4.24}$$

Thus, Eq. A.4.21 is reformulated as

$$u(\omega) = \frac{2u_0(\omega)}{\alpha^2} \sum_{i=0}^{\infty} \frac{(-1)^i}{\lambda_i^2 H^{(2)}_{\lambda_i}(a_0)} H^{(2)}_{\lambda_i}\left(\frac{\omega r}{c_s}\right) \sin \lambda_i \theta \tag{A.4.25}$$

The shear stress amplitude $\tau_{zr}(\omega)$ follows from Eqs. A.4.25, A.4.1a and A.4.2a as

$$\tau_{zr}(\omega) = \frac{2Gu_0(\omega)}{\alpha^2 r} \sum_{i=0}^{\infty} \frac{(-1)^i}{\lambda_i^2 H^{(2)}_{\lambda_i}(a_0)} \left(\frac{\omega r}{c_s} H^{(2)}_{\lambda_i-1}\left(\frac{\omega r}{c_s}\right) - \lambda_i H^{(2)}_{\lambda_i}\left(\frac{\omega r}{c_s}\right)\right) \sin \lambda_i \theta \tag{A.4.26}$$

where

$$H^{(2)}_{\lambda_i}\left(\frac{\omega r}{c_s}\right)_{,r} = \frac{1}{r}\left(\frac{\omega r}{c_s} H^{(2)}_{\lambda_i-1}\left(\frac{\omega r}{c_s}\right) - \lambda_i H^{(2)}_{\lambda_i}\left(\frac{\omega r}{c_s}\right)\right) \tag{A.4.27}$$

is substituted. Based on virtual work considerations, the interaction force amplitude $R(\omega)$ is calculated as

$$R(\omega) = -\int_0^\alpha \tau_{zr}(\omega, r = r_0)\frac{\theta}{\alpha}d\theta \qquad (A.4.28)$$

Substituting Eq. A.4.26 in Eq. A.4.28 yields

$$R(\omega) = \frac{2G}{\alpha^3}\sum_{i=0}^\infty \frac{1}{\lambda_i^3}\left(1 - \frac{a_0 H^{(2)}_{\lambda_i-1}(a_0)}{\lambda_i H^{(2)}_{\lambda_i}(a_0)}\right)u_0(\omega) \qquad (A.4.29)$$

The ratio of $R(\omega)$ to $u_0(\omega)$ is equal to the dynamic-stiffness coefficient

$$S^\infty(a_0) = \frac{2G}{\alpha^3}\sum_{i=0}^\infty \frac{1}{\lambda_i^3}\left(1 - \frac{a_0 H^{(2)}_{\lambda_i-1}(a_0)}{\lambda_i H^{(2)}_{\lambda_i}(a_0)}\right) \qquad (A.4.30)$$

Setting $a_0 = 0$ in Eq. A.4.30 results in the static-stiffness coefficient

$$K^\infty = \frac{2G}{\alpha^3}\sum_{i=0}^\infty \frac{1}{\lambda_i^3} \qquad (A.4.31)$$

For $\alpha = 30°$, $K^\infty = 0.5427G$ results. The decomposition of Eq. A.0.3 yields the dimensionless spring coefficient $k(a_0)$ and damping coefficient $c(a_0)$. They are plotted for $\alpha = 30°$ in Figure A-12.

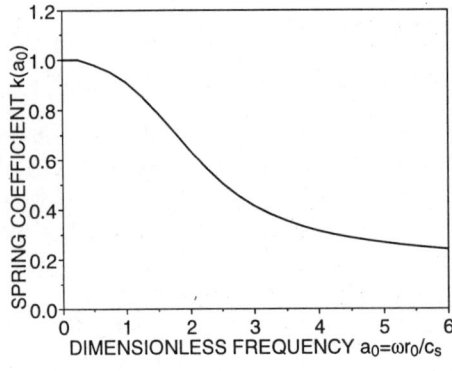
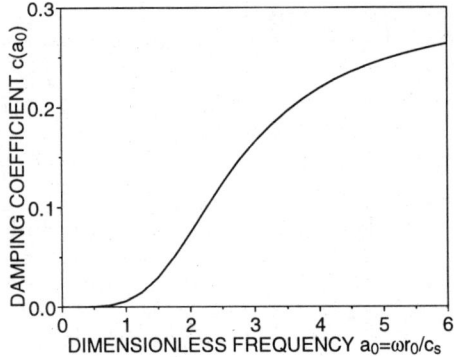

Figure A-12 Dynamic-stiffness coefficient in frequency domain of out-of-plane motion of semi-infinite wedge

A.5 IN-PLANE MOTION OF CIRCULAR CAVITY

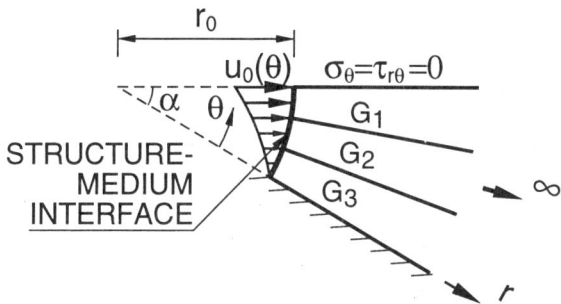

Figure A-13 In-plane motion of inhomogeneous semi-infinite wedge with prescribed horizontal displacement

A.4.2 In-plane Motion

As a two-dimensional wave propagation problem the in-plane motion of a semi-infinite wedge is addressed (Figure A-13). One of the boundaries extending to infinity is fixed, and the other is a free surface. The structure-medium interface is an arc of radius r_0 with an opening angle $\alpha = 30°$. Inhomogeneity compatible with similarity exists. 3 domains of equal opening angle are present with the shear moduli G_1, G_2 and G_3, whereby Poisson's ratio $\nu = 0.25$ and the mass density ρ do not vary. On the structure-medium interface a linear function in the circumferential direction of the horizontal motion $u_0(\theta)$ and zero vertical motion are prescribed.

A.5 IN-PLANE MOTION OF CIRCULAR CAVITY EMBEDDED IN FULL-PLANE

As a two-dimensional in-plane wave propagation problem with an analytical solution available, the circular cavity embedded in a full-plane is discussed (Figure A-14). On its rigid wall, the structure-medium interface, a constant horizontal displacement u_0 is enforced. The radius of the cavity is r_0, and the material properties of the full-plane are chosen as the Lamé parameters λ and G, the shear modulus ($\lambda = 2G\nu/(1-2\nu)$ with Poisson's ratio ν). The shear- and dilatational-wave velocities equal

$$c_s = \sqrt{\frac{G}{\rho}} \qquad \text{(A.5.1a)}$$

$$c_p = \sqrt{\frac{\lambda + 2G}{\rho}} = \sqrt{\frac{2(1-\nu)}{1-2\nu}} c_s \qquad \text{(A.5.1b)}$$

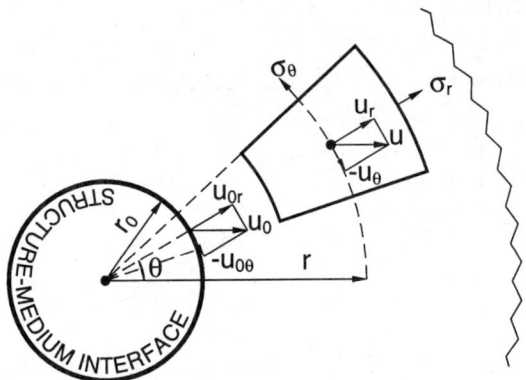

Figure A-14 Circular cavity embedded in full-plane with prescribed translational motion

The in-plane wave equation in polar coordinates r, θ for the displacement in the radial direction $u_r = u_r(r,\theta,t)$ and the displacement in the circumferential direction u_θ is derived. The radial strain ε_r, the circumferential strain ε_θ and the shear strain $\gamma_{r\theta}$ are specified as

$$\varepsilon_r = u_{r,r} \tag{A.5.2a}$$

$$\varepsilon_\theta = \frac{u_r}{r} + \frac{1}{r} u_{\theta,\theta} \tag{A.5.2b}$$

$$\gamma_{r\theta} = \frac{1}{r} u_{r,\theta} + u_{\theta,r} - \frac{1}{r} u_\theta \tag{A.5.2c}$$

The corresponding stresses follow from Hooke's law as

$$\begin{Bmatrix} \sigma_r \\ \sigma_\theta \\ \tau_{r\theta} \end{Bmatrix} = \begin{bmatrix} \lambda+2G & \lambda & 0 \\ \lambda & \lambda+2G & 0 \\ 0 & 0 & G \end{bmatrix} \begin{Bmatrix} \varepsilon_r \\ \varepsilon_\theta \\ \gamma_{r\theta} \end{Bmatrix} \tag{A.5.3}$$

The dynamic equilibrium equations are formulated as

$$\sigma_{r,r} + \frac{1}{r}\tau_{r\theta,\theta} + \frac{1}{r}(\sigma_r - \sigma_\theta) - \rho \ddot{u}_r = 0 \tag{A.5.4a}$$

$$\tau_{r\theta,r} + \frac{1}{r}\sigma_{\theta,\theta} + \frac{2}{r}\tau_{r\theta} - \rho \ddot{u}_\theta = 0 \tag{A.5.4b}$$

The boundary conditions on the structure-medium interface are written as

$$u_r(r=r_0) = u_0 \cos\theta \tag{A.5.5a}$$

$$u_\theta(r=r_0) = -u_0 \sin\theta \tag{A.5.5b}$$

A.5 IN-PLANE MOTION OF CIRCULAR CAVITY

The displacement dynamic-stiffness coefficient in the frequency domain $S^\infty(\omega)$ is discussed, which relates the displacement amplitude $u_0(\omega)$ to the corresponding interaction force amplitude

$$R(\omega) = -\int_0^{2\pi} (\sigma_r(\omega, r = r_0) \cos\theta - \tau_{r\theta}(\omega, r = r_0) \sin\theta) r_0 d\theta \tag{A.5.6}$$

as

$$R(\omega) = S^\infty(\omega) u_0(\omega) \tag{A.5.7}$$

Introducing the Lamé potentials $\phi(\omega)$ and $\psi(\omega)$ the displacement amplitudes follow

$$u_r(\omega) = \phi(\omega)_{,r} + \frac{1}{r}\psi(\omega)_{,\theta} \tag{A.5.8a}$$

$$u_r(\omega) = \frac{1}{r}\phi(\omega)_{,\theta} - \frac{1}{r}\psi(\omega)_{,r} \tag{A.5.8b}$$

Using Eqs. A.5.2 to A.5.4 and Eq. A.5.8 yields the wave equations in the potentials

$$\phi(\omega)_{,rr} + \frac{1}{r}\phi(\omega)_{,r} + \frac{1}{r^2}\phi(\omega)_{,\theta\theta} + \frac{\omega^2}{c_p^2}\phi(\omega) = 0 \tag{A.5.9a}$$

$$\psi(\omega)_{,rr} + \frac{1}{r}\psi(\omega)_{,r} + \frac{1}{r^2}\psi(\omega)_{,\theta\theta} + \frac{\omega^2}{c_s^2}\psi(\omega) = 0 \tag{A.5.9b}$$

Consistent with the boundary condition of Eq. A.5.5

$$\phi(\omega) = \Phi(r) \cos\theta \tag{A.5.10a}$$

$$\psi(\omega) = \Psi(r) \sin\theta \tag{A.5.10b}$$

is formulated with the frequency-dependent functions $\Phi(r)$ and $\Psi(r)$. Substituting in Eq. A.5.9 results in the Bessel equations

$$r^2 \Phi(r)_{,rr} + r\Phi(r)_{,r} + \left(\left(\frac{\omega r}{c_p}\right)^2 - 1\right) \Phi(r) = 0 \tag{A.5.11a}$$

$$r^2 \Psi(r)_{,rr} + r\Psi(r)_{,r} + \left(\left(\frac{\omega r}{c_s}\right)^2 - 1\right) \Psi(r) = 0 \tag{A.5.11b}$$

The general solution of Eq. A.5.11a equals

$$\Phi(r) = c_1 H_1^{(1)}\left(\frac{\omega r}{c_p}\right) + c_2 H_1^{(2)}\left(\frac{\omega r}{c_p}\right) \tag{A.5.12}$$

with the first and second kind Hankel functions of the first order and the integration constants c_1, c_2. To introduce the radiation condition, the asymptotic behaviour for $r \to \infty$ is examined

$$H_1^{(1)}\left(\frac{\omega r}{c_p}\right) \approx \sqrt{\frac{2\,c_p}{\pi\,\omega r}}\,e^{+i\left(\frac{\omega r}{c_p} - \frac{3\pi}{4}\right)} \quad \text{(A.5.13a)}$$

$$H_1^{(2)}\left(\frac{\omega r}{c_p}\right) \approx \sqrt{\frac{2\,c_p}{\pi\,\omega r}}\,e^{-i\left(\frac{\omega r}{c_p} - \frac{3\pi}{4}\right)} \quad \text{(A.5.13b)}$$

Following the argument presented in connection with Eq. A.1.14, $c_1 = 0$ is obtained. This yields

$$\Phi(r) = c H_1^{(2)}\left(\frac{\omega r}{c_p}\right) \quad \text{(A.5.14a)}$$

where the subscript 2 in the integration constant is omitted. Analogously, from Eq. A.5.11b

$$\Psi(r) = d H_1^{(2)}\left(\frac{\omega r}{c_s}\right) \quad \text{(A.5.14b)}$$

is deduced with the integration constant d.

The boundary conditions in Eq. A.5.5 determine c and d. Using Eqs. A.5.8, A.5.10 and A.5.14 leads to

$$u_r(\omega) = \left(c H_1^{(2)}\left(\frac{\omega r}{c_p}\right)_{,r} + d\frac{1}{r} H_1^{(2)}\left(\frac{\omega r}{c_s}\right)\right)\cos\theta \quad \text{(A.5.15a)}$$

$$u_\theta(\omega) = -\left(c\frac{1}{r} H_1^{(2)}\left(\frac{\omega r}{c_p}\right) + d H_1^{(2)}\left(\frac{\omega r}{c_s}\right)_{,r}\right)\sin\theta \quad \text{(A.5.15b)}$$

Eliminating the derivatives of the Hankel functions

$$H_1^{(2)}\left(\frac{\omega r}{c_p}\right)_{,r} = \frac{1}{r}\left(\frac{\omega r}{c_p} H_0^{(2)}\left(\frac{\omega r}{c_p}\right) - H_1^{(2)}\left(\frac{\omega r}{c_p}\right)\right) \quad \text{(A.5.16)}$$

and substituting Eq. A.5.15 in A.5.5 results in

$$c\left(\frac{\omega r_0}{c_p} H_0^{(2)}\left(\frac{\omega r_0}{c_p}\right) - H_1^{(2)}\left(\frac{\omega r_0}{c_p}\right)\right) + d H_1^{(2)}\left(\frac{\omega r_0}{c_s}\right) = r_0 u_0(\omega) \quad \text{(A.5.17a)}$$

$$c H_1^{(2)}\left(\frac{\omega r_0}{c_p}\right) + d\left(\frac{\omega r_0}{c_s} H_0^{(2)}\left(\frac{\omega r_0}{c_s}\right) - H_1^{(2)}\left(\frac{\omega r_0}{c_s}\right)\right) = r_0 u_0(\omega) \quad \text{(A.5.17b)}$$

Introducing the dimensionless frequency

$$a_0 = \frac{\omega r_0}{c_s} \quad \text{(A.5.18)}$$

A.5 IN-PLANE MOTION OF CIRCULAR CAVITY

and the abbreviations

$$\alpha = a_0 \frac{H_0^{(2)}(a_0)}{H_1^{(2)}(a_0)} \tag{A.5.19a}$$

$$\beta = \frac{c_s}{c_p} a_0 \frac{H_0^{(2)}\left(\frac{c_s}{c_p} a_0\right)}{H_1^{(2)}\left(\frac{c_s}{c_p} a_0\right)} \tag{A.5.19b}$$

c and d follow from Eq. A.5.17

$$c = \frac{\alpha - 2}{\alpha\beta - \alpha - \beta} \frac{r_0 u_0(\omega)}{H_1^{(2)}\left(\frac{c_s}{c_p} a_0\right)} \tag{A.5.20a}$$

$$d = \frac{\beta - 2}{\alpha\beta - \alpha - \beta} \frac{r_0 u_0(\omega)}{H_1^{(2)}(a_0)} \tag{A.5.20b}$$

Using Eqs. A.5.20, A.5.15, A.5.2, A.5.3 with Eq. A.5.11 and substituting in Eq. A.5.6 yields

$$R(\omega) = \pi G a_0^2 \frac{\alpha + \beta - 4}{\alpha\beta - \alpha - \beta} u_0(\omega) \tag{A.5.21}$$

Comparing with Eq. A.5.7, the dynamic-stiffness coefficient follows as

$$S^\infty(a_0) = \pi G a_0^2 \frac{\alpha + \beta - 4}{\alpha\beta - \alpha - \beta} \tag{A.5.22}$$

To be able to determine the static-stiffness coefficient, the asymptotic behaviour of α for $a_0 \to 0$ is studied. For $a_0 \to 0$

$$H_0^{(2)}(a_0) \approx -i\frac{2}{\pi} \ln\left(\frac{a_0}{2}\right) \tag{A.5.23a}$$

$$H_1^{(2)}(a_0) \approx i\frac{2}{\pi a_0} \tag{A.5.23b}$$

apply. Substituting Eq. A.5.23 in Eq. A.5.19a

$$\alpha \approx a_0 \frac{-i\frac{2}{\pi} \ln\left(\frac{a_0}{2}\right)}{i\frac{2}{\pi a_0}} = -a_0^2 \ln\left(\frac{a_0}{2}\right) \tag{A.5.24}$$

results. Analogously, for $a_0 \to 0$ (Eq. A.5.19b)

$$\beta \approx -\frac{c_s^2}{c_p^2} a_0^2 \ln\left(\frac{c_s}{c_p} \frac{a_0}{2}\right) \tag{A.5.25}$$

is obtained. Substituting Eqs. A.5.24 and A.5.25 in Eq. A.5.22 yields for $a_0 \to 0$

$$S^\infty(a_0) = -4\pi G \frac{1}{\ln\left(\frac{a_0}{2}\right) + \frac{c_s^2}{c_p^2}\ln\left(\frac{c_s}{c_p}\frac{a_0}{2}\right)} \quad (A.5.26)$$

In the limit $K^\infty = S^\infty(a_0 = 0) = 0$, as expected.

The high-frequency behaviour is addressed. For $a_0 \to \infty$,

$$H_0^{(2)}(a_0) \approx \sqrt{\frac{2}{\pi a_0}} e^{-i\left(a_0 - \frac{\pi}{4}\right)} \left(1 - \frac{1}{8ia_0}\right) \quad (A.5.27a)$$

$$H_1^{(2)}(a_0) \approx \sqrt{\frac{2}{\pi a_0}} e^{-i\left(a_0 - \frac{3\pi}{4}\right)} \left(1 + \frac{3}{8ia_0}\right) \quad (A.5.27b)$$

holds. Substituting Eq. A.5.27 in Eq. A.5.19 results in

$$\alpha \approx -ia_0 + \frac{1}{2} \quad (A.5.28a)$$

$$\beta \approx -i\frac{c_s}{c_p}a_0 + \frac{1}{2} \quad (A.5.28b)$$

Substituting Eq. A.5.28 in Eq. A.5.22 yields for $a_0 \to \infty$

$$S^\infty(\omega) = \pi r_0 \rho (c_s + c_p) i\omega + \frac{\pi}{2} G \left(-\left(\frac{c_s}{c_p}\right)^2 + 4\frac{c_s}{c_p} - 1\right) \quad (A.5.29)$$

The second term on the right-hand side denotes a spring coefficient and the coefficient of $i\omega$ in the first term a dashpot coefficient. The latter can also be derived on physical grounds. In the radial and circumferential directions of the structure-medium interface the impedances are ρc_p and ρc_s, respectively. For the enforced horizontal velocity amplitude $i\omega u_0(\omega)$, the components in the radial and circumferential directions are $\cos\theta i\omega u_0(\omega)$ and $-\sin\theta i\omega u_0(\omega)$ (Eq. A.5.5). The corresponding stress amplitudes follow as

$$\sigma_r(\omega, r = r_0) = -\rho c_p \cos\theta i\omega u_0(\omega) \quad (A.5.30a)$$

$$\tau_{r\theta}(\omega, r = r_0) = +\rho c_s \sin\theta i\omega u_0(\omega) \quad (A.5.30b)$$

Substituting Eq. A.5.30 in Eq. A.5.6 yields as coefficient of $i\omega u_0(\omega)$ the dashpot coefficient $\pi r_0 \rho (c_s + c_p)$.

The decomposition of Eq. A.0.3 with G instead of K^∞ applied to Eq. A.5.22 leads to the dimensionless spring coefficient $k(a_0)$ and damping coefficient $c(a_0)$. They are plotted for $\nu = 1/3$ in Figure A-15.

A.6 IN-PLANE MOTION OF EMBEDDED STRIP FOUNDATION

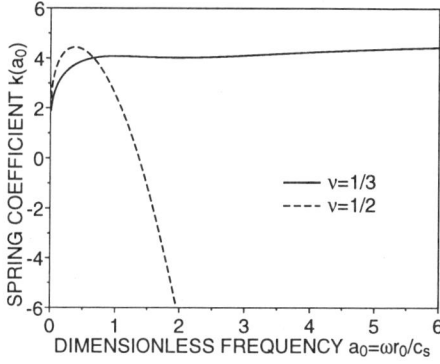
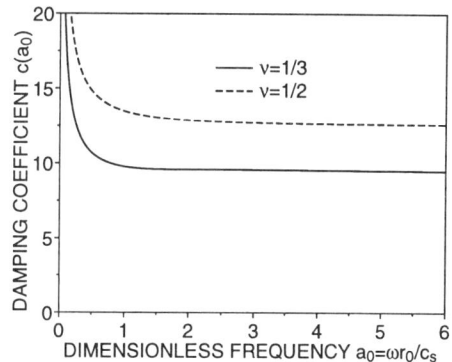

Figure A-15 Dynamic-stiffness coefficients in frequency domain of circular cavity embedded in full-plane

For the incompressible case $v \to 1/2$, $c_p \to \infty$ which results in (Eq. A.5.19b)

$$\beta = \lim_{c_p \to \infty} \frac{c_s}{c_p} a_0 \frac{-i\frac{2}{\pi} \ln\left(\frac{c_s}{c_p} \frac{a_0}{2}\right)}{i\frac{c_p}{c_s} \frac{2}{\pi a_0}} = 0 \qquad (A.5.31)$$

Eq. A.5.22 using Eq. A.5.19a and A.5.31 equals

$$S^\infty(\omega) = -\pi r_0^2 \rho \omega^2 + 4\pi G a_0 \frac{H_1^{(2)}(a_0)}{H_0^{(2)}(a_0)} \qquad (A.5.32)$$

Using Eq. A.5.27, the high-frequency limit of the dynamic-stiffness coefficient in the incompressible case equals

$$S^\infty(\omega) = -\pi r_0^2 \rho \omega^2 + 4\pi r_0 \rho c_s i\omega + 2\pi G \qquad (A.5.33)$$

The third term on the right-hand side denotes a spring coefficient. The coefficient of $i\omega$ in the second term is a dashpot coefficient. It is interesting to note that the physical interpretation discussed in connection with the dashpot coefficient of Eq. A.5.29 still holds, when c_p is replaced by $3c_s$. The coefficient of $-\omega^2$ in the first term represents a mass equal to the area of the circular cavity multiplied by the mass density.

The spring coefficient $k(a_0)$ and damping coefficient $c(a_0)$ corresponding to Eq. A.5.32 are plotted in Figure A-15.

A.6 IN-PLANE MOTION OF STRIP FOUNDATION WITH RECTANGULAR CROSS-SECTION EMBEDDED IN HALF-PLANE

A strip foundation with a rectangular cross-section of width $2b$ and depth e embedded in an inhomogeneous half-plane is addressed (Figure A-16). The embedment ratio e/b is equal to 1. The inhomogeneity consisting of 3 zones with different material properties is compatible with similarity. Poisson's ratio ν and the mass density ρ do not vary. Boundary-element results for the horizontal, vertical and rocking dynamic-stiffness coefficients enforcing a rigid structure-medium interface for the homogeneous case are available [56].

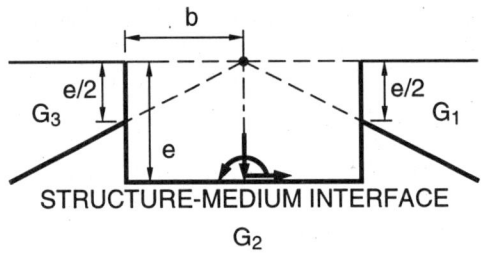

Figure A-16 Strip foundation with rectangular cross-section embedded in inhomogeneous half-plane

A.7 PRISM EMBEDDED IN HALF-SPACE FOR VECTOR WAVE EQUATION

As a truly three-dimensional problem, a square prism of length $2b$ embedded with depth e in an inhomogeneous half-space is addressed (Figure A-17). The inhomogeneity consisting of two zones is compatible with similarity. Zones 1 and 2 are adjacent to the walls and the base, respectively. The mass density ρ does not vary. Boundary-element results for the horizontal, vertical, rocking and torsional dynamic-stiffness coefficients enforcing a rigid structure-medium interface for the homogeneous case are available [15, 35].

A.8 TRANSVERSELY ISOTROPIC MATERIAL

Some of the benchmark examples are calculated for a simple anisotropic material behaviour, the transversely isotropic case with 5 material constants. The plane of

A.8 TRANSVERSELY ISOTROPIC MATERIAL

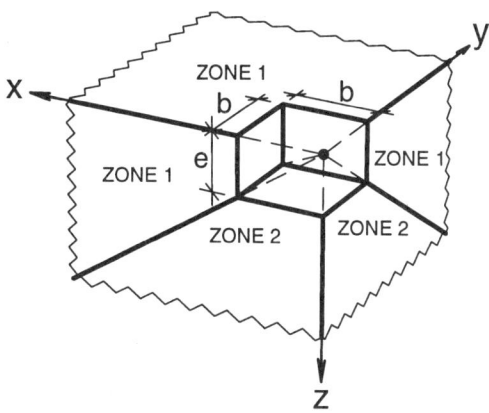

Figure A-17 One quarter of square prism embedded in inhomogeneous half-space

isotropy lies in the x-y plane (subscript h). The axis of elastic symmetry is the z-axis (subscript v)

Using engineering elastic moduli, the strain-stress relationship is formulated as

$$\begin{Bmatrix} \varepsilon_x \\ \varepsilon_y \\ \varepsilon_z \\ \gamma_{yz} \\ \gamma_{xz} \\ \gamma_{xy} \end{Bmatrix} = \begin{bmatrix} \frac{1}{E_{hh}} & -\frac{\nu_{hh}}{E_{hh}} & -\frac{\nu_{hv}}{E_{hv}} & 0 & 0 & 0 \\ -\frac{\nu_{hh}}{E_{hh}} & \frac{1}{E_{hh}} & -\frac{\nu_{hv}}{E_{hv}} & 0 & 0 & 0 \\ -\frac{\nu_{hv}}{E_{hv}} & -\frac{\nu_{hv}}{E_{hv}} & \frac{1}{E_{hv}} & 0 & 0 & 0 \\ 0 & 0 & 0 & \frac{1}{G_{hv}} & 0 & 0 \\ 0 & 0 & 0 & 0 & \frac{1}{G_{hv}} & 0 \\ 0 & 0 & 0 & 0 & 0 & \frac{1}{G_{hh}} \end{bmatrix} \begin{Bmatrix} \sigma_x \\ \sigma_y \\ \sigma_z \\ \tau_{yz} \\ \tau_{xz} \\ \tau_{xy} \end{Bmatrix} \quad (A.8.1)$$

with the Young's moduli in the plane of isotropy E_{hh} and perpendicular to it E_{hv}, the shear modulus in a plane perpendicular to the plane of isotropy G_{hv} and the Poisson's ratios in the plane of isotropy ν_{hh} and perpendicular to it ν_{hv}. The shear modulus in the plane of isotropy is calculated as $G_{hh} = E_{hh}/(2(1+\nu_{hh}))$. Inverting Eq. A.8.1 yields

$$\begin{Bmatrix} \sigma_x \\ \sigma_y \\ \sigma_z \\ \tau_{yz} \\ \tau_{xz} \\ \tau_{xy} \end{Bmatrix} = \begin{bmatrix} d_{11} & d_{12} & d_{13} & 0 & 0 & 0 \\ d_{12} & d_{11} & d_{13} & 0 & 0 & 0 \\ d_{13} & d_{13} & d_{33} & 0 & 0 & 0 \\ 0 & 0 & 0 & d_{44} & 0 & 0 \\ 0 & 0 & 0 & 0 & d_{44} & 0 \\ 0 & 0 & 0 & 0 & 0 & 2(d_{11}-d_{12}) \end{bmatrix} \begin{Bmatrix} \varepsilon_x \\ \varepsilon_y \\ \varepsilon_z \\ \gamma_{yz} \\ \gamma_{xz} \\ \gamma_{xy} \end{Bmatrix} \quad (A.8.2)$$

with the coefficients

$$d_{11} = \frac{E_{hh}(E_{hv} - E_{hh}v_{hv}^2)}{(1+v_{hh})(E_{hv}(1-v_{hv}) - 2E_{hh}v_{hv}^2)} \tag{A.8.3a}$$

$$d_{12} = \frac{E_{hh}(E_{hv}v_{hh} + E_{hh}v_{hv}^2)}{(1+v_{hh})(E_{hv}(1-v_{hv}) - 2E_{hh}v_{hv}^2)} \tag{A.8.3b}$$

$$d_{13} = \frac{E_{hh}E_{hv}v_{hv}}{E_{hv}(1-v_{hv}) - 2E_{hh}v_{hv}^2} \tag{A.8.3c}$$

$$d_{33} = \frac{E_{hv}^2(1-v_{hh})}{E_{hv}(1-v_{hv}) - 2E_{hh}v_{hv}^2} \tag{A.8.3d}$$

$$d_{44} = G_{hv} \tag{A.8.3e}$$

The coefficient matrix of Eq. A.8.2 is the elasticity matrix $[D]$ used in calculating the static-stiffness matrix and the coefficient matrices of the consistent infinitesimal finite-element cell method.

Appendix B

DESCRIPTION OF COMPUTER PROGRAMME *SIMILAR* FOR CONSISTENT INFINITESIMAL FINITE-ELEMENT CELL METHOD

The largest part of the book is devoted to the consistent infinitesimal finite-element cell method, which is based on finite elements. As the limit of the infinitesimal cell width is performed analytically, only the boundary is discretized. The corresponding coefficient matrices are calculated similarly to the standard static-stiffness and mass matrices of finite elements. It is thus appropriate to present a computer programme called *SIMILAR* for the conSistent Infinitesimal finite-element cell Method - a fInite-eLement boundARy for modelling unbounded and bounded media.

This programme can help the reader to understand the implementation and to evaluate the method. The examples presented in the book were calculated using this programme. More complicated cases can straightforwardly be analysed, whereby the software for the pre- and post-processing must be provided by the user. The subroutines of the programme can be incorporated in a general finite-element programme to model, as a super-element, an unbounded medium or a bounded medium without discretizing the interior. They also serve as the basis for developing more advanced versions of the programme as the source code is provided. Emphasis is placed on readability of the code at the expense of computational efficiency.

In principle, the programme covers all cases discussed in Chapters 5 to 10. Two- and three-dimensional vector and scalar wave equations and diffusion equations for an unbounded medium are solved in the time and frequency domains. Statics is included as a special case. General anisotropy and inhomogeneity (compatible with similarity) of the material is incorporated. Incompressible elasticity is addressed. A bounded medium can also be analysed.

A brief description of the input follows, which consists of 4 items:

1. Type of problem with corresponding parameters of algorithm

Governing differential equation (e.g. three-dimensional vector wave equation for unbounded medium), domain of analysis and corresponding parameters (time domain with size of time step Δt and number of time steps or frequency domain with starting frequency ω_h, order of asymptotic expansion and initial frequency step $\Delta\omega$)
2. Similarity
   Coordinates of similarity centre
3. Mesh of structure-medium interface
   Nodal coordinates, element data and boundary condition (free or fixed)
4. Material properties of medium
   Elasticity matrix $[D]$ and mass density ρ

The output consists of one of the following items according to the type of problem:

1. Unit-impulse response matrix for time-domain analysis of unbounded medium (e.g. acceleration unit-impulse response matrix $[M^\infty(t)]$ for three-dimensional vector wave equation)
2. Dynamic-stiffness matrix for frequency-domain analysis of unbounded medium (e.g. dynamic-stiffness matrix $[S^\infty(\omega)]$ for three-dimensional vector wave equation)
3. Static-stiffness matrix and eigenvalues and eigenvectors of Riccati equation for statics of unbounded medium (e.g. static-stiffness matrix $[K^\infty]$ and eigenvalues $\lfloor\lambda\rfloor$ and eigenvectors $[\Phi]$ to calculate internal displacements and stresses for three-dimensional elasticity)
4. Static-stiffness and mass matrices, eigenvalues and eigenvectors of Riccati equation of bounded medium (e.g. static-stiffness matrix $[K^b]$ and mass matrix $[M^b]$ for dynamic analysis)

A detailed user's manual with examples and the source code written in FORTRAN 77 is available on disk, free of charge, from:

>Customer Services
>John Wiley & Sons, Ltd.
>Distribution Centre
>Southern Cross Trading Estate
>Shripney Road
>Bognor Regis
>West Sussex PO22 9SA
>U.K.
>Phone: +44 1243 843 294

0471 96879 X   Disk   Free of Charge

REFERENCES

[1] Aliabadi, M. H. (1994) "Application of the boundary element method to fracture mechanics", *Static and Dynamic Fracture Mechanics*, Aliabadi, M. H., Brebbia C. A. and Parton, V. Z., ed., Computational Mechanics Publications, Southampton, Chapter 3, 113–175.
[2] Anderson, E., Bai, Z. *et al.* (1992) *LAPACK Users' Guide*, Society for Industrial and Applied Mathematics, Philadelphia, PA.
[3] Apsel, R. J. and Luco, J. E. (1987) "Impedance functions for foundations embedded in a layered medium: an integral equation approach", *Earthquake Engineering and Structural Dynamics*, Vol. 15, 213–231.
[4] Banerjee, P. K. (1994) *The Boundary Element Methods in Engineering*, McGraw-Hill, London.
[5] Bartels, R. H. and Stewart, J. L. (1972) "Solution of the matrix equation $AX + XB = C$", *Communications of the ACM*, Vol. 15, 820–826.
[6] Beskos, D. E. (1987) "Boundary element methods in dynamic analysis", *Applied Mechanics Reviews*, Vol. 40, 1–23.
[7] Bettess, P. (1992) *Infinite Elements*, Penshaw Press, Sunderland, UK.
[8] Bougacha, S., Tassoulas, J. L. and Roësset, J. M. (1993) "Analysis of foundations on fluid-filled poroelastic stratum", *Journal of Engineering Mechanics*, ASCE, Vol. 119, 1632–1648.
[9] Brebbia, C. A., Telles, J. C. F. and Wrobel, L. C. (1984) *Boundary Element Techniques*, Springer, Berlin.
[10] Campbell, G. A. and Foster, R. M. (1967) *Fourier Integrals for Practical Applications*, Van Nostrand, Princeton, NJ.
[11] Civelek, M. B. and Erdogan, F. (1982) "Crack problems for a rectangular plate and an infinite strip", *International Journal of Fracture*, Vol. 19, 139–159.
[12] Clayton, R. and Engquist, B. (1977) "Absorbing boundary conditions for acoustic and elastic wave equations", *Bulletin of the Seismological Society of America*, Vol. 67, 1529–1540.
[13] Cundall, P. A., Kunar, R. R., Carpenter, P. C. and Marti, J. (1974) "Solution of infinite dynamic problems by finite modelling in the time domain",

Proceedings of the 2nd International Conference on Applied Numerical Modelling, Madrid, 339–351.

[14] Dasgupta, G. (1982) "A finite element formulation for unbounded homogeneous continua", *Journal of Applied Mechanics*, ASME, Vol. 49, 136–140.

[15] Dominguez, J. (1993) *Boundary Elements in Dynamics*, Computational Mechanics Publications, Southampton.

[16] Engquist, B. and Majda, A. (1977) "Absorbing boundary conditions for the numerical simulation of waves", *Mathematics of Computation*. Vol. 31, 629–651.

[17] Givoli, D. (1992) *Numerical Methods for Problems in Infinite Domains*, Elsevier, Amsterdam.

[18] Gurtin, M. E. (1972) *The Linear Theory of Elasticity*, Section 74, *Mechanics of Solids II*, C. Truesdell, ed., in *Encyclopedia of Physics*, Vol. VIa/2, Springer, Berlin, 257–264.

[19] Higdon, R. L. (1986) "Absorbing boundary conditions for difference approximations to the multi-dimensional wave equation", *Mathematics of Computation*, Vol. 176, 437–459.

[20] Higdon, R. L. (1992) "Absorbing boundary conditions for acoustic and elastic waves in stratified media", *Journal of Computational Physics*, Vol. 101, 386–418.

[21] Holley, R.F. and Hibbert, P.D. (1966) "Bounding plane stress solutions by finite elements", *Journal of the Structural Division*, ASCE, Vol. 92, 39-48.

[22] Kausel, E. (1988) "Local transmitting boundaries", *Journal of Engineering Mechanics*, ASCE, 1011–1027.

[23] Kausel, E. and Roësset, J. M. (1975) "Dynamic stiffness of circular foundations", *Journal of Engineering Mechanics Division*, ASCE, Vol. 101, 771-785.

[24] Kausel, E., Roësset, J. M. and Waas, G. (1975) "Dynamic analysis of footings on layered media", *Journal of Engineering Mechanics*, ASCE, Vol. 101, 679–693.

[25] Keys, R. G. (1985) "Absorbing boundary conditions for acoustic media", *Geophysics*, Vol. 50, 892–902.

[26] Laub, A. J. (1979) "A Schur method for solving algebraic Riccati equations", *IEEE Transactions on Automatic Control*, Vol. AC-24, 913-921.

[27] Liao, Z. P. and Wong, H. L. (1984) "A transmitting boundary for the numerical simulation of elastic wave propagation", *Soil Dynamics and Earthquake Engineering*, Vol. 3, 174–183.

[28] Luco, J. E. (1976) "Torsional response of structures for SH-waves: the case of hemispherical foundations," *Bulletin of the Seismological Society of America*, Vol. 66, 109-124.

REFERENCES

[29] Luco, J. E. (1982) "Linear soil-structure interaction: A review", *Earthquake Ground Motion and Its Effects on Structures*, Datta, S. K., ed., Applied Mechanics Division, ASME, Vol. 53, 41–57.

[30] Lysmer, J. (1970) "Lumped mass method for Rayleigh waves", *Bulletin of the Seismological Society of America*, Vol. 60, 89–104.

[31] Lysmer, J. (1978) "Analytical procedures in soil dynamics", *Proceedings Specialty Conference on Earthquake Engineering and Soil Dynamics*, Geotechnical Engineering Division, ASCE, Pasadena, CA, Vol.III, 1267–1316.

[32] Lysmer, J. and Kuhlemeyer, R. L. (1969) "Finite dynamic model for infinite media", *Journal of Engineering Mechanics Division*, ASCE, Vol. 95, 859–877.

[33] Lysmer, J. and Waas, G. (1972) "Shear waves in plane infinite structures", *Journal of Engineering Mechanics Division*, ASCE, Vol. 98, 85–105.

[34] Manolis, G. D. and Beskos, D. E. (1988) *Boundary Element Methods in Elastodynamics*, Unwin Hyman, London.

[35] Mita, A. and Luco, J. E. (1989) "Impedance functions and input motions for embedded square foundations", *Journal of Geotechnical Engineering*, ASCE, Vol. 115, 491–503.

[36] Nardini, D. and Brebbia, C.A. (1984) "Boundary intergral formulation of mass matrices for dynamic analysis", *Topics in Boundary Element Research*, Brebbia, C.A., ed., Vol. 2, Chapter 7, Springer, Berlin, 191-208.

[37] Nicolas-Vullierme, B. (1991) "A contribution to doubly asymptotic approximation: an operator top-down approach", *Journal of Vibration and Acoustics*, ASME, Vol. 113, 409-415.

[38] Paronesso, A. and Wolf, J.P. (1995) "Global lumped-parameter model with physical representation for unbounded medium", *Earthquake Engineering and Structural Dynamics*, Vol. 24, 637–654.

[39] Potter, J. E. (1966) "Matrix quadratic solutions", *SIAM Journal on Applied Mathematics*, Society for Industrial and Applied Mathematics, Vol. 14, 496–501.

[40] Press, W. H., Flannery, B. P., Teukolsky, S. A. and Vetterling, W. T. (1988) *Numerical Recipes*, Chapter 15, Cambridge University Press, Cambridge.

[41] Reissner, E. (1936) "Stationäre, axialsymmetrische, durch eine schüttelnde Masse erregte Schwingungen eines homogenen elastischen Halbraumes", *Ingenieur-Archiv*, Vol. 7, 381–396.

[42] Rice, J. R. (1988) "Elastic fracture mechanics concepts for interfacial cracks", *Journal of Applied Mechanics*, ASME, Vol. 55, 98–103.

[43] Rossët, J. M. (1980) "A review of soil-structure interaction", *Soil-Structure Interaction: The Status of Current Analysis Methods and Research, Report NUREG/CR-1780 UCRL-53011*, Johnson, J. J., ed., Lawrence Livermore

National Laboratory, Livermore, CA.

[44] Sezawa, K. and Kanai, K. (1935) "Decay in the seismic vibration of a simple or tall structure by dissipation of their energy into the ground", *Bulletin of the Earthquake Research Institute*, University of Tokyo, XIII, Part 3, 681-697.

[45] Sezawa, K. and Kanai, K. (1936) "Improved theory of energy dissipation in seismic vibrations on a structure", *Bulletin of the Earthquake Research Institute*, University of Tokyo, XIV, Part 2, 164-168.

[46] Smith, W. D. (1974) "A nonreflecting plane boundary for wave propagation problems", *Journal of Computational Physics*, Vol. 15, 492–503.

[47] Sommerfeld, A. (1949) *Partial Differential Equations in Physics*, Chapter 28, Academic Press, New York, NY.

[48] Song, Ch. and Wolf, J. P. (1994) "Dynamic stiffness of unbounded medium based on damping-solvent extraction" *Earthquake Engineering and Structural Dynamics*, Vol. 23, 169–181.

[49] Song, Ch. and Wolf, J. P. (1995) "Unit-impulse response matrix of unbounded medium by finite-element based forecasting", *International Journal for Numerical Methods in Engineering*, Vol. 38, 1073–1086.

[50] Song, Ch. and Wolf, J. P. (1995) "Consistent infinitesimal finite-element–cell method: out-of-plane motion", *Journal of Engineering Mechanics*, ASCE, Vol. 121, 613–619.

[51] Song, Ch. and Wolf, J. P. (1996) "Consistent infinitesimal finite-element cell method: three-dimensional vector wave equation", *International Journal for Numerical Methods in Engineering*, (in press).

[52] Song, Ch. and Wolf, J. P. (1996) "Consistent infinitesimal finite-element cell method for diffusion equation in unbounded medium",*Computer Methods in Applied Mechanics and Engineering*, (in press).

[53] Song, Ch. and Wolf, J. P. (1996) "Consistent infinitesimal finite-element cell method for incompressible unbounded medium", submitted for review and possible publication to *International Journal for Numerical Methods in Engineering*.

[54] Underwood, P., and Geers, T. L. (1981) "Doubly asymptotic boundary-element analysis of dynamic soil-structure interaction", *International Journal of Solids and Structures*, Vol. 17, 687–697.

[55] Waas, G. (1972) *Linear Two-Dimensional Analysis of Soil Dynamics Problems in Semi-Infinite Layered Media*, PhD dissertation, University of California, Berkeley, CA.

[56] Wang, Y. and Rajapakse, R. K. N. D. (1991) "Dynamics of rigid strip foundations embedded in orthotropic elastic soils", *Earthquake Engineering and Structural Dynamics*, Vol. 20, 927–947.

[57] Wolf, J. P. (1985) *Dynamic Soil-Structure-Interaction Analysis*, Prentice-Hall, Englewood Cliffs, NJ.

[58] Wolf, J. P. (1988) *Soil-Structure-Interaction Analysis in Time Domain*, Prentice-Hall, Englewood Cliffs, NJ.

[59] Wolf, J. P. (1991) "Classification of analysis methods for dynamic soil-structure interaction", *Proceedings of the Second International Conference on Recent Advances in Geotechnical Earthquake Engineering and Soil Dynamics*, Prakash, S., ed., St. Louis, MO, Vol. II, 1821–1832.

[60] Wolf, J.P. and Motosaka, M. (1989) "Recursive evaluation of interaction forces of unbounded soil in the time domain", *Earthquake Engineering and Structural Dynamics*, Vol. 18, 345–363.

[61] Wolf, J.P. and Motosaka, M. (1989) "Recursive evaluation of interaction forces of unbounded soil in the time domain from dynamic-stiffness coefficients in the freqency domain", *Earthquake Engineering and Structural Dynamics*, Vol. 18, 365–376.

[62] Wolf, J.P. and Song, Ch. (1993) "Dynamic finite-element analysis in time domain of unbounded medium based on damping-solvent extraction", *Computational Mechanics*, ASME Winter Annual Meeting, New Orleans, LA, AD-Vol. 39, 57-70

[63] Wolf, J. P. and Song, Ch. (1994) "Dynamic-stiffness matrix of unbounded soil by finite-element multi-cell cloning", *Earthquake Engineering and Structural Dynamics*, Vol. 23, 233–250.

[64] Wolf, J.P. and Song, Ch. (1994) "Insight on representation of unbounded medium using damping-solvent extraction", *Second International Conference on Earthquake Resistant Construction and Design*, Savidis, S.A., ed., Berlin, 315–320.

[65] Wolf, J. P. and Song, Ch. (1994) "Dynamic-stiffness matrix in time domain of unbounded medium by infinitesimal finite-element cell method", *Earthquake Engineering and Structural Dynamics*, Vol. 23, 1181–1198.

[66] Wolf, J. P. and Song, Ch. (1995) "Unit-impulse response matrix of unbounded medium by infinitesimal finite-element cell method", *Computer Methods in Applied Mechanics and Engineering*, Vol. 122, 251–272.

[67] Wolf, J. P. and Song, Ch. (1995) "Doubly asymptotic multi-directional transmitting boundary for dynamic unbounded medium-structure–interaction analysis", *Earthquake Engineering and Structural Dynamics*, Vol. 24, 175–188.

[68] Wolf, J. P. and Song, Ch. (1995) "Consistent infinitesimal finite-element cell method: in-plane motion", *Computer Methods in Applied Mechanics and Engineering*, Vol. 123, 355-370.

[69] Wolf, J. P. and Song, Ch. (1996) "Static stiffness of unbounded soil by finite-element method", *Journal of Geotechnical Engineering*, ASCE, Vol. 122, 267–273.

[70] Wolf, J. P. and Song, Ch. (1996) "Consistent infinitesimal finite-element

cell method: three-dimensional scalar wave equation", *Journal of Applied Mechanics*, ASME, (in press).

[71] Wolf, J. P. and Weber, B. (1982) "On calculating the dynamic-stiffness matrix of the unbounded soil by cloning", *International Symposium on Numerical Models in Geomechanics*, Dungar, R., *et al*, eds., Zurich, 486–494.

[72] Wonham, W. M. (1968) "On a matrix Riccati equation of stochastic control", *SIAM Journal on Applied Mathematics*, Society for Industrial and Applied Mathematics, Vol. 6, 681–697.

As indicated in the main text, certain parts of the book are based on some of the references mentioned above. The permission to include the material is gratefully acknowledged.

INDEX

Acoustics, 1
Actual structure, 1
Artificial boundary, 5, 263
Artificial damping, 17, 223
Asymptotic expansion
  early time, 155
  high frequency, 152
    diffusion, 190
  low frequency, 207

Benchmark, 281
Body wave, 2
Boundary condition
  approximate, 6
  consistent infinitesimal finite-element cell method in frequency domain, 152
  rigorous, 5
Boundary finite-element method, 13, 201, 205
Boundary-element method, 8
Boundary-integral equation, 8
Bounded medium, 1, 201, *see also* Consistent infinitesimal finite-element cell method
  accuracy, 208
  assemblage, 202
  computer programme *SIMILAR*, 317
  concept, 202
  consistent infinitesimal finite-element cell equation, 204
  displacement at internal point, 206, **207**
  mass matrix, 207, **208**
  similarity, 204
  static-stiffness matrix, 204, **205**
  strain at internal point, 206
Bounded structure, 1

Cantilever, 212
Characteristic length, **11**, 43
Circular cavity embedded in full-plane for in-plane motion, 307
  analytical solution, 311, 313
  consistent infinitesimal finite-element cell method, 121
  frequency domain, 162
  incompressibe elasticity, 147
Circular cavity with varying shear modulus in radial direction for out-of-plane motion
  consistent infinitesimal finite-element cell method, 114
Cloning, 12, *see also* Generalized cloning
Coefficient matrix
  axisymmetric scalar wave equation, 87, **88, 89**
  axisymmetric vector wave equation, 81, **85, 86**
  incompressible elasticity, 138, **139, 142**

three-dimensional diffusion, **184, 185**
three-dimensional scalar wave equation, 76, **77**
three-dimensional vector wave equation, 68, **73, 75**
two-dimensional diffusion, **185**
two-dimensional layer, **97, 98**
two-dimensional scalar wave equation, 80
two-dimensional vector wave equation, 78
Computer programme *SIMILAR*, 317
Concentrated mass, 143, 144, 147, 313
Conservation of momentum, 3
Consistent boundary, 9
Consistent infinitesimal finite-element cell method, 10, 13, 65
  accuracy, 107
  assemblage, 89
  bounded medium, *see* Bounded medium
  coefficient matrix, *see* Coefficient matrix
  computer programme *SIMILAR*, 317
  concept, 65
  diffusion, *see* Diffusion
  equation in frequency domain, 91, **92**, 151
    bounded medium, 204
    diffusion, **186**
    incompressible elasticity, **142**
    two-dimensional layer, **98**
    variation of material properties in radial direction, 102
  equation in statics, 166
  equation in statics for bounded medium, 204
  equation in time domain, 93, **93**
    diffusion, **187**
    incompressible elasticity, **143**
    two-dimensional layer, **98**
    variation of material properties in radial direction, 102
  features, 14
  frequency domain, *see* Frequency domain
  incompressible elasticity, *see* Incompressible elasticity
  SIMILAR, 317
  spatial discretization, 68
  statics, *see* Statics
  time discretization, 104
    first time step, 105
    nth time step, 106
    two-dimensional layer, 94
    variation of material properties in radial direction, 99
    dimensionless frequency, 99
  wave propagation, 65
Continuous-time formulation, 40, 42
Convolution, 6, 7, 32
Correspondence principle, 226, 231
Coupling
  global in space and time, **6**, 7
  local in space and time, **7**, 9, 275
  partially global in space, 276
Cut-off frequency, 4, 104, 114, 119, 292, 295
Cylinder embedded in half-space for vector wave equation, 128

Damping-solvent extraction method, 10, 16, 223
  accuracy
    frequency domain, 246
    time domain, 251
  analytical verification, 237
  bounded medium, 223
  concept, 223
  extraction of damping, 228
    frequency domain, **233**
    time domain, **234**

flexibility formulation, 235
implementation in frequency domain, 231
implementation in time domain, 233
Dashpot matrix, 32, 154, 157
Decision flowchart, 276
Diffusion, 1, 179, *see also* Consistent infinitesimal finite-element cell method
  accuracy, 190
  assemblage, 185
  asymptotic expansion for high frequency, 190
  coefficient matrix, 184, 185
  computer programme *SIMILAR*, 317
  consistent infinitesimal finite-element cell equation in frequency domain, **186**, 189
  consistent infinitesimal finite-element cell equation in time domain, **187**
  dimensional analysis, 182
  dynamic-stiffness matrix, 181, 189
  time discretization, 188
    first time step, 188
    nth time step, 188
  unit-impulse response matrix, 181, 182
  unit-step response matrix, 182
Dilatational-wave velocity, 3
Dimensional analysis
  frequency domain, 44, 182
  time domain, 46
Dimensionless frequency, 44
  diffusion, 183
  variation of material properties in radial direction, 99
Dimensionless time, 47
Direct method, *see also* Transmitting boundary
  definition, **6**

displacement-force relationship, 7
state of the art, 9
Discrete-time formulation, 41, 42
Domain of influence, 3
Doubly-asymptotic boundary, 9, 263, 266, 273–275
Doubly-asymptotic multi-directional transmitting boundary, 10, 19
  accuracy, 271
  concept, 263
  decision flowchart, 276
  numerical implementation, 267
Dynamic condensation, 156
Dynamic condensation equation, 90
Dynamic unbounded medium-structure interaction, 1
Dynamic-stiffness matrix, 4, 6, 151
  acceleration, 34
  diffusion, 189
  displacement, 31, 181
  effect of damping, 225
  extraction of damping, 228
  velocity, 38

Earthquake, 1, 7
Edge-cracked plate, 213
Elastic restoring force, 4, 103
Electromagnetism, 1
Energy sink, 4
Energy source, 4
Equation of motion, 7
  finite-element region, 53
Extended mesh, 3, 282
Exterior mass-proportional dashpots, 225
Extrapolation boundary, 10, 269, 273, 274

Finite-element boundary, 317
Fluid-structure interaction, 1
Forecasting method, 10, 12, 49
  accuracy, 58

concept, **49**
features, 14
fundamental equation, 52
numerical algorithm, **54**
time discretization, 54
Fracture mechanics, 213
Free-field motion, 7
Frequency domain, 151, *see also* Consistent infinitesimal finite-element cell method
 accuracy, 158
 asymptotic expansion
  early time, 155
  high frequency, 152
 computer programme *SIMILAR*, 317
 consistent infinitesimal finite-element cell equation, 151
 dynamic condensation, 155
 substructure deletion, 155
Fundamental solution, 8, 13

Generalized cloning, 12
Generalized structure, 1
Generalized structure-medium interface, 1
Geometric spreading, 2
Geophysics, 1
Global coupling, *see* Coupling

Heat conduction, 1, 179, 180, 190, 194, 198
Hemispherical foundation embedded in half-space, 173

Impact, 2
Impedance, **3**
Incompressible elasticity, 137, *see also* Consistent infinitesimal finite-element cell method
 accuracy, 144
 coefficient matrix, 138, **139**, **142**
 computer programme *SIMILAR*, 317

concentrated mass, 143, 144, 147, 313
consistent infinitesimal finite-element cell equation in frequency domain, **142**
consistent infinitesimal finite-element cell equation in time domain, **143**
statics, 167
Inertial force, 4, 103
Infinite element, 10
Infinite medium, 1
Infinitesimal cell width, 13, 65
 analytical limit, 91
 dimensionless, 67
Infinity, 2–4, 12
Initial value theorem, 155
Instantaneous response, 33, 143, 179
Interaction
 fluid-structure, 1
 medium-structure, 1
 soil-structure, 1
Interaction force-acceleration relationship, 34, 41, 52
Interaction force-displacement relationship, 6, 7, 31, 40
Interaction force-velocity relationship, 37, 41
Interaction horizon, 5
Irregular bounded medium, 1

LAPACK, 105
Linear, 1, 2
Linear system theory, 40
Lingering response, 33
Local, *see* Coupling
Lyapunov equation, **106**, 188, 208

Machine, 2
Material damping, 225
Medium, 1
 infinite, 1

irregular bounded, 1
semi-infinite, 1
unbounded, 1
Medium-structure interaction, 1
Medium-structure interface, *see* Structure-medium interface
Modelling
approximate, 9
rigorous, 8
Motion
free-field, 7
scattered, 7
Multi-directional boundary, 10, 263, **265**, 273–275

Natural frequency, 213
Nonlinear, 1, 2, 4

Outgoing plane wave, 9, 264

Paraxial boundary, 9, 265, 273
Prism embedded in half-space for diffusion, 198
Prism embedded in half-space for scalar wave equation, 129
Prism embedded in half-space for vector wave equation, 314
consistent infinitesimal finite-element cell method, 132
frequency domain, 162
incompressible elasticity, 147
statics, 173

Radiation condition, 2, **2**, 4, 8, 12
Radiation damping, 2–4
insight, 102, 118
Rational approximation, 40
Realization, 40
Recursive equation, 41
Regular part
displacement, 32
velocity, 38

Review paper, 8
Riccati equation, **105**, 166, 168, 188, 205

Scattered motion, 7
Schur factorization, 105
Semi-infinite layer for in-plane motion, 301
doubly-asymptotic multi-directional transmitting boundary, 273
incompressible elasticity, 145
Semi-infinite layer for out-of-plane motion, 294
analytical solution, 298, 300
consistent infinitesimal finite-element cell method, 120
damping-solvent extraction method, 246
Semi-infinite medium, 1
Semi-infinite rod on elastic foundation, 289
analytical solution, 292, 293
damping-solvent extraction method, 237
doubly-asymptotic multi-directional transmitting boundary, 271
forecasting method, 58
Semi-infinite wedge for diffusion, 194
Semi-infinite wedge for in-plane motion, 307
consistent infinitesimal finite-element cell method, 121
damping-solvent extraction method, 248
doubly-asymptotic multi-directional transmitting boundary, 274
forecasting method, 60
Semi-infinite wedge for out-of-plane motion, 302
analytical solution, 306
consistent infinitesimal finite-element cell method, 121

forecasting method, 59
statics, 172
Settlement, 1
Shear-wave velocity, 3
SIMILAR, 317
Similarity, **11**
  bounded medium, 204
  consistent infinitesimal finite-element cell method, 65
  diffusion, 182
  dynamic-stiffness matrix, 44, **46**, **184**
  forecasting method, 49
  statics, 166
  unit-impulse response matrix, 46, **47**
  variation of material properties in radial direction, 101
Singular part
  acceleration, 34
  displacement, 32
  velocity, 38
Soil-structure interaction, 1, **1**, 2
Solid sphere, 208
Solvent, 223
Spherical cavity embedded in full-space, 282
  analytical solution, 285, 286, 288, 289
  consistent infinitesimal finite-element cell method, 107
  diffusion, 190
  frequency domain, 158
  incompressible elasticity, 144
  statics, 171
Spherical cavity with varying material properties in radial direction, 112
Spring matrix, 32, 154
State variable, 41
Statics, 1, 165, *see also* Consistent infinitesimal finite-element cell method
  accuracy, 171

bounded medium, 204
computer programme *SIMILAR*, 317
consistent infinitesimal finite-element cell equation, 166
displacement at internal point, 167, **170**
incompressible elasticity, 167
similarity, 166
strain at internal point, 167, **170**
Stress intensity factor, 214
Strip foundation embedded in half-plane for in-plane motion, 314
  consistent infinitesimal finite-element cell method, 125
  damping-solvent extraction method
    frequency domain, 250
    time domain, 251
  forecasting method, 61
Strip foundation embedded in half-plane for out-of-plane motion, 124
Structure
  actual, 1
  bounded, 1
  generalized, 1
Structure-fluid interaction, *see* Fluid-structure interaction
Structure-medium interaction, *see* Medium-structure interaction
Structure-medium interface, **1**, 2–5, 7
  generalized, 1
Structure-soil interaction, *see* Soil-structure interaction
Substructure deletion, 158
Substructure method
  definition, **5**
  interaction force-acceleration relationship, 34
  interaction force-displacement relationship, 7, 31
  interaction force-velocity relationship, 37

state of the art, 8
Superposition boundary, 9, 273
Surface wave, 2

Temporal coupling, 6
Thin-layer method, 9
Time delay, 13, 49
Time discretization, 104, 188
  first time step, 105, 188
  nth time step, 106, 188
Transmitting boundary, 7, 9
  doubly-asymptotic boundary, 9, 273–275
  doubly-asymptotic multi-directional boundary, 10, 271
  extrapolation boundary, 10, 273, 274
  multi-directional boundary, 10, 273–275
  paraxial boundary, 9, 273
  superposition boundary, 9, 273
  viscous boundary, 9, 273–275
Transversely isotropic material, 125, 132, 176, 314
Tunnel in inhomogeneous transversely isotropic unbounded rock, 176

Unbounded medium, 1
Unbounded medium-structure interaction, 1
Underground explosion, 1
Uniqueness
  frequency domain, 4
  time domain, 3
Unit-impulse response matrix, 6
  acceleration, **34**
  displacement, **32**, 181
  forecasted, 51
  velocity, **38**, 182
Unit-step response matrix, 182

Vehicle, 1, 2
Viscous boundary, 9, 273–275

Wave front, 2, 3
Wave propagation, 1, 2, 49